# CHEMICAL MÖSSBAUER SPECTROSCOPY

# CHEMICAL MÖSSBAUER SPECTROSCOPY

Edited by

## R.H. Herber

Rutgers University
New Brunswick, New Jersey

PLENUM PRESS • NEW YORK AND LONDON

Library of Congress Cataloging in Publication Data

Main entry under title:

Chemical Mössbauer spectroscopy.

  Proceedings of a symposium held at the 187th Meeting of the American Chemical
Society, St. Louis, Mo., 4/8–13/1984.
  Includes bibliographies and index.
  1. Mössbauer spectroscopy—Congresses. I. Herber, Rolfe H. II. American Chemical
Society. Meeting (187th: 1984: St. Louis, Mo.)
QD96.M6C48   1985                         543 .08586                         84-20657

ISBN-13: 978-1-4612-9479-5          e-ISBN-13: 978-1-4613-2431-7
DOI: 10.1007/978-1-4613-2431-7

Proceedings of a symposium entitled Twenty-Five Years of Chemical
Mössbauer Spectroscopy, held April 8–13, 1984, in St. Louis, Missouri

This volume is dedicated to Professor Rudolf L. Mössbauer of the Technische Universität München, as a measure of respect and admiration on the occasion of the anniversary marking twenty-five years of the use of nuclear gamma-ray resonance spectroscopy for the solution of problems in chemistry and associated sciences.

# PREFACE

The past twenty five years - roughly the period from 1960 to 1985 - have been by all measures among the most exciting and challenging times of our science. The increasing sensitivity of chemical instrumentation, the introduction of the routine use of computers for data reduction and of microprocessors for instrumental control, the wide-spread utilization of lasers, and the disappearance of traditional disciplinary boundaries between scientific fields are but a few of the examples one could cite to support the introductory contention.

Almost all of these developments have had their impact on the development of Mössbauer Effect Spectroscopy into a technique par excellence for the elucidation of problems in all areas of chemistry and its associated sister sciences. Indeed, because this spectroscopy is based on fundamental phenomena in nuclear physics, is described in terms of the theory of the solid state and structural chemistry, is useful in the understanding of chemical reactivity and biological phenomena, and can serve to supplement information developed by many other experimental techniques, it has provided an unparalleled opportunity for the exchange of ideas among practitioners of a very wide variety of subfields of the physical and biological sciences. The present collection of contributions is the direct result of such an interaction.

It has been especially appropriate to mark the first twenty five years of the use of recoilless gamma ray resonance spectros-copy to the application of the solution of chemical problems by a special Symposium of the American Chemical Society. This Symposium was held at the 187th Meeting of the Society in St. Louis, Missouri, from 8 to 13 April 1984, under the aegis of the Division of Nuclear Chemistry and Technology, and with the co-sponsorship of the Division of Inorganic Chemistry and the Division of Physical Chemistry of the Society. Fifteen invited speakers participated in four half-day sessions, presenting both an overview of the application of Mössbauer Spectroscopy to their particular areas of interest, as well as a description of new results and future prospects. Through the generosity of the donors of the Petroleum Research Fund, it was possible to invite a number of participants from outside the United States, and this happy circumstance has made it possible to include in the discussions – as well as in the present volume – a wide variety of viewpoints and perspectives. The organizer wishes to express his gratitude to the PRF, as well as to the Division of Nuclear Chemistry and Technology of the ACS, for this generous support.

St. Louis, Missouri and
New Brunswick, New Jersey
April 1984

R.H. Herber

# CONTENTS

# ASPECTS OF ORGANOIRON MÖSSBAUER SPECTROSCOPY

Barrie A. Sosinsky

Olin Corporation, Research Center
275 South Winchester Avenue
New Haven, Connecticut 06511

## INTRODUCTION

In this the silver anniversary year of the discovery of the Mössbauer effect its role in the characterization of the organometallic chemistry of iron will be considered. Within the framework of this discussion we begin by examining the range of parameters normally encountered in low valent iron Mössbauer spectroscopy and the structure of the experiment for this application. Mössbauer spectroscopy studies the interaction of nuclear spin transitions with the electronic and magnetic environment. This provides information regarding electronic configuration or oxidation state, stereochemistry and the bonding capabilities of ligands. Some of these arguments have impacted historically on the field of organoiron chemistry. More detailed expositions of the theory and practice of Mössbauer spectroscopy may be found published elsewhere[1-13] and also accompany this review. Previous reviews of this subject have also appeared.[1,2,10]

For our purposes this discussion includes only those iron complexes in low oxidation states with direct metal to carbon bonding. This encompasses iron carbonyls, $\sigma$- and $\pi$-complexes, cyanides and some related complexes. As a general rule these are all strong field ligands and diamagnetic environments which limits the power of the Mössbauer technique. Exceptions will be noted. Molecular orbital theory provides a basis for rationalizing the trends noted for these complexes as a function of geometry and metal ligand bonding. Mössbauer spectroscopy may be seen to be very valuable in studying trends in organoiron chemistry such as covalency <u>vs</u>. ionicity of bonds, synergistic bonding, thru-space and thru-bond interactions and subtle differences in complex stereochemistry and these will be highlighted.

1

## THE MÖSSBAUER EXPERIMENT

The iron (57) Mössbauer spectrum involves the $I_g$ = 1/2 ——>
$I_e$ = 3/2 (g=ground, e=excited state) transition which has an energy of 14.41 keV and a natural line-width $\Gamma_r$ (at 1/2 height) of
0.192 mm s$^{-1}$. In the usual configuration a cobalt (57) source
with a half life of 270 days is embedded in an inert matrix such
as palladium. Cobalt (57) decays by electron capture (99.84%
conversion $I_g$ = 7/2 ————> $I_e$ = 5/2) to $^{57}$ Fe nucleus (Figure 1)
yielding a γ-ray of 136.32 keV. Source iron (57) $I_e$ = 5/2 further
decays to either I = 3/2 or I = 1/2. It is the subsequent decay
of iron $I_e$ = 3/2 ——> $I_g$ = 1/2 that is used to excite iron (57) in
the absorber (material to be studied). This is shown schematically
in Figure 1. Iron (57) has a low natural abundance of 2.17% and a
high internal conversion coefficient by electron emission (α =
8.17) which degrades the signal intensity but an extremely high
γ-ray absorption cross section $\sigma_o$ = 23.5 x 10$^{-19}$ cm$^{-1}$. Iron is
in fact one of the most easily studied nuclei for this technique.
The Mössbauer condition has such high intrinsic resolution (one
part in 10$^{13}$) that Doppler shifting the absorber velocities on the
order of a couple of mm s$^{-1}$ relative to a reference matrix (generally iron metal or sodium nitroprusside) will include all the iron
compounds one could study. This velocity parameter commonly called
isomer shift or chemical shift given the symbol δ measures the
coulombic interaction between the electronic configuration and the
nucleus and is discussed in the next section. The reader is referred elsewhere for more detailed presentation of this material.[1-3,9]

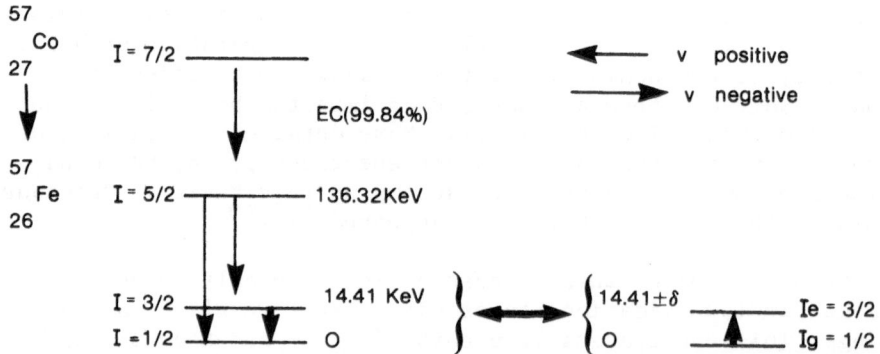

Figure 1.   Source decay scheme and resonant absorption (indicated
by double arrow). The source is Doppler shifted by ± δ
where δ is much smaller than $E_\gamma$. For $^{57}$Fe every one mm
s$^{-1}$ of velocity, $E_\gamma$ cosθ or 4.8 x 10$^{-8}$ eV (1.1 x 10$^{-3}$
cal mol$^{-1}$). This corresponds to a frequency of 11.6 MHz,
a very small amount compared to the energy of the transition (∿2 parts in 10$^{12}$).

## Isomer Shift

The isomer shift (or chemical shift) $\delta$ is the amount of velocity of the souce either towards the absorber or away from it. This Doppler shifts the source so that resonant absorption of the $\gamma$-ray may occur. By convention positive velocities mean relative movement towards one another. This is illustrated in Figures 1 and 2. Whereas for many analyses of coulombic interaction between the electronic state and the nucleus the latter may be treated as a point charge, this is not the case here. Electron density <u>within</u> the nucleus has a predominant effect in establishing this interaction with s-orbital configuration the largest part of this. Interpenetration shielding of other orbitals are also important. This point will be discussed in more detail in later sections when we consider the effect of electronic occupancy and stereochemistry on the Mössbauer parameters $\delta$ and $\Delta E_Q$ (quadrupole splitting, discussed below). A further complication is that the nucleus changes size on de-excitation with $\delta R/R \sim 10^{-4}$. For $^{57}Fe$ the $I_e$ = 3/2 state is smaller than $I_g$ = 1/2 and negative differences in isomer shifts means enhanced coulombic interaction. This is commonly interpreted as more s-electron density at the nucleus or a lower oxidation state. For $^{119}Sn$ the situation is exactly reversed. A common approximation of the isomer shift is given by <u>1</u> where $|\Psi_s(0)|^2$ = the non-relativistic Schrödinger wave-

$$\delta = \frac{1}{5\varepsilon_o} Ze^2R^2 \frac{\Delta R}{R} \left( |\psi_s(0)_{absorber}|^2 - |\psi_s(0)_{source}|^2 \right) \qquad \underline{1}$$

function at r=0, Ze=nuclear charge, -e=electron charge, $\Delta R$=change in nuclear radius, and $\varepsilon_o$ = the permittivity of a vacuum. In most instances <u>1</u> is replaced by a simplified form <u>2</u>, a product of a nuclear term and a chemical term. The former normally is considered a constant for a given transition.

$$\delta = \text{constant} \times \left( |\Psi_s(0)_A|^2 - |\Psi_s(0)_B|^2 \right) \qquad \underline{2}$$

Sometimes equations similar to <u>1</u> are given with R as a mean-square value to account for deviation of the nucleus from spherical symmetry.

Although it is convenient to discuss $\delta$ in terms of $|\Psi_s(0)|^2$ and describe the electronic occupancy as an s-electron interaction such descriptions must always be tempered by the fact that p- and d-orbital occupancy contributes significantly to $\delta$ through interpenetration shielding. Whereas these effects are implied in the analysis of $\delta$ they are so obviously important in determining $\Delta E_Q$, the quadrupole splitting, that they are normally not emphasized in a discussion of $\delta$. We will return to this point when we consider oxidation state ranges. In attempting to establish isomer shift scales for compounds measured in various studies second order Doppler shifts occur because of zero point motion in the lattice. Low temperature spectra minimize this effect. It is worthwhile noting that $\delta$ and $\Delta E_Q$ are really intimately related and that one of the significant achievements of organoiron Mössbauer spectroscopy has been the development of schemes for assigning contributing values to different orbitals and to ligands in various geometries. These schemes have worked particularly well for six coordination where this field is symmetry related to the cubic field imposed by the quadrupolar state $I_e = 3/2$.

## Electric Quadrupole Interactions

It is a great simplification in the analysis of organoiron Mössbauer spectroscopy that rarely do we need consider the interaction of the nuclear (magnetic moment $\mu$) spin I with either intrinsic magnetic fields caused by unpaired electrons or extrinsic fields caused by magnetic domains or applied magnetic field. Hyperfine splitting rarely occurs because in the strong crystal fields of organoiron complexes paramagnetic configurations spin-pair or the complex oligomerizes. The lack of magnetic hyperfine splittings which in an $I_g = 1/2 \longrightarrow I_e = 3/2$ transition leads to a six line pattern (as shown in Figure 2) and the relative temperature independence of organoiron Mössbauer spectra substantially diminishes the amount of information available about the symmetry of the electronic field. The energy levels of this splitting are defined by 3 and 4 where $\mathcal{H}$ = Hamiltonian, $\mu_N$ =

$$\mathcal{H} = \mu \cdot B = -g\mu_N I \cdot B \qquad \qquad \underline{3}$$

$$E_m = \frac{\mu B}{I} m_z = -g\mu_N B m_z \qquad \qquad \underline{4}$$

nuclear magnetron, g = nuclear g-factor (g = $\mu/(I\mu_N)$, $m_z$ = magnetic quantum number with all integer values I to $-$I. Experiments necessary to ascertain symmetry parameters in organoiron chemistry have applied magnetic fields to split the spectrum. The electric quadrupole interaction occurs when a nuclear state with I>1/2 splits due to the influence of a non-spherically symmetric field. For $^{57}$Fe $I_e$ splits into two non-degenerate states with X-ray transitions $I_g = \pm 1/2 \longrightarrow I_e = \pm 1/2$ and $I_g = \pm 1/2 \longrightarrow I_e = \pm 3/2$ populated as shown in Figure 2. This gives rise to the symmetrical doublet pattern normally seen for organoiron complexes. In general, changes in $\Delta E_Q$ can be quite large for a given series of organoiron compounds where changes in $\delta$ are small. We will not consider further the combination of hyperfine splitting and quadrupole splitting which can give rise to unsymmetrical patterns.[1-5,9]

Quadrupole splitting ($\Delta E_Q$) often represented as $\Delta$ for $^{57}$Fe is commonly defined as 5 where eQ= nuclear quadrupole moment, e=charge of a proton, eq=principle component of the electric field gradient, and $\eta$=assymetry parameter which measures departure from

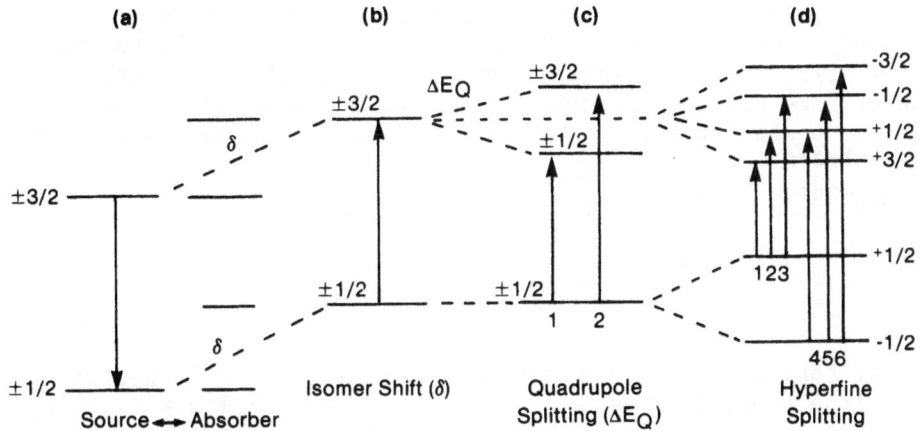

Figure 2. (a) Source γ-ray emission. (b) Absorber γ-ray absorption. (c) Quadrupole split doublet ($\Delta E_Q$). (d) Six line hyperfine pattern for a polycrystalline absorber. For (c) and (d) the splittings are much smaller than the γ-ray transitions.

axial symmetry [$\eta=(V_{xx}-V_{yy})/V_{zz}$], V is the potential of the electric field tensor, z is the principle axis).[1-3,9] For $I_g=1/2 \longrightarrow I_e = 3/2$

$$\Delta E_Q = 1/2 \ e^2qQ(1 + \eta^2/3)^{1/2} \hspace{3cm} \underline{5}$$

the transitions take place according to the selection rule $\Delta M_z = 0, \pm1$ (M1 dipolar radiation), giving rise to a Lorenzian shaped symmetrical doublet with $2\Delta E_Q$ splitting. $\Delta E_Q$ is then the deviation from cubic symmetry arising from the electric quadrupole interaction between the nuclear quadrupole moment and the local electric field gradient tensor at the nucleus. When $\eta=0$ (axial symmetry), $\Delta E_Q=e^2qQ/2$ or the quadrupole coupling constant is half the peak separation as defined in nuclear quadrupole resonance (NQR) spectroscopy. The difference is that whereas in Mössbauer spectroscopy we measure γ-ray absorptions between multiplets in NQR we measure radio frequency transitions within the multiplets. It is not possible to measure $\eta$ or $e^2qQ$ directly from a powder Mössbauer spectrum. Single-crystal measurements of the angular dependency of line intensities or the dependence of the Mössbauer spectrum upon large external magnetic fields can, however, establish these parameters. It is the latter technique that has been most widely used to elucidate the field gradients in several important organo-iron complexes.

An approximate value for the field gradient may be obtained by adding contributing values for each orbital. Conversely, knowledge of the field gradient can lead one to assign the electronic state. For various orbitals the following q (=$V_{zz}/e$) and $\eta$ may be given: $P_z(q = 4/5 <r^{-3}>$ , $\eta=0)$, $P_x(+2/5,-3)$, $P_y(+2/5, +3)$, $d_{x2-y2}(+4/7, 0)$, $d_{z2}$ $(-4/7, 0)$, $d_{xy}(+4/7, 0)$, $d_{xz}$ $(-2/7, +3)$ and $d_{yz}$ $(q= -2/7$, $\eta=3$, respectively) where $<r^{-3}>$ is the expectation value of $1/r^3$ for the appropriate orbital function and the orbital functions 6 and 7 evaluate the matrix elements 9 from polar coordinate analyses.

$$q_{zz} = q = - < \Psi \mid \frac{3\cos^2 \theta -1}{r^3} \mid \Psi > \qquad \underline{6}$$

$$\eta q = < \Psi \mid \frac{\sin^2\theta - \cos 2\phi}{r^3} \mid \Psi > \qquad \underline{7}$$

For ferrocene with a ground state $a_{1g}^2(d_z^2)$, $e_{2g}^4$ $(d_{xy}^2$, $d_{x2-y2}^2)$ the prediction is that $q_{zz} = 2(-4/7 <r^{-3}> ) + 4(4/7 <r^{-3}>)$ = $8/7 <r^{-3}>$ . The field gradient should be large and positive and this is what is observed. Generally $q_{valence}$ may be expressed as in 8 where we neglect the much smaller 4p-electron imbalance.

$$q_{valence} = K_d \, [-N_{d_z2} + N_{d_x2-y2} + N_{dxy}-1/2(N_{dxz}+N_{dxy})] \qquad \underline{8}$$

where $K_d$ is a constant for a given d-orbital in a given atom.

## OXIDATION STATE AND BONDING

Organoiron complexes occupy an historic place in the field of transition metal chemistry.[14-16] Iron pentacarbonyl was the second metal carbonyl to be discovered by Monde very early in this century and the iron carbonyl complexes were recognized early on as essential zero valent complexes, that is ones where the metal is very nearly in a native metallic state. Vibrational spectra of iron carbonyls[17-19] confirm this notion. The Mössbauer parameter $\delta$ reinforced this concept with values of diamagnetic covalent iron complexes narrowly clustered about that for metallic iron (referenced as $\delta$ = 0 mm s$^{-1}$ within a range defined by $d^{10}$ - $Fe^{-2}$ $Na_2[Fe(CO)_4]$ $(\delta= -0.18)$[20] to $d^5-Fe^{3+}$ ferricenium $[(C_5H_5)_2Fe]$ $BPh_4$ $(\delta= + 0.58)$.[21] The development of the Mössbauer technique in the late fifties and early sixties coincided with the discovery of ferrocene as the first recognized $\pi$-sandwich complexes and the refinement of the Duncanson-Dewar-Chatt model of synergistic metal to ligand bonding. Iron Mössbauer spectroscopy impacted conceptually on both of these developments.

Although it is inappropriate to develop the complete bonding treatments for organoiron complexes within this text and the reader is referred to standard texts[14,15] for this presentation some general statements concerning organoiron complexes are in order. Nearly all stable organoiron complexes follow the effective atomic number rule, are strong field diamagnetic complexes and fall into formal oxidation states of +2 or 0 with +1 or +3 being much less common. By formal oxidation state we mean considering anionic ligands as $\sigma$-donating point changes without considering $\pi$-backbonding effects. Ferrocene would therefore have formal $Fe^{+2}$, however Mössbauer $\delta$ parameters indicate it is less and much closer to $Fe^{+1}$ [22] Thus covalency and synergistic bonding gives complexes with intermediate values throughout the range. Not only is the formal oxidation state range small but only three idealized coordination numbers need be considered 6, 5, and 4-coordination and are discussed below individually. When finally the combinations of oxidation state and coordination number are considered they are greatly restricted. For example, $Fe^{+2}$ and $Fe^{-2}$ octahedral dominates 6-coordination as does $Fe^{-2}$ tetrahedral for 4-coordination. This is seen to be the obvious extention of low spin system for a $d^8$-metal system.

Six Coordination

Six coordinate organoiron complexes are octahedral with only slight z-in or z-out tetragonal distortions observed. The potential energy barriers separating $O_h$ geometry from other potential idealized forms such as trigonal prism ($D_{3h}$) or hexagonal (trigonal antiprism, $D_{3d}$) are not normally surmountable in these systems and need not be considered. Formal oxidation states can be either $\pm 2$ depending on whether one places the two additional electrons on the metal as in the $[Fe(CN)_6]^{-2}$ ion or on the ligand as in $[Fe(Co)_4Br_2]$. Synergistic bonding L ———>M $\sigma$-bonds counterbalanced by M ———>L $\pi$-backbonding clouds the issue and in reality examples exist throughout this range. Undoubtedly the greatest contribution six coordinate organoiron Mössbauer studies provided arises from the fact that in an octahedral field the symmetry is appropriate so that a point charge approximation may be made for the ligand-nuclear coulombic interaction. The three L ——> M $T_{2g}$ $\pi$-orbitals at $(-2/5$ $\Delta_0$ ($\Delta_0$=octahedral crystal field splitting) are offset exactly by two M ——> L $E_g$ $\sigma$-orbitals at $+3/5$ $\Delta_0$ and more importantly both these orbital sets have the spherical symmetry elements of the cubic field.[22] It was obvious to early workers in this field that for organoiron compounds Mössbauer parameters for different isomers like cis- or trans- $FeA_2B_4$ had $\Delta E_Q$'s related by nearly simple whole numbers [23-26] whereas $\delta$ values remained constant.[28,29] Ligands treated as point charges in an octahedral field give individual effects that are additive. This led Herber and co-workers to advance the concept of partial isomer shifts (PIS) [28,29] and others notably Bancroft [30] to define the partial quadrupole splitting (PQS) parameter for each ligand thereby providing a direct measure of the relative importance of $\sigma$- and $\pi$-bonding. Only infrared spectroscopy ($\nu_{co}$ stretches) provide this kind of data but in this particular case Mössbauer spectroscopy gives a more consistent story.

The concept of additive values for ligand $\delta$ and $\Delta E_Q$ contributions work well in 6-coordination and was extended to other metals isoelectronic with $d^6 - Fe^{2+}$ $(t_{2g}{}^6)$ namely $Co^{+3}$, $Ru^{+2}$ and $Mn^+$ [29,31,32] and also to $d^{10} -Fe^{-2}$.[33] In general PQS values are more consistent than PIS values, and the correlation which has been so widely used in octahedral coordination tends to break down as symmetry is lost ($MA_6 \longrightarrow MA_5B \longrightarrow MA_4BC \longrightarrow$ etc.) and does not work at all in other geometries. A study confirming the signs of $V_{zz}$ and $e^2qQ$ as measured by the magnetic perturbation method for low spin $Fe^{+2}$ has appeared[25] and confirms those predicted by the point charge model.

The reader is referred to the article by Bancroft for the derivation of the relationship for $\Delta E_Q$ and ligands PQS values for different complexes.[30] More germane to this discussion are the trends noted in ligands as a function of their σ- and π-bonding. PQS values increase with enhanced π-backbonding and decrease with enhanced σ-bonding and q lattice parameters. Equations 9 and 10 qualitatively summarize the nearly linear relationships[25] that are seen for δ vs. Δ plots.

$$PIS = -k \ (\sigma + \pi) \qquad\qquad\qquad \underline{9}$$
$$PQS = q_{lattice} + C \ (\sigma' - \pi') \qquad\qquad \underline{10}$$

Small hard ligands such as $H^-$, $Cl^-$ and $CN^-$ which are better approximations for point charges work well, ligands such as CO, $N_2$, CNR, and $P(OR)_3$ are not particularly good actors, and their deviation off the linear relationship emphasizes their enhanced π-bonding capabilities. Deviations occur when certain geometries give strong ligand-ligand interactions such as trans-effects. Since PQS values are synergistically related (σ-π) they tend to occur over a narrow range compared to PIS values. Clearly σ-bonding involves orbital overlap with more s-character so this interaction tends to be more dominant in determining the Mössbauer parameters.

## Five Coordination

In five coordination organoiron complexes are usually formally $d^8$ and zerovalent with large $\Delta E_Q$ due to an unsymmetrical d-shell. Two geometries dominate five coordination, the trigonal bipyramid ($D_{3h}$, "TBP") and the square pyramid ($C_{4v}$, "SP") with, in general, low barriers to interconversion through intermediate unstable $C_{2v}$ forms. In their classic treatment of five coordination Rossi and Hoffmann have calculated[34] the preferred geometry as a function of electronic occupancy, the barriers to interconversion, and preferred ligand substitution patterns. For 5-coordination the following geometries are favored: $d^0-d^2$, $C_{4v}$; $d^3-d^4$, $D_{3h}$; $d^5-d^7$, $C_{4v}$; $d^8-d^{10}$ $D_{3h}$. The major effect on the electric field gradient in $D_{3h}$ is the vacancy of the LUMO $d_{z^2}$ orbital. For $d^8-D_{3h}$ stronger σ-donors substitute axially; the weaker σ-donors, equatorially. In $d^0-d^4$ and $d^{10}-D_{3h}$ the situation is reversed.

In 1963, Collins and Pettit reported[35] that for a plot of $\Delta E_Q$ vs δ pentacoordinate organoiron complexes follow with a high degree of correlation a linear function (with negative slope) shown in Figure 3. This is really a remarkable relationship for three reasons. First assuming synergistic bonding operates in these cases (as well as it should) Mössbauer parameters are being dominated by $\sigma$-effects alone. Also, there is a seeming insensitivity to the site of substitution with both axial and equatorial substitution noted. In the order of increasing $\delta$ (and decreasing $\Delta E_Q$) we have (a) $[Fe(CO)_5]$, (b) $[Fe(CO)_4$ (phosphine or phosphite)$]$,$[Fe(CO)_4-(P(NMe_2)_2)]^+AlCl_4^-$, (c) $[Fe(CO)_4(olefin)]$, (d) $[Fe(CO)_4$ (allyl)$]^+$, and even (e) $[Fe(CO)_4I_2]$ (cis) and $[Fe_2CO_9]$. Stereochemistry in this series changes from (a) TBP, (b) axially substituted TBP (c) equatorially substituted TBP, to (d) $C_{2v}$, and to (e) formal six coordinate $Fe^{+2}$. It was suggested initially that in (e) iodine was bound as a diatomic.[35]

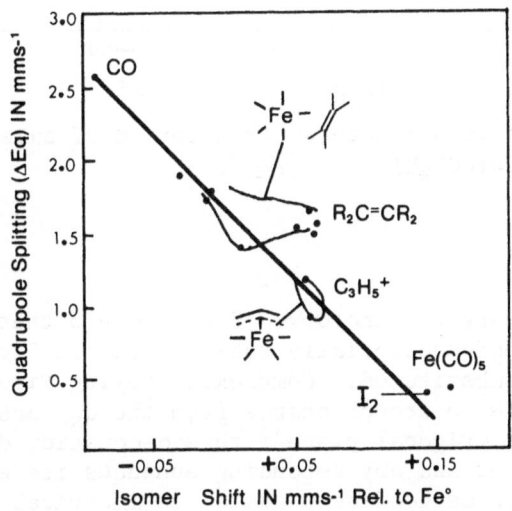

Figure 3. The Collins-Pettit plot for 5-coordinate equatorially substituted $D_{3h}$ $[Fe(CO)_4L]$ and pseudo- 5 coordinate tetracarbonyl ferrates.

Subsequent work by other authors on a series of closely related axially substituted $D_{3h}$-$Fe(CO)_4L$ (L = phosphines, phosphates, arsines and stibenes) and disubstituted $C_{3v}$-$Fe(CO)_3L_2$ established[36,37,38] the $\Delta E_Q$ vs. $\delta$ plot of these complexes lie on a line intersecting Figure 3 perpendicularly as in Figure 4. This relationship with the expected positive slope established that for axial substitution $\sigma$-donation is offset by $\pi$-acceptance (synergistic bonding).

In this author's study of the Mössbauer parameters in 5-coordinate tetracarbonyl ferrates we advanced[39] the following explana-

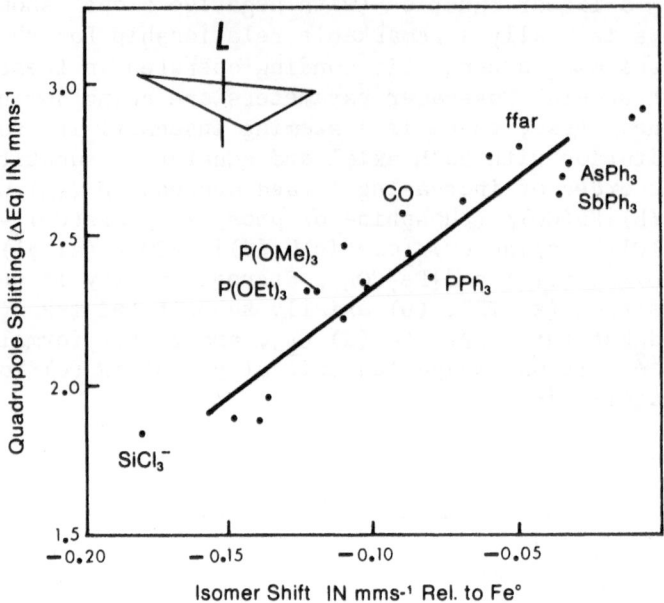

Figure 4. Mössbauer parameters for a series of axially substituted $D_{3h}$-[Fe(CO)$_4$L]

tion for this apparent dichotomy. For Figure 3 these complexes are equatorial or pseudoequatorially substituted, in Figure 4 they are clearly axially substituted. Complexes obeying Figure 3 like olefins are unable to accept charge from the $d_{xz}$ orbital because they align in the trigonal plane[34] thus σ-donation dominates bonding in the xy plane and any π-bonding enhances its effect. When σ-donation becomes as good as CO, axial (symmetrical) substitution occurs. Complexes obeying Figure 4 lie and both sides of [Fe(CO)$_5$]. Decreased donation into the $d_{z^2}$ orbital increases the positive field gradient as would back donation from the $d_{xz}$ and $d_{yz}$ orbitals. The reader is referred to the article by Elian and Hoffman on metal carbonyl fragment analysis.[40] The major difference between both classes of complexes is that in axially substituted systems σ donation leads to a "formally" spherically symmetric d-shell and π-bonds which have less s-character can then show effects. When σ-effects dominate as in Figure 3, δ and $\Delta E_Q$ will both decrease in the order [Fe$^0$(CO)$_4$L$^0$] >[Fe$^+$(CO)$_4$L$^-$]> [Fe$^{+2}$(CO)$_4$L$^{-2}$]. Distortions such as having two cis I$^-$ ligands cannot be differentiated from 5-coordination. Furthermore the Mössbauer parameters of 5-coordination flow naturally towards those of low spin Fe$^{+2}$. In a situation where there is a nonspherical d$^n$-configuration there tends to be much less sensitivity to geometry. The reader is referred elsewhere for a fuller discussion of the spectra of [Fe(CO)$_4$I$_2$].[35,41-44]

Two papers have appeared describing experimental and theoretically calculated field gradients in isoelectronic pentacoordinate $[M(CO)_{5-x}(phosphine)_x]^n$ (M=Fe[45,46], Mn[45] and Co[45]) systems. One would expect a large positive field gradient in $[Fe(CO)_5]$ ($q_{zz} = 2d_{xy} + 2d_{xz} + 2d_{yz} + 2d_{x^2-y^2} = 2(4/7) + 2(-2/7) + 2(-2/7) + 2(4/7) = 8/7 <r^{-3}>$) and it has been found that $\Delta E_Q$ is positive.[47] Phosphine substitution in $[Fe(CO)_4L]$ complexes shows[45,48] that for axial substitution ($\eta=0$) there is little effect on either the magnitude or the size of the electric field gradient. However, both $\eta$ and the sign of the quadrupole splitting are very sensitive to breaking the threefold symmetry by substitution whereas the overall magnitude $|\Delta E_Q|$ is not.

## Four Coordination

Four coordination is a restricted geometry in organoiron chemistry. Complexes are normally $d^{10}$-Fe$^{-2}$ tetrahedral (T$_d$) with high spherical symmetry as was the case in $d^6$-Fe$^{+2}$. $\Delta E_Q$ tends to be small and representative of ligand inequivalence. Point charge approximations have been applied to these systems.[33] Isomer shifts are found at the low (-0.2 to -0.05) end of the diamagnetic covalent series consistent with their enhanced low valent state with $[Fe(CO)_4^{-2}]$ ($\delta = -.0.18$, $\Delta E_Q=0$)[20] and its substitution derivatives $[Fe(CO)_3NO^{-1}]$ (-0.10, 0.38)[49] and $[Fe(CO)_2NO_2]$ (-0.08, 0.36)[49] being prototypical. For $[Fe(CO)_3NO^{-1}]$ the field gradient is positive[49], CO being a better $\sigma$-donor and/or worse $\pi$-acceptor than NO$^+$.

## ORGANOIRON COMPLEXES

### Binary Iron Carbonyls

The binary iron carbonyls as a class show low $\delta$ with a restricted range. For a given series $\Delta$ values generally vary over a greater range and are more indicative of stereochemistries. We discussed aspects of this in previous sections on coordination numbers and will return to this point later in this section. Iron carbonyl complexes exhibit distinct trends. Buildup of anionic charge as in the sequence neutral to anionic to dianionic lowers the isomer shift as does going from high coordination numbers to lower coordination numbers. The iron carbonyls are noted for their very good correlation of $\delta$ with formal oxidation state.

Iron pentacarbonyl has been well studied in a variety of ways by Mössbauer spectroscopy. Its parameters were discussed in the section on five coordination. A theoretical study based on extended Hückel calculations has appeared for $[Fe(CO)_5]$ (as well as $[Fe(CN)_6]^{-3}$, $[Fe(CN)_5NO]^{-2}$, and $[Fe(CN)_6]^{-4}$) which interprets the Mössbauer parameters as a function of their orbital occupancy. Since $[Fe(CO)_5]$ is a liquid with a freezing point near -21°C, frozen polycrystalline samples gives a Mössbauer spectrum that is an

11

asymmetric doublet. This asymmetry has been ascribed[51] to a preferential orientation within polycrystalline $[Fe(CO)_5]$. A study of a
wide variety of frozen solutions of $[Fe(CO)_5]$ has appeared.[51] In
solutions the asymmetry disappears. Some solvents, those with
carbonylic CO groups interact with $[Fe(CO)_5]$ to slightly lower the
$\Delta$ in a manner which appears to be temperature independent. This
interaction was not considered to be a dipole-dipole interaction
because of its specificity. A weak bond was postulated and these
interactions are really quite common when one studies metal carbonyl chemistry as a class.

The iron carbonyl $[Fe_2(CO)_9]$ and its substitution compounds
have been studied by Mössbauer spectroscopy. In $[Fe_2CO_8H]^{-1}$ a
single quadrupole split doublet argues for one type of iron, and in
conjunction with other spectroscopic techniques it can be seen to
have two $Fe(CO)_3$ groups bridged by two carbonyls and a hydride.
Hydride, a good $\sigma$-donor and a non-existant $\pi$-acceptor lowers $\delta$
and increases $\Delta E_Q$. A similar effect is found in complexes such as
$[Fe_3(CO)_{12}]$ and $[Fe_3(CO)_{11}H]^{-1}$. For systematic studies of the iron
carbonyls and its substitution compounds the reader is referred
elsewhere.[1,10,20]

The story of Fe(57) Mössbauers' role in the elucidation of the
structure of $Fe_3(CO)_{12}$ was an important one. Initially the structure of $[Fe_3(CO)_{12}]$ was believed to be an equilateral triangle of
$Fe(CO)_4$ groups, instead of its true structure with 2 CO's bridging
one edge. Complications due to a crystal disorder left the solution
incompletely refined. The Mössbauer spectrum of $[Fe_3(CO)_{12}]$ is a
three peak pattern, a doublet around a singlet suggest two iron
environments in a 2:1 ratio. The central singlet is a more symmetrical iron with no resolvable $\Delta$. Here Mössbauer $\Delta$ values are
very sensitive to small changes in geometry. The related
$[Fe_3(CO)_{11}H^-]^{-1}$ has a similar spectrum with a large $\Delta$ for the
doublet, $H^-$ has replaced a bridging CO. Dahl and co-workers were
finally able to resolve the structure of $[Fe_3(CO)_{12}]$ and the story
of its resolution makes interesting reading.[54]

The use of Mössbauer spectroscopy to assign iron atoms in higher iron carbonyl clusters is not without its difficulties. The
pyramidal anion $[Fe_4(CO)_{13}]^{-2}$ gives a simple doublet spectrum with
no resolution of the unique iron atom from the three basal atoms.[20]
As the number of iron atoms increase all with similar $\delta$ values
signals tend to be broadened and hard to resolve. This author
studied a series of iron carbido carbonyl clusters and observed in
general broad and unresolved signals.[55] Other workers subsequently
reported spectra of these complexes which were better resolved.[56]
Since resolving the spectra changes the $\delta$ position our values are
similar but not exact. In some of the cases studied notably
$[Fe_5C(CO)_{15}]$ ($\delta$=0.46, 0.47; $\Delta$=0.98, 0.45; area 1:4 (apical:basal),
respectively) and $[Fe_5N(CO)_{14}H]$ the authors used aligned samples and

the spectra are resolved.[56] The other spectra studied are complex $[Fe_5C(CO)_{14}]^{-2}$ or a broadened doublet (of doublets?) assigned from the more resolved spectra. The complex $[Fe_6C(CO)_{16}]^{-2}$ with an interstitial carbon atom gives a symmetrical doublet ($\delta=0.49$, $\Delta=0.56$). It is claimed that the ability to form cluster bonds is of the order $Fe(CO)_3^t$ apical$\sim Fe(CO)_2^t(CO)_2^b$ >$Fe(CO)_2^t(CO)^b H \sim Fe(CO)_2^t(CO)^{semi-b}$ >$Fe(CO)_3 basal$>$Fe(CO)_2^t Co^b$ (t=terminal, b=bridging) based on a correlation between $\Delta E_Q$ and the approach of the iron site to axial symmetry.[56] The larger the deviation from axial symmetry the smaller $\Delta$. What is clear from this discussion is that it is only the differences in $\Delta$ that allow any assignments to be made in some very similar iron environments.

## Mixed Metal Tetracarbonyl Ferrates

We have studied two systems of mixed metal tetracarbonyl ferrates which demonstrate the power of Mössbauer spectroscopy in organoiron chemistry. The complexes $[M'Fe(CO)_4]_x$ (M' = Zn, Cd, and Hg) undergo reduction in the presence of Lewis bases to give complexes formulated as $(Na\{THF\})_2^+[M'(Fe(CO)_4)_2]^{-2}$ with the structures of M'= Zn and Hg shown in Figure 5.[57,58] Structures of both the dianions (5A) and associated cation structures (5B) are shown. Formally these complexes should contain two axially substituted $C_{3v}$ $d^8$- $Fe(CO)_4$ fragments. With M' = Hg (5A) this is nearly what is seen. Good $\sigma$-donors and poor $\pi$-acceptors substitute axially (see the section on 5-coordination). There is a small umbrella bending of the three equatorial carbonyl ligands inward towards Hg as would be expected from theoretical considerations[34] for slight ionicity in the Hg-Fe bond. The cation structure containing 5-coordinate $Na^+$ (5B) is regular and imposes no constraints on the dianion structure. The Zn complex has a radically different structure. For M' = Zn the dianion structure (6A) has iron in a geometry approaching $T_d$ - $Fe(CO)_4^{-2}$. In fact Zn can be seen to be intermediate between a face capping and edge capping position. A true ionic complex such as $Na_2Fe(CO)_4$ would have $T_d$-$Fe(CO)_4$ with sodium associating with carbonyl oxygens. For M'=Zn which has a near ionic description for the Zn-Fe bonds, association of zinc with carbonyl ligands can be seen. The cation structure (5B) is highly associated with the dianion and the complex forming an infinite polymeric array.

Our attention was drawn to these complexes by the progressive complexity of the M'=Hg, Cd, and Zn $\nu_{co}$ infrared patterns as shown in Figure 6. This structure is not imposed by any lattice considerations, as it is maintained in solution. What we have here is an example of electronic crossover $d^8$-$Fe^o$ (Hg) to $d^9$-$Fe^{-1}$ (Zn) and the Mössbauer parameters show this clearly. For M'=Zn, $\delta=0.1.56$, $\Delta=1.429$ mm s$^{-1}$. For $[Fe_2(CO)_8]^{-2}$ $\delta=0.16$ mm s$^{-1}$ so that Fe is in a near -1 oxidation state. The complex M'=Cd gave a resolved three-peak Mössbauer spectrum which could be interpreted as three singlets ($\delta= -0.064$, 1.048 and 1.856), two doublets ($\delta=0.492$, 1.452; $\Delta=1.112$,

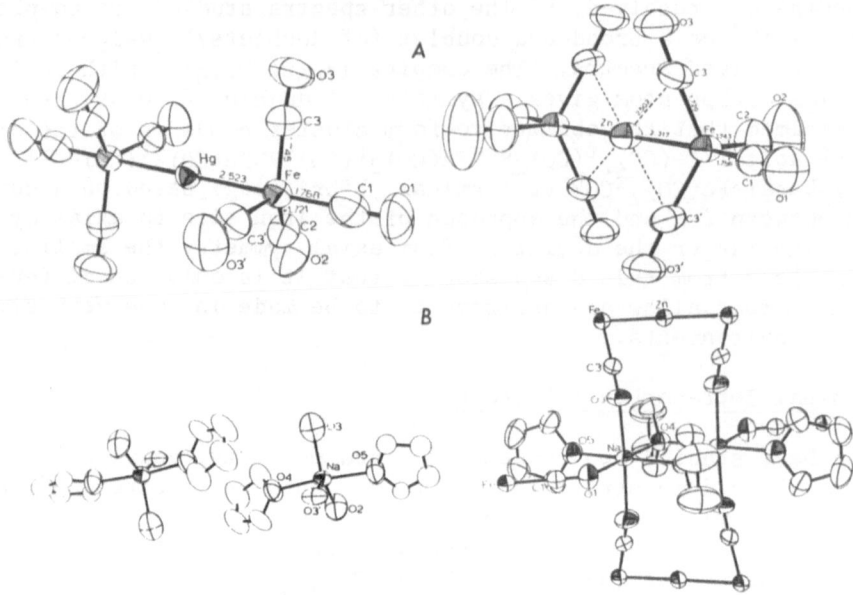

Figure 5. Structures of the $(Na\{THF\}_2)_2^+[M'Fe(CO)_4)_2]^{-2}$ (M' = Hg, and Zn) compounds: (A) dianion structure; (B) cation structure.

0.808) and a doublet around a singlet ($\delta$=1.048, 0.896; $\Delta$= small, 1.921). We felt the two-doublet scheme to be most consistent with the known structure of M'=Hg (whose IR spectrum is very similar). The formula M'Fe$_2$ argues against three singlets as does the high $\delta$ values. The location and symmetry of the peaks which are not in a 1:1 ratio to our minds argues against the doublet-singlet scheme. Here again we see that small changes in geometry can often lead to large changes in Mössbauer parameters when the bonding description changes.

What we have here is an example of intersystem crossing where the d$^n$- occupation number (and not just the symmetry of the state) is changing. Intersystem crossing is an area in which iron Mössbauer spectroscopy has played a very important role.[1,14,59] The area of iron containing metalloenzymes spin transitions is the classic case[14,60]. Professor Gutlich, the next speaker in this symposium will describe some of his work in this area. Previous studies have established[59] that spin crossover high spin $\rightleftharpoons$ low spin transitions are affected by crystal packings and involve complex anion-cation associations. Cycling through the transition point leads to hysteresis due to microdestruction of the crystal. The strain arises in changing molecular size and shape which

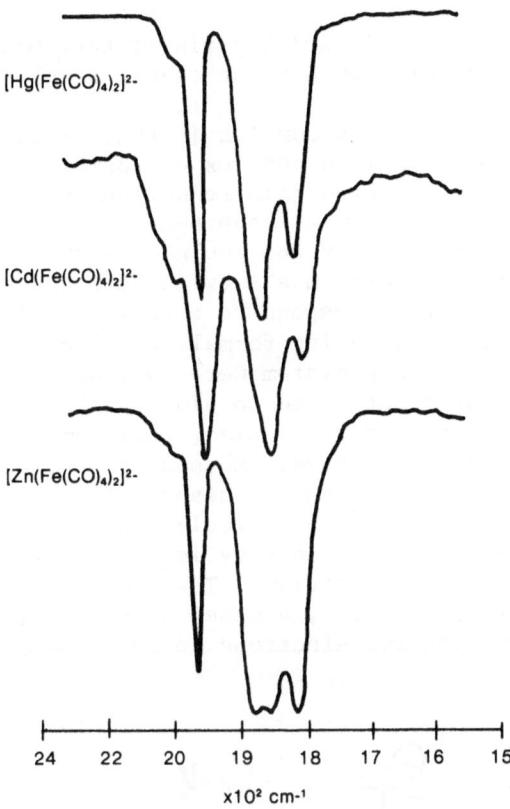

Figure 6. Solution IR spectra of $[M'(Fe(CO)_4)_2]^{2-}$ in the $\nu_{co}$ region.

stresses the crystal. For the molecules $[M'(Fe(CO)_4)_2]^{-2}$ particularly M'=Zn cation-anion associations play a large role in stabilizing the complex.

Several related complexes of Zn, Cd and Hg tetracarbonyl ferrates have appeared of the type $[B_nM'Fe(CO)_4]_x$.[61-63] Their Mössbauer spectra are well resolved into the mononuclear $B_3M'Fe(CO)_4$ species or species of the type $[B_2M'Fe(CO)_4]_x$ with x $\alpha$ B.[64,65] The latter complexes are composed of rings with Mössbauer parameters of iron in traditional 6-coordinate environments and their structures[63] were later confirmed as such.

Complexes containing both iron and tin atoms should allow study of how the Mössbauer parameters of one varies of a function of the other as bonding is changed. Mössbauer studies of the following types of complexes have appeared: $[Fe(CO)_4(SnR_3)_{2-n}X_n]$ (n=1 or 0)[66,67], $[Fe(CO)_4(SnR_2)_2]$[68], $[Fe(CO)_4)_2SnR_2]X$ (X=R or Sn)[68], and

$[(CpFe(CO)_2)_nSnX_{4-n}]$.[69,70] Invariably tin is tetrahedrally bonded as Sn IV and tin may be treated by a point charge model as may iron.[70]

We have reported that the complex $[Fe(CO)_4(SnR_2)]_2$ may be reduced in Lewis bases to yield a complex we formulated $\tilde{N}a_2[Fe(CO)_4-(SnR_2)]$.[71] We initially viewed this reduction as populating a nonbonding $sp^3$ lone-pair on tin with some $\pi$-backbonding from tin to iron. Mössbauer parameters for the two nuclei are (Fe: $\delta=0.061$, $\Delta=1.498$; Sn: $\delta=0.169$, $\Delta=1.498$ mm s$^{-1}$ (with Sn referenced to SnO$_2$)). when B=4-MeC$_5$H$_4$NO. These correspond to a region of iron (-I) and tin IV with each metal undergoing formal one election reduction. A number of features of this system were curious. It was hard to reconcile the electron count based on simple EAN considerations with the formal oxidation states. Also, this complex is tight ion paired and highly base associated. Mössbauer parameters as well as compound colors change with changing bases. In a subsequent paper we carried out an extended Hückel calculation of the complex $[Fe(CO)_4-(SnH_2)]^{-2}$.[72] The most stable geometry is 10 with TBP-Fe and a highly pyramidal SnH$_2$ at an axial position. The two electrons are localized on a lone pair at tin. However, another stable geometry 11 has a trigonal SnH$_2$ group with two electrons in an a$_2$ equatorial carbonyl based orbital.

10          11

It is instructive to consider just how deviation from $d^8$-Fe$^0$ 5-coordination shows up in the Mössbauer parameters. In Figure 7 we have plotted $\Delta$ vs. $\delta$ for tetracarbonyl ferrates. Lines obtained in Figure 4 for axial substitution and in Figure 5 for equatorial (and pseudo equatorial) substitution are drawn in Figure 7 with positive and negative slopes, respectively. A number of mixed metal complexes, iron carbonyl anions, and carbonyl hydrides can be seen to fall substantially off the lines in quadrants labelled I and II.

Deviation from the $d^8$ configuration by the tetracarbonyl ferrate moiety can be analyzed by a metal carbonyl fragment analysis.[40] For $d^8$- and $d^9$-Fe(CO)$_4$ the lowest form is D$_{2d}$. At $d^{10}$-Fe(CO)$_4$ T$_d$ geometry is preferred and completion of the d-shell leads to ligand-ligand repulsions becoming dominant in setting the geometry. For reasons discussed previously when we have a formally spherically symmetric d-shell ligand inequalities are the major cause of the size (and sign) of $\Delta$. $[Fe(CO)_4]^{-2}$ lies in Quadrant 1 at low $\delta$ with

16

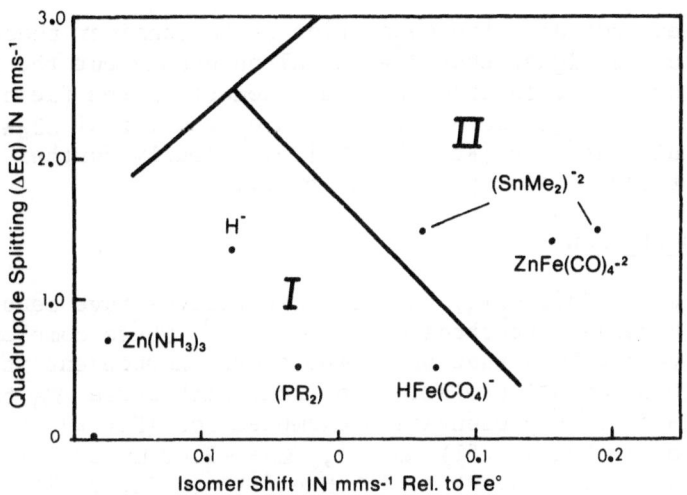

Figure 7. Mössbauer parameters for mixed-metal tetracarbonyl ferrates. See reference 39.

$\Delta = 0$. Other complexes found in this quadrant include $[Fe(CO_4H]^{-1}$ and $[Fe_2(CO)_8H]^{-1}$ and structural distortions may be seen[73] which leads us to believe that these complexes distort towards the structure[74] of $[Fe(CO)_4]^{-2}$.

Ligands which are very good σ-donors/poor π-acceptors will also distort iron towards $d^{10}$-$T_d$ if there is sufficient ionicity in the bond. For $Zn(NH_3)_3$ found in Quadrant 1 this is the case and we would be inclined to consider zinc as $Zn^{+2}$. A similar complex $[Fe(CO)_4SiCl_3]^{-1}$ falls nicely on the line for axial substitution arguing for a covalent bond description with the $SiCl_3^-$ ligand not being a sufficiently strong σ-donor to change the $d^n$-occupancy. It is worth noting that $[Fe(CO)_4Zn(NH_3)_3]$ and $[Fe(CO)_4(SiCl_3)]^{-1}$ lie at almost identical δ values. Lastly we find the complex $[Fe(CO)_4-(PR_2\cdot)]$ in Quadrant 1 which we believe is consistent with the unpaired electron residing on iron.[39,72]

It is important to note that what we are describing here are effects occurring <u>at the nucleus</u> and that valid electronic descriptions are not geared to consider this isolated aspect.

Complexes to be found in Quadrant II include $[Fe(CO)_4SnR_2]^{2-}$ and $[(Fe(CO)_4)_2Zn]^{-2}$. The former complex is found at two places with the same $\Delta E_Q$ values but differing δ values. These correspond to counterions of $(Na(THF)_x)^+$ and $Na(4-MeC_5H_4NO)_{1.4}^+$ (higher δ) and the importance of these interactions has been discussed before. The causes that lead to complexes being found in Quadrant II are to our minds less certain. The temptation is to consider these complexes as having higher than $d^8$-orbital occupancy but the large $\Delta E_Q$ values argue for considerable enhancement of the field gradient. This may be a consequence of population of equatorial $CO_{\pi*}$ orbitals as was postulated[72] for $[Fe(CO)_4SnH_2]^{-2}$. Clearly further examples are needed to ascertain the trends involved.

## π-Complexes of Iron

Ferrocene $[Fe(C_5H_5)_2]$ and its derivatives have been closely studied by Mössbauer spectroscopy.[1,10,75-80] These complexes are found to give $\delta$ values that are temperature independent in the range 0.40-0.55 mm $s^{-1}$ (relative to iron) with large $\Delta E_Q$ between 2.05-2.42 mm $s^{-1}$. Representative examples for $[Fe(C_5H_5)(C_5H_4R)]$ are as follows: R=H($\delta$=0.51) mm $s^{-1}$, $\Delta E_Q$= +2.40 mm $s^{-1}$)[76], Cl (0.51, 2.42)[77], $CH_2OH$ (0.51, 3.39)[77], $[CH_2NMe_3]^+I^-$ (0.51, 2.38)[77], $C_6H_5$ (0.42, 2.34)[77], $CO_2Na$ (0.48, 2.34)[77], $C(CH_3)_3$ (0.44, 2.32)[77], CN (0.41, 2.30)[76], $COCH_3$ (0.45, 2.27)[77], $NO_2$ (0.43, 2.27)[79], and $CO_2H$ (0.44, 2.16).[79] The $\Delta E_Q$ is positive and large on the basis of $d^n$-orbital occupancy (discussed in the section on $\Delta E_Q$). Collins showed[81] that the field gradient of ferrocene was positive and corresponded to the molecular orbital schemes advanced by other workers.[82,83] The bonding in ferrocene is treated in standard texts.[14,15] Cyclopentadienyl being a five atom $6\pi$ aromatic system has five orbitals in a two ($e_2$, two node) over two ($e_1$, one node) over one ($a_1$, no node) pattern. The lowest orbital of a ($\delta$) symmetry has a match with $4s$, $4p_z$ and $4d_{z^2}$ but is too low lying to interact much it being essentially bonding orbitals localized on cyclopentadiene. The two highest $e_2(\pi)$ orbitals similarly match with $3d_{xy}$ and $3d_{x^2-y^2}$ ($\delta$ bonds) but $e_2$ is too high lying to interact much and is essentially $\pi^*$ orbitals localized on cyclopentadiene. However the one node $e_{1g}$ ring orbitals overlap strongly with $3d_{xz}$ and $3d_{yz}$ to form two strong $\pi$-bonds. Thus low $\Delta E_Q$ values are found for electron withdrawing substituent $CO_2H$, $COCH_3$, and $NO_2$ because they favor back donation by lowering ring orbital energies and reduce the field gradient along the z-axis as defined by the centroid of the $C_5H_5$ rings.[83b] The isoelectronic benzene complex $[Fe(C_5H_5)-(C_6H_5R)]^+PF_6^-$ and its substituents may be seen to give totally analogous results.[78]

Ferrocene may be reduced by removal of one electron to give the ferrocenium ion $d^5$-$Fe^{III}$ $[Fe(C_5H_5)]^+$. The Mössbauer spectra of these species consists of a singlet with the following parameters for $[Fe(C_5H_5)(C_5H_4R)]^+X^-$: R=H, X=$BF_4$ ($\delta$=0.58 mm $s^{-1}$, $\Delta$=0 {$\Gamma$=0.61} mm $s^{-1}$); H, Cl (0.54, 0 {$\Gamma$= 0.70}); and H, $BPh_4$ (0.58, 0{$\Gamma$=0.77}).[21]. A variety of complexes based on ferrocene units such as biferrocenyl,[84] bisdicarbollyl iron,[21,75] and mixed valence complexes have been studied.[84-88] Broadening in the ferricenium singlet may be unresolved quadrupole splitting (ligand and not d-orbitally induced) in concert with a magnetic relaxation effect.[21] $\Gamma$ represents peak width at half height.

Mixed cyclopentadienyl iron complexes based on the $[(C_5H_5)-Fe(CO)_2]$ group and its derivatives have been widely studied. For example $[Fe(C_5H_5)(CO)_2]_2$ interconverts through rotamers to cis-trans-isomers. Mössbauer spectroscopy is almost unable to differentiate the two (cis: $\delta=0.21$, $\Delta=1.92$; trans: 0.21, 1.90 mm s$^{-1}$).[89] Mössbauer parameters for representative examples of the class $[Fe(C_5H_5)(CO)LX]$ are as follows: L=PMe$_2$, X=Fe(C$_5$H$_5$(CO)PMe$_2$ (cis - $\delta=0.14$, $\Delta=1.61$; trans = $\delta=0.16$, $\Delta=1.64$ mm s$^{-1}$)[52]; L=PPh$_2$, X=Fe(C$_5$H$_5$)-(CO)PPh$_2$ (cis = 0.17, 1.60; trans = 0.17, 1.66)[28]; CS, [Fe(C$_5$H$_5$)CO-(CS)]$^-$PF$_6^+$ (-0.05, 1.89)[91]; CO, CO]$^+$PF$_6^-$ (0.01, 1.78{note this abbreviates the complex $[Fe(C_5H_5)(CO)_3]^+PF_6^-$})[91]; CO, CN(0.04, 1.96)[91]; CO, COCH$_3$ (0.04, 1.68)[27]; CO, PPh$_3$]$^+$Cl$^-$ (0.05, 1.92)[91]; CO, CONEt (0.06, 1.73)[27]; CO, SnPh$_3$ (0.10, 1.75)[10]; CO, SnCl$_3$ (0.15, 1.80)[10]; CO, py]$^+$PF$_6^-$ (0.15, 1.86)[67]; CO, C$_2$H$_4$]$^+$PF$_6^-$ (0.17, 1.77)[67]; CO, I (0.22, 1.86)[10]; CO, Br (0.24, 1.77)[10] and CO, Cl (0.24, 1.89).[10] Other examples may be found elsewhere[90,92-97] as well as additional examples in the references just quoted. The series of complexes based on $[Fe(C_5H_5)L_1L_2L_3]$ units have been examined[10,93,97] by the point charge model to ascertain $\sigma/\pi$ ligand characteristics. The complexes $[Fe(C_5H_5)(CO)_2X]$ show systematic dependence of $\delta$ with ligands with $\Delta E_Q$ variable unlike other organo-iron complexes where the situation is reversed. Cyclopentadienyl ligand must be instrumental in establishing the field gradient in this case. This might be expected for a ligand which is formally three coordinate. It is noteworthy in this regard that the sign of the field gradient in butadiene iron tricarbonyl is opposite that found for cyclobutadiene iron tricarbonyl.

It is really beyond the scope of this presentation to discuss more than a representative set of examples of organo $\pi$-complexes of iron. The number of complexes studied seem almost limitless including studies of allyl complexes,[98,99] carbene complexes,[100] imidizole derivatives,[101] and other extended $\pi$-systems such as polyolefin (as well as their fluxional aspects) complexes.[1,10] There is a registry kept on computer file at the Mössbauer Effect Data Center at the University of North Carolina (USA) and for iron Mössbauer results to which the reader is referred.

Iron Cyanide Complexes

The last topic we will consider is the application of Mössbauer spectroscopy to cyano-complexes of iron. These complexes fulfill the requirements stated previously in that they are dia-magnetic covalent complexes in a low oxidation state with an iron to carbon bond. This class of complexes has been widely studied and is noted for the ability to form linkage isomers, that is cyanide bound to the metal either at carbon or nitrogen. In general this means that four types of iron atoms are to be found: low spin Fe$^{+2}$ or Fe$^{+3}$ coordinated to carbon and high spin Fe$^{+2}$ or Fe$^{+3}$ coordinated to nitrogen. High spin Fe$^{+2}$ and Fe$^{+3}$ are easy to distinguish and have unique $\delta$ values (0.75 to 1.55 and 0.25 to 0.60 mm s$^{-1}$, respectively) separating one from the other. Low spin $Fe^{+2}$ and $Fe^{+3}$ have nearly identical $\delta$ ranges (-0.20 to 0.40 and -0.30 to 0.40, respectively) and both low spin states overlap entirely with the general range of diamagnetic covalent range (-0.30 to 0.55).[1] This makes Mössbauer characterization often-times difficult. A simplification arises because $d^6$-$Fe^{II}$ [Fe(CN)$_6$]$^{-2}$ is usually a single line spectrum while $d^5$-$Fe^{III}$ [Fe(CN)$_6$]$^{-3}$ shows a large $\Delta E_Q$. In complex high spin-low spin cations of the type [$Fe^{+3}$($Fe^{III}$(CN)$_6$)]$^{-3}$ the complex shape may be resolved by selective isotope $^{57}$Fe enrichment and assignments may be made.[102]

Mössbauer has successfully resolved the question concerning the composition of "Prussian Blue" made from $Fe^{+3}$ and [Fe(CN)$_6$]$^{-4}$ and "Turnbulls Blue" made from $Fe^{+2}$ and [Fe(CN)$_6$]$^{-3}$. Among a variety of possibilities $Fe_4^{+3}$[$Fe^{II}$(CN)$_6$]$_3$, $Fe_3^{+2}Fe^{+3}$[$Fe^{III}$(CN)$_6$]$_3$, etc., or the case of fast electron transfer averaging the environments both spectra are identical and through selective enrichment both may be persuasively argued to be $Fe_4^{+3}$[$Fe^{II}$(CN)$_6$]$_3$.[103]

The cyanides are well treated by a point charge approximation.[1,10,30] Low spin $d^6$-$Fe^{+2}$ [Fe(CN)$_6$]$^{-2}$ salts have $\delta$ values between -0.17 to -0.03 with $\Delta E_Q=0$ with the following as representative examples: H$_4$[Fe(CN)$_6$] ($\delta$ = - 0.14 mm s$^{-1}$, relative to $Fe^0$ at 293°K)[104], Mg[Fe(CN)$_6$] (-0.10)[104], Zr[Fe(CN)$_6$] (-0.16)[104], K$_4$[Fe(CN)$_6$] (-0.03)[105], [Fe(CNH)$_4$(CNBF$_3$)$_2$] (-0.12)[105], and [Fe(CNMe)$_6$] (HSO$_4$)$_2$ (-0.11).[106] The $\delta$ for these complexes increases (decreasing $|\psi_s(0)|$ or oxidation state) as the polarizability of the cation increases. This trend suggest a large "soft" cation enhances delocalization of non-bonding 3d electrons by polarizing CN$^-$ ligands. Removal of O$_h$ symmetry such as is the case in the nitroprusside ion [Fe(CN)$_5$NO]$^{-2}$ with C$_{4v}$ symmetry leads to a large positive field gradient.[107] The point charge model quite successfully predicts that $\pi$-bonding is dominant in setting the value of $\Delta E_Q$ for this class of molecules (low spin $Fe^{+2}$).[1,10,30] For low spin $d^5$-$Fe^{+3}$ the asymmetry of the electronic configuration complicates the analysis. Related isonitrile complexes have also been studied.[106]

ACKNOWLEDGEMENTS

The work described was carried out at Rice University, Department of Chemistry, Houston, Texas 77251 and was supported by the Research Corporation, the Robert A. Welch Foundation, and Dow Corporation. Dr. Robert Shong, Dr. Javan Shelly, Nathan Norem, and Carol O'Rourke were responsible for the experimental work described. During the course of these investigations the author was privileged to have as collaborators Professor Cortlandt Pierpont (University of Colorado, Department of Chemistry), who determined the structures of our complexes, Professor Thomas Albright (University of Houston, Department of Chemistry) who performed theoretical calculations described, and Professor Lon Wilson (Rice University) who provided the facilities each providing meaning criticism along the way. Vice-President John Margrave (Rice University, Advanced Studies and Research) provided computing funds. The author especially wishes to acknowledge his present employer, the Olin Corporation for supplying the support and time necessary to complete this work, and to Sonny Pane for typing the manuscript.

REFERENCES

1. T. C. Gibb, "Principles of Mössbauer Spectroscopy", Wiley, New York (1976).
2. T. C. Gibb and N. N. Greenwood, "Mössbauer Spectroscopy", Chapman and Hall, London (1971).
3. G. M. Bancroft, "Mössbauer Spectroscopy - An Introduction for Inorganic Chemists and Geochemists", McGraw-Hill, London (1973).
4. V. I. Gol'danskii and R. H. Herber, (Eds.),"Chemical Applications of Mössbauer Spectroscopy", Academic Press, New York (1968).
5. R. H. Herber, Scientific American, 225:86 (1971).
6. L. May (Ed.), "Introduction to Mössbauer Spectroscopy", Plenum Press, New York (1971).
7. R. L. Mössbauer, Angew.Chem.Int.Ed., 10:462 (1971).
8. J. G. Stevens and G. K. Shenoy, "Mössbauer Spectroscopy and Its Chemical Applications" in Advances in Chemistry Series, Vol. 194; American Chemical Society, Washington, D.C. (1981).
9. R. S. Drago, "Physical Methods in Chemistry", W. B. Saunders Company, Philadelphia (1977), Chapter 15.
10. R. V. Parish, Mössbauer Spectroscopy, in: "The Organic Chemistry of Iron, Volume 1", Avademic Press, New York (1978).
11. Richard L. Cohen (Ed.), "Applications of Mössbauer Spectroscopy-Vol. 1", Academic Press, Inc., New York (1976); Vol. 2 (1976).
12. U. Gonser (Ed.) "Mössbauer Spectroscopy II", in Topics in Current Physics, Springer-Verlag, New York (1981) and also Vol. 1.
13. I. J. Gruverman and C. W. Seidel (Eds.) "Mössbauer Effect Methodology-Vol. 9", Plenum Press, New York (1974) and all preceding volumes based on these symposia.
14. F. A. Cotton and G. Wilkinson, "Advanced Inorganic Chemistry", 4th Edition, J. Wiley & Sons, New York (1980).
15. J. E. Huheey, "Inorganic Chemistry", 2nd Edition, Harper and Row, New York (1978).
16. E. W. Abel and F.G.A. Stone, Q. Rev., 24:498 (1970).
17. P. S. Braterman, Struct. Bonding, 26:1 (1976).
18. P. S. Braterman, "Metal Carbonyl Spectra", Academic Press, New York (1975).
19. S.F.A. Kettle, Topics in Current Chem., 71:111 (1977).
20. K. Farmery, M. Kilner, R. Greatrex and N. N. Greenwood, J. Chem. Soc. A, (1969): 2339.
21. T. Birchall and I. Drummond, Inorg. Chem., 10:399 (1971).

22.  M. G. Clark, Mol. Phys., 20:257 (1971).

23.  G. M. Bancroft, M. J. Mays and B. E. Prater, Discuss. Faraday
     Soc., 47:136 (1969); J. Chem. Soc. A, (1970): 956.

24.  G. M. Bancroft, R.E.B. Garrod, A. G. Maddock, M. J. Mays, and
     B. E. Prater, J. Amer. Chem. Soc., 94:647 (1972).

25.  G. M. Bancroft, R.E.B. Garrod, and A. G. Maddock, J. Chem. Soc.
     (A), (1971): 3165.

26.  G. M. Bancroft and E. T. Libbey, J.C.S. Dalton, (1973): 2103.

27.  R. Herber, R. B. King and G. K. Wertheim, Inorg. Chem., 3:101
     (1964).

28.  R. Herber and R. G. Hayter, J. Am. Chem. Soc., 86:301 (1964).

29.  G. M. Bancroft, Chem. Phys. Letters, 10:449 (1971).

30.  G. M. Bancroft, Coord. Chem. Rev., 11:247 (1973) and references
     therein.

31.  G. M. Bancroft, K. D. Butler and E. T. Libbey, J.C.S. Dalton,
     (1972): 2643.

32.  G. M. Bancroft, H. C. Clark, R. G. Kidd, A. T. Rake, and H. G.
     Spinney, Inorg. Chem., 12:728 (1973).

33.  R. A. Mazak and R. L. Collins, J. Chem. Phys., 51:3220 (1969).

34.  A. R. Rossi and R. Hoffmann, Inorg. Chem., 14:365 (1975).

35.  R. L. Collins and R. Pettit, J. Chem. Phys., 39:3433 (1963).

36.  W. E. Carrol, F. A. Deeney, J. A. Delaney, and F. J. Lalor,
     J. Chem. Soc. (Dalton), (1973): 718.

37.  J. Ensling, P. Gutlich, and L. Rösch, Z. Naturforsch, 30b: 850
     (1975).

38.  H. Mosbaek, Acta.Chem.Scand.A, 29:957 (1975).

39.  B. A. Sosinsky, N. Norem, and R. G. Shong, Inorg. Chem., 21:
     4229 (1982).

40.  M. Elian and R. Hoffmann, Inorg. Chem., 14:1058 (1975).

41.  R. H. Herber, W. R. Kingston, and G. K. Wertheim, Inorg. Chem.,
     2:153 (1963).

42.  P. Vasudev and C.H.W. Jones, Can. J. Chem., 51:405 (1973).

43.  R. Robinette and R. L. Collins, J. Chem. Phys., 57:4319 (1972).

44.  G. M. Bancroft and E. T. Libbey, J. Chem. Soc. (Dalton), (1973):
     2103.

45.  C. G. Pribula, T. L. Brown, and E. Munck, J. Am. Chem. Soc.,
     96:4149 (1974).

46.  K. Welgehausen, M. L. Rudee, and R. B. McLellan, Acta Metall.,
     21:589 (1973).

47.  P. Kienle, Phys. Verh., 3:33 (1963).

48.  M. G. Clark, W. R. Cullen, R.E.B. Garrod, A. G. Maddock, and
     J. R. Sams, Inorg. Chem., 12:1045 (1973).

49.  R. A. Mazak and R. L. Collins, J. Chem. Phys., 51:3220 (1969).

50.  A. Trautwein and F. E. Harris, Theoret. Chim. Acta., 30:45
     (1978).

51.  P. Kuhn, U. Hauser, and W. Neuwirth, Z. Physik, 264:287 (1973).

52.  T. C. Gibb, R. Greatrex, N. N. Greenwood, and D. T. Thompson,
     J. Chem. Soc. (A), (1967): 1663.

53.  L. F. Dahl and R. E. Rundle, J. Chem. Phys., 26:1751 (1957).

54.  C. H. Wei and L. F. Dahl, J. Am. Chem. Soc., 88:1821 (1966).

55. B. A. Sosinsky, N. Norem and J. Shelly, Inorg. Chem., 21:348 (1982).

56. R. P. Brint, K. O'Cuill, T. R. Spalding and F. A. Deeney, J. Organometal. Chem., 247:61 (1983).

57. C. G. Pierpont, B. A. Sosinsky, and R. G. Shong, Inorg. Chem., 21:3248 (1982).

58. B. A. Sosinsky, R. G. Shong, B. J. Fitzgerald, N. Norem, and C. O'Rourke, Inorg. Chem., 22:3124 (1983).

59. See Reference 8, Chapter 19, page 405, by P. Gutlich; references contained therein.

60. E. I. Ochiai, "Bioinorganic Chemistry - An Introduction", Allyn and Bacon, Inc., Boston (1977).

61. A.T.T. Hsieh and M. J. Mays, MTP Int. Rev. Sci. Inorg. Chem., Series One, 6:43 (1972).

62. R. J. Neustadt, T. H. Cymbaluk, R. D. Ernst, and F. W. Cagle, Inorg. Chem., 19:2375 (1980).

63. R. D. Ernst, T. J. Marks, and J. A. Ibers, J. Am. Chem. Soc., 99:2090 (1977). ibid 2098.

64. A.T.T. Hsieh, M. J. Mays and R. H. Platt, J. Chem. Soc. (A), (1971): 3296.

65. T. Takano and Y. Sasaki, Bull. Chem. Soc. Japan, 44:431 (1971).

66. N. Dominelli, E. Wood, P. Vasudev and C.H.W. Jones, Inorg. Nucl. Chem. Lett., 8:1077 (1972).

67. G. M. Bancroft, K. D. Butler, L. E. Manzer, A. Shaver, and J.E.H. Ward, Canad. J. Chem., 52:782 (1974).

68. M. T. Jones, Inorg. Chem., 6:1249 (1967).

69. V. I. Gol'danskii, B. V. Borshagovskii, E. F. Makarov, R. A. Stukan, K. N. Anisimov, N. E. Kolobova, and V. V. Skripkin, Theor. Eksp. Khim., 2:126 (1966).

70. R. J. Dickinson, R. V. Parish, P. J. Rowbotham, A. R. Manning, P. Hackett, J.C.S. Dalton, (1975): 424.

71. B. A. Sosinsky, J. Shelly and R. Shong, Inorg. Chem., 20:1370 (1981).

72. J. Silvestre, T. A. Albright, and B. A. Sosinsky, Inorg. Chem., 20:3937 (1981).

73. R. Bau (Ed.), "Transition Metal Hydrides", Adv. Chem. Ser., American Chemical Society, Washington, D.C. (1978).

74. H. B. Chin and R. Bau, J. Am. Chem. Soc., 98:2434 (1976).

75. R. H. Herber, Inorg. Chem., 8:174 (1969).

76. A. V. Lesikar, J. Chem. Phys., 40:2746 (1964).

77. R. A. Stukan, S. P. Gubin, A. N. Nesmeyanov, V. I. Gol'danskii and E. F. Makarov, Theor. Eksp. Khim., 7:486 (1971); Theor. Exper. Chem., 2:581 (1966).

78. K. I. Turta, R. A. Stukan, V. I. Gol'danskii, N. A. Vol'Kenan, E. I. Sirotkina, I. N. Isaeva, and N. N. Nesmeyanov, Theor. Eksp. Khim., 7:486 (1971); Theor. Exper. Chem., 7:401 (1971).

79. G. K. Wertheim and R. H. Herber, J. Chem. Phys., 38:2106 (1963).

80. U. Zahn, P. Kienle and H. Eicher, Z. Phys., 166:220 (1962).

81. R. L. Collins, J. Chem. Phys., 42:1072 (1965).

82. L. F. Dahl and C. F. Baulhausen, Mat. Fys. Medd. Dan. Vid. Selsk., 33:5 (1961).

83. E. M. Shustorovich and Dyatkina, Dokl. Akad. Nauk. SSSR, 128: 1234 (1959).

83b. D. O. Cowan, R. L. Collins, G. A. Landela, U. T. Muelter-Westerhoef, and P. Eilbracht, J.C.S. Chem. Comm., (1973): 329.

84. D. O. Cowan, R. L. Collins, and F. Kaufman, J. Phys. Chem., 75: 2025 (1971).

85. W. H. Morrison, Jr. and D. N. Hendrickson, Chem. Phys. Letts., 22:119 (1973).

86. W. H. Morrison, Jr. and D. N. Hendrickson, J. Chem. Phys., 59: 380 (1973).

87. L. A. Aliev, T. P. Vishnakov, Ya. M. Paushkin, A. A. Pendin, T. A. Sokolinskaya, and R. A. Stukan, Izv. Akad. Nauk. SSSR, Ser. Khim., (1970): 306; Bull. Acad. Sci. USSR, Div. Chem. Sci., (1970): 256.

88. S. G. Baxter, R. L. Collins, A. H. Cowley, and S. F. Sena, J. Am. Chem. Soc., 103:714 (1981).

89. R. F. Bryan, P. T. Greene, M. J. Newlands, and D. S. Fields, J. Chem. Soc. (A), (1970): 3068.

90. A. J. Carty, A. Efraty, T. W. Ng and T. Birchall, Inorg. Chem., 9:1263 (1970).

91. K. Burger, L. Korecz, P. Mag, U. Belluco and L. Busetto, Inorg. Chim. Acta., 5:362 (1971).

92. R. B. King, L. M. Epstein, and E. W. Crowling, J. Inorg. Nucl. Chem., 32:441 (1970).

93. B. Johnson, P. J. Ouseph, J. S. Hsieh, A. L. Steinmetz, and J. E. Shade, Inorg. Chem., 18:1796 (1979).

94. G. L. Long, D. G. Alway and K. W. Barnett, Inorg. Chem., 17: 486 (1978).

95. M. J. Mays and P. L. Sears, J.C.S. Dalton, (1973): 1973.

96. A. J. Carty, T. W. Ng, W. Carter, G. J. Palenik, and T. Birchall, Chem. Comm., (1969): 1101.

97. R. H. Herber, Prog. Inorg. Chem., 8:1 (1967).

98. H. L. Clark and N. J. Fitzpatrick, J. Organomet. Chem., 66:119 (1974).

99. B. C. Parakkat, P. J. Ouseph, R. L. Vonnahme, and D. H. Gibson, J. Inorg. Nucl. Chem., 37:2342 (1975).

100. J. Pebler and W. Petz, Z. Naturforsch, 29b:658 (1974).

101. F. Seel, R. Lehnert, E. Bill, and A. Trautwein, Z. Naturforsch, 35b:631 (1980).

102. K. Maer, M. L. Beasley, R. L. Collins and W. O. Milligan, J. Am. Chem. Soc., 90:3201 (1968).

103. W. Zerler, W. Neuwirth, E. Fluck, P. Kuhn, and B. Zimmermann, Z. Physik, 173:321 (1963).

104. B. V. Borshagovskii, V. I. Gol'danskii, G. B. Seifer, and R. A. Stukan, Izv. Akad. Nauk. SSSR, Ser. Khim., (1968): 87; Bull. Acad. Sci. USSR, Div. Chem. Sci., (1968): 81.

105. D. Hall, J. H. Slater, B. W. Fitzsimmons and K. Wade, J. Chem. Soc. (A), (1966): 162.

106. R. R. Berrett and B. W. Fitzsimmons, <u>J. Chem. Soc. (A)</u>, (1967): 525.
107. W. Kerler, <u>Z. Phys.</u>, 167:194 (1962).
108. G. M. Bancroft and K. D. Butler, <u>J.C.S. Dalton</u>, (1972), 1209.
109. G. M. Bancroft, M. J. Mays and B. E. Prater, <u>J. Chem. Soc. (A)</u>, (1970): 956.

# SPIN TRANSITION IN IRON COMPOUNDS

Philipp Gütlich

Institut für Anorganische und Analytische Chemie
Johannes-Gutenberg-Universität
D-6500 Mainz, West Germany

## 1.    INTRODUCTION

First-row transition metal complexes with $d^4$ up to $d^7$ electron configuration in octahedral ligand fields (and $d^8$ electron configuration in six-coordinate complexes of lower symmetry) may undergo temperature dependent high spin (HS) $\rightleftharpoons$ low spin (LS) transition, provided the ligand field strength ($\Delta$), including low symmetry contributions, becomes comparable in magnitude with the mean spin pairing energy (P). At a critical field strength $\Delta_{crit}$ = P, the energy levels of the two spin states cross. This is illustrated in Fig. 1, where a simplified Tanabe-Sugano type energy level diagram for $d^6$ systems (e.g. $Fe^{2+}$, $Co^{3+}$) is shown as an example. For weak ligand fields, $\Delta < \Delta_{crit}$, the HS state $^5T_{2g}(O_h)$ is the ground state; for strong ligand fields, $\Delta > \Delta_{crit}$, the LS state $^1A_{1g}(O_h)$ becomes more stable. If the difference $|\Delta-P|$, or, in terms of thermodynamics, the difference in Gibbs free energy $\Delta G=G(HS)-G(LS)$, is on the order of thermal energy kT, both spin states HS and LS may be populated in thermal equilibrium.

The phenomenon of thermally induced spin transition (spin crossover, spin state equilibrium) has first been discovered by Cambi et al.[1] in ferric tris(dithiocarbamato) complexes. Later on it has mostly been observed for complex compounds of $Fe^{3+}(d^5)$, $Fe^{2+}(d^6)$, $Co^{2+}(d^7)$, and to a lesser extent of $Ni^{2+}(d^8)$. The work has been reviewed in a number of articles; see e.g. refs.2-5. In fewer cases temperature dependent spin transition has been reported for complexes of $Mn^{3+}(d^4)$,[6] $Mn^{2+}(d^5)$,[7] $Co^{3+}(d^6)$,[8,9] for inorganic compounds of $Fe^{2+}$ [10] or $Co^{3+}$,[11] and for niobium cluster compounds.[12]

27

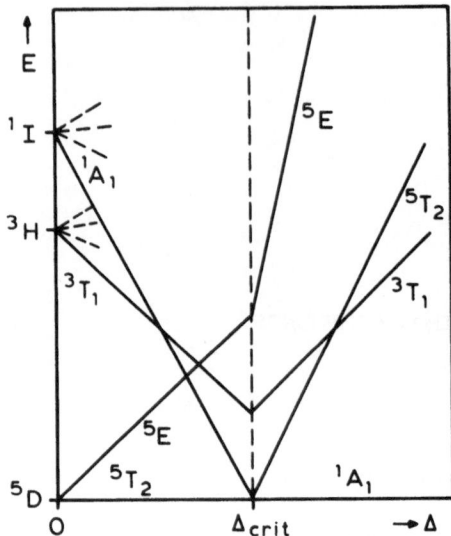

Fig. 1. Simplified (Tanabe-Sugano type) energy level diagram for
$d^6$ ions in octahedral ligand fields. Only the low-energy
Russell Saunders terms and their schematic splittings under $O_h$ symmetry are shown. The electronic ground state is
$^5T_2(O_h)$ for ligand field strengths $\Delta < \Delta_{crit}$, and $^1A_1(O_h)$
for $\Delta > \Delta_{crit}$. Spin transition (spin crossover) between
the high spin (HS) and the low spin (LS) states,
$^5T_2(O_h) \rightleftharpoons {}^1A_1(O_h)$, occurs if the critical field strength
$\Delta_{crit}$ equals the mean spin pairing energy P.


Most of the spin crossover studies have so far been done on
ferrous complexes; the results reported up to 1981 have been reviewed in ref.5. The vast majority of ferrous spin crossover systems
are six-coordinate complexes with nitrogen donor ligands, mostly bis-
and trisdentate but also mono-, tetra- and hexadentate ligands. In
ferric spin crossover complexes, where the contribution to the ligand field strength from the central ferric ion is a priori larger
than in case of ferrous complexes,[13] the total critical field strength
$\Delta_{crit}$ is set up by somewhat weaker ligands forming, e.g., $FeS_6$ cores
as in tris(dithiocarbamates)[1] or $FeN_4O_2$ cores as reported in refs.
14-16.

Temperature dependent spin transition takes place in solids as
well as in the liquid state. In both cases, spin state changes are
easily detected by following the magnetic susceptibility as a function of temperature, $\chi(T)$. The effective magnetic moment, $\mu_{eff}(T)$,
or the fraction of molecules in one spin state, say $\gamma_{HS}(T)$, derived
from $\chi(T)$ for spin transition in solution is generally gradual and
follows a Boltzmann distribution over the vibronic levels of the LS

and the HS states. The change of spin multiplicity results from an intramolecular electron transfer between the metal-localized valence orbitals (in $O_h$ symmetry),

$$(t_{2g})^n (e_g^*)^m \;\rightleftharpoons\; (t_{2g})^{n-x} (e_g^*)^{n+x} \qquad (1)$$

(LS state)          (HS state)

The spin transition behaviour in solution is generally not greatly affected by interactions with the solvent.[9] In solids, however, interactions with the lattice appear to be of great importance, which is the reason for the fact that the spin transition curves $\mu_{eff}(T)$ or $\gamma_{HS}(T)$ come in a variety of types: discontinuous or gradual, with or without hysteresis, complete or incomplete spin state conversion with residual fractions of HS molecules at low temperatures and of LS molecules at high temperatures, respectively. To study the origin of such interactions has been one of our primary goals in recent years.

The molecular properties such as bond lengths, bond forces, vibrational frequencies, magnetism and thermodynamic quantities change spontaneously with an intramolecular electron transfer according to (1) and thereby going from one spin state to another. This is schematically illustrated in Fig. 2. The double-potential-well is composed of the parabolic functions (harmonic approximation) for a characteristic stretching vibrational mode in the LS state (of effective frequency $\omega_{LS}$) and in the HS state ($\omega_{HS}$), respectively. Electron transfer from $t_{2g}$ to antibonding $e_g^*$ orbitals weakens the metal-ligand bond strength and, along with it, lengthens the bond distance. Therefore, $r_{LS} < r_{HS}$ and $\omega_{LS} > \omega_{HS}$, which has been confirmed for several spin crossover systems by single crystal X-ray diffraction and farinfrared spectroscopy at variable temperatures; see e.g. the compilation in ref. 5. The electronic energies in the two spin states are also different, $E_{LS} < E_{HS}$, which can be stated by optical spectroscopy in connection with ligand field theory.[13] The difference between the zero-point vibronic energies is $\Delta E_O = E_{HS} - E_{LS} + \frac{1}{2}\hbar\,\Sigma(\omega_{HS}^i - \omega_{LS}^i)$ and corresponds, at T = 0 K, to the enthalpy difference $\Delta H_O = H_{HS} - H_{LS} > 0$. At T > 0, the entropy term $T\Delta S$ in the Gibbs free energy difference $\Delta G = G_{HS} - G_{LS} = \Delta H - T\Delta S$ becomes an increasingly important factor, and it can be shown that the spin state conversion from LS to HS is mainly entropy driven, because both the magnetic contribution $S_{mag} = k\,\ln(2S+1)$ and the vibrational contribution $S_{vib}$ to the total entropy of a complex molecule are larger in the HS state than in the LS state. Precise heat capacity measurements have been successfully applied to spin crossover complexes in order to evaluate the changes in thermodynamic parameters as well as to detect eventually whether there is latent heat on passing through the transition temperature $T_c$ in case of first-order nature transitions.[17]

Spin transitions in iron complexes are most elegantly followed by Mössbauer spectroscopy. The different electronic structures of

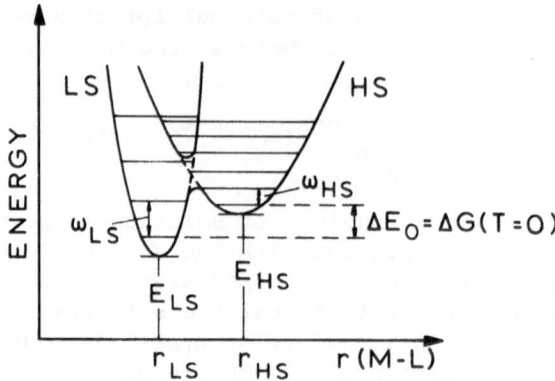

Fig. 2. Schematic potential energy diagram for a spin crossover
compound. The potential energy is shown for the HS state
and the LS state as a function of the metal-ligand bond
length r(M-L). Note that $r_{LS} < r_{HS}$, the electronic energies
$E_{LS} < E_{HS}$, and the stretching vibrational frequencies
$\omega_{LS} > \omega_{HS}$. $\Delta E_O = \Delta G(T=0)$ is the difference of the zero-point
vibronic energies or the difference of the Gibbs free ener-
gies $G^O_{HS} - G^O_{LS}$ of the two spin states.

the HS and LS states cause different electric field gradients and
electron densities at the iron nucleus; therefore, the Mössbauer
spectra of HS and LS spin states (in a certain oxidation state)
differ significantly in the quadrupole splitting and the isomer shift.
As an example, Fig. 3 shows some representative variable-temperature Mössbauer spectra of $[Fe(phen)_2(NCS)_2]$ (phen = 1,10-phenan-
throline), which is the first ferrous spin crossover complex reported in the literature.[18] Of course, resolved spectra characteristic
of different spin states as shown in Fig. 3 are only observable if
the spin state relaxation times $\tau_{H,L}$ (the reciprocal of the spin
state interconversion rates) are longer than the quadrupolar precession time $\tau_Q$ in the nuclear excited state. This is the case in
nearly all ferrous spin crossover systems reported so far; an
exception is $Fe_xTa_{1-x}S_2$ studied by Eibschütz et al.,[19] where
$\tau_{H,L} < \tau_Q$ and only one Mössbauer resonance is observed arising from
a population-averaged electronic structure. In most ferric spin
crossover compounds examined so far the spin state interconversion
is fast ($\tau_{H,L} < \tau_Q$), as e.g. in the tris(dithiocarbamato) complexes,
which has been ascribed to strong spin-orbit coupling effects, and
separate Mössbauer spectra cannot be observed. An early example of
a slowly relaxing ferric spin crossover compound with $FeS_3O_3$ chromophore was published by Cox et al..[20] Hendrickson et al.[163] have recently measured Mössbauer and EPR spectra of a ferric spin crossover
compound with $N_4O_2$ core, which, depending on the preparation method,
shows spin state relaxation times extending from $\sim 10^{-7}$s (with separate HS and LS Mössbauer spectra) up to $\sim 10^{-10}$s (derived from separate EPR signals). Spin state relaxation times ranging from $10^{-6}$ to

30

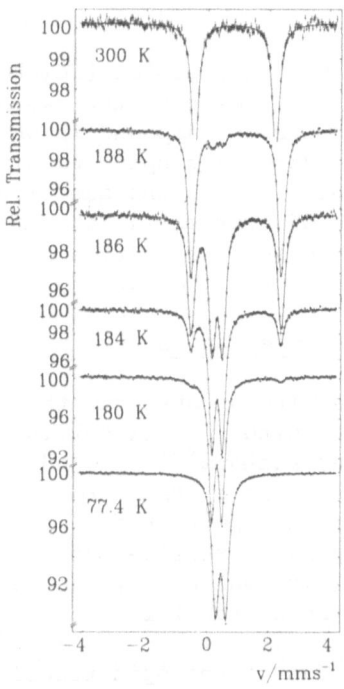

Fig. 3. Mössbauer spectra of [Fe(phen)$_2$(NCS)$_2$] as a function of temperature. The quadrupole doublet with the large splitting, which dominates in the high temperature region, refers to the HS state, $^5T_2(O_h)$; the doublet with the small splitting energy, whose intensity increases with decreasing temperature, refers to the LS state, $^1A_1(O_h)$, of iron(II).[19]

$10^{-8}$ s have been measured in solution using relaxation techniques (laser Raman T-jump[21,22] and pulsed ultrasonic perturbation,[23] and it has been found that both structural and electronic factors are likely to influence the relaxation times. Sutin[24] has shown that spin state interconversion rates may be evaluated taking a nuclear, and electronic, and a frequency factor into account.

Mössbauer spectroscopy has proven to be an exceptionally powerful tool to study the various influences on the spin transition behaviour in iron complexes, e.g. the effect of intraligand substitution, of ligand replacement, of anion replacement, of metal dilution, of H/D or other isotopic exchange, of applying pressure or magnetic fields. Examples are given in the review of ref. 5.

In the following we shall report on more recent studies of spin transitions in iron complexes, mainly employing Mössbauer spectroscopy and magnetic susceptibility measurements. Emphasis will be placed on experiments which have been conducted to learn more about the mechanism and the interactions involved in spin transitions in solids.

## 2. $[Fe(phen)_2(NCS)_2]$

The neutral six-coordinate complex $[Fe(phen)_2(NCS)_2]$ with six nitrogen donor atoms in $C_{2v}$ symmetry (the two $NCS^-$ groups in cis-positions) is the first ferrous compound in which Baker and Bobonich[18] discovered a temperature dependent spin transition twenty years ago. Special problems, however, such as the question about structural phase change, occurrence of hysteresis, the effect of metal dilution and crystal quality on the spin transition behaviour have led us to reexamine this system with great care.

### 2.1 Spin Transition Characteristics

Baker et al.[18] have already observed that the spin transition is very abrupt and occurs within a few Kelvin around 175 K, but these, and later on other authors[25,26] found no indication for the occurrence of a hysteresis and a structural phase change. Only recently one has seen different peak profile patterns in X-ray diffraction measurements on a polycrystalline material above and below the transition temperature $T_c$,[27] which calls for a structural phase change accompanying the spin transition.

The first-order nature of the spin transition in $[Fe(phen)_2(NCS)_2]$, which has been established previously by heat capacity measurements,[17] has been further confirmed by the detection of a hysteresis in the $\mu_{eff}(T)$ function of only ca. 0.2 K width;[28] cf. Fig. 5.

The quadrupole splitting QS(T) and the resonance line width $\Gamma(T)$ as a function of temperature evaluated from the Mössbauer resonance doublet of the HS state of $[Fe(phen)_2(NCS)_2]$ show discontinuities near the transition temperature $T_c$ (s. Fig. 4); this is another indication for a structural change going along with the spin transition.[27]

### 2.2 Effect of Crystal Quality

Casey et al.[26] have already observed that the spin transition characteristics, as reflected by the $\mu_{eff}(T)$ curve, may change significantly with the sample preparation method. In a later study of the effect of crystal quality on the spin transition behaviour in $[Fe(phen)_2(NCS)_2]$ we could confirm Casey's observation, cf. Fig. 5. A (slow) extraction method yielded relatively good crystals with a sharp spin transition and no HS fraction at low temperatures. A (rapid) precipitation method resulted in a poor quality material with much smaller particles; the spin transition is still rather abrupt, but shows a considerable HS fraction at low temperatures. Grinding the good material from the extraction method yields a $\mu_{eff}(T)$ curve with transition features (steepness and residual HS fraction) somewhat in between the extracted and the precipitated samples.

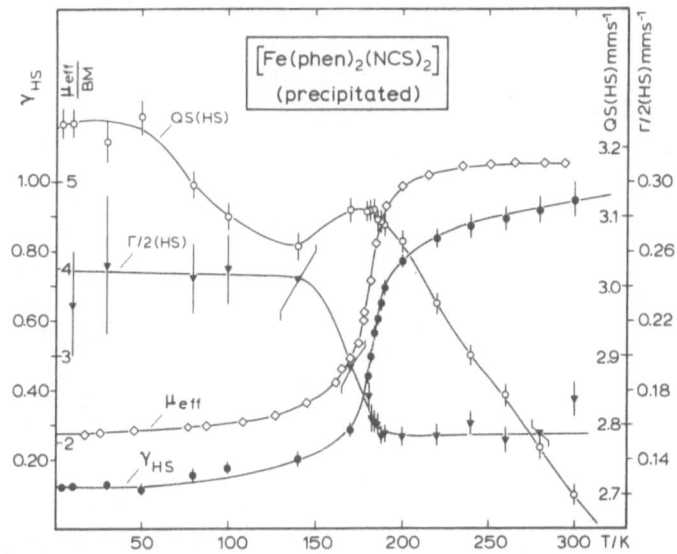

Fig. 4. Quadrupole splitting (QS, O), linewidth ($\Gamma/2$,▼), area fraction ($x_{HS}$, ●), and effective magnetic moment ($\mu_{eff}$,◇) of the high spin state in $[Fe(phen)_2(NCS)_2]$ (precipitated) as a function of temperature.[27]

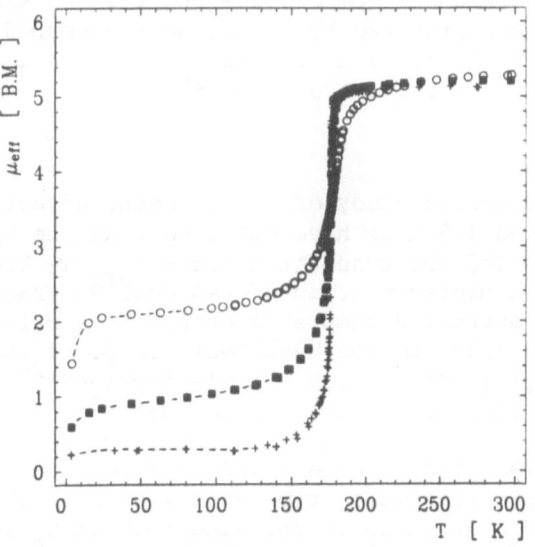

Fig. 5. Effective magnetic moment of different samples of $[Fe(phen)_2(NCS)_2]$ as a function of temperature: + "extracted" no treatment; ■ "extracted", 6 h ball mill; o "precipitated".[28]

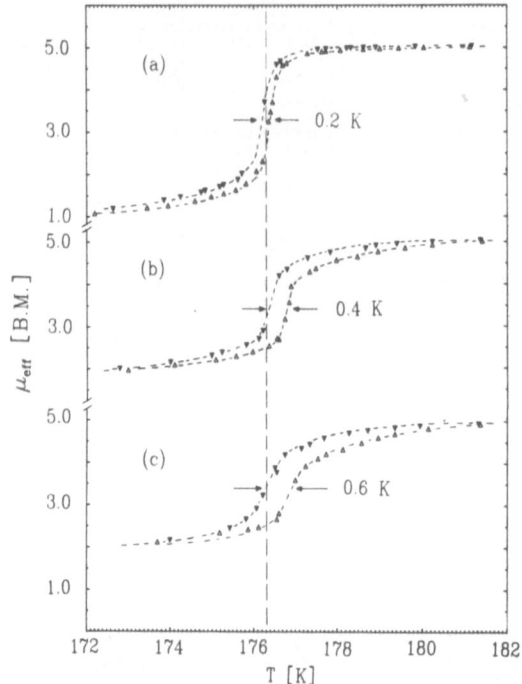

Fig. 6. Effective magnetic moment $\mu_{eff}(T)$ of $[Fe(phen)_2(NCS)_2]$ as a function of temperature and various sample treatments: (a) sample prepared by an extraction method,[33] no grinding; (b) grinding for 1 h, and (c) for 6 h in a ball mill (petroleum ether suspension).[28]

In a very detailed study of the grinding effect in the cross-over region around 175 K we have observed that the hysteresis (s. Fig. 6) broadens and the transition curve $\mu_{eff}(T)$ becomes more gradual and merges into a higher plateau of residual HS fraction at low tempe-ratures, if the extracted sample is ground in a ball mill. Similar grinding effects occur to the hysteresis loops of the spin crossover systems $[Fe(bipy)_2(NCS)_2]$ (bipy = $\alpha,\alpha'$-bipyridyl)[28] and $[Fe(bt)_2(NCS)_2]$ (bt = 2,2'-bi-2-thiazoline).[29]

In an earlier study of the grinding effect on a ferric complex of the abrupt spin transition type Hendrickson et al.[30] have already observed a decrease in the steepness at $T_c$ and an increase of the residual HS fraction upon grinding, and it appears that this is a common phenomenon with discontinuous spin transition systems. It is believed, and there is support from X-ray studies,[31] that the grinding effect is predominantly due to the incorporation of defects into the crystallites and to a lesser extent to a reduction of the particle size. These findings, together with results from studies of the metal dilution effect, have led Hendrickson et al.[30]

to suggest that the "nucleation and growth" mechanism[32] may be considered for spin transition in solids.

## 2.3  Effect of Metal Dilution

The fact that the molecular dimension changes abruptly with the spin transition (the change in the metal ligand bond length is $r_{HS} - r_{LS} \lesssim 0.2$ Å in ferrous spin crossover compounds[34]) has led us to conclude that a spin state transition in a solid should induce drastic changes in the vibrational and elastic properties, and we have conducted several series of experiments to prove this. Among these are the studies of the isostructural metal dilution in $[Fe_x M_{1-x}(phen)_2(NCS)_2]$[27] and other ferrous mixed-crystal systems (s. below).

Some representative $^{57}$Fe Mössbauer spectra of the systems $[Fe_x M_{1-x}(phen)_2(NCS)_2]$ with M = Mn, Co are shown in Figs. 7 and 8, respectively. The spectra demonstrate clearly that, in the host lattice with M = Mn, the ferrous HS state is favored with decreasing iron concentration, and in the case of M = Co the LS state stabilizes more and more with decreasing x. It could be shown,[27] that the spin transition tends to disappear totally in the limit of infinite dilution, with all the isolated $[Fe(phen)_2(NCS)_2]$ molecules embedded in the manganese host stabilizing in the $^5T_{2g}$ $(O_h)$ ground state, and those embedded in the cobalt host stabilizing in the $^1A_{1g}$ $(O_h)$ state. These observations may be qualitatively understood on the basis of the different ionic radii: $Mn^{2+}_{HS}(0.82$ Å$)$ > $Fe^{2+}_{HS}(0.78$ Å$)$ > $Co^{2+}_{HS}(0.74$ Å$)$ > $Fe^{2+}_{LS}(0.61$ Å$)$.[35] If $r(M^{2+})_{HS} > r(Fe^{2+})_{HS}$ a kind of negative lattice pressure acts on the $[FeN_6]$ core and causes the iron-nitrogen bond lengths to increase abruptly and the crystal field strength to decrease following the rule $\Delta \sim (1/r)^5$,[13] which increases the tendency to stabilize the HS state. If $r(M^{2+}) < r(Fe^{2+}_{HS})$, a positive pressure acting on the $[FeN_6]$ core forces the Fe-N bond length to decrease, and the concomitant increase in the crystal field strength favors the LS state.

We are not yet in the position to describe the metal dilution effect in $[Fe_x M_{1-x}(phen)_2(NCS)_2]$ quantitatively.

## 2.4  Effect of a Magnetic Field on the Spin Transition

Only one report has been published so far on the influence of a magnetic field on the spin transition.[36] $[Fe(phen)_2(NCS)_2]$ was placed in magnetic fields of 1 T and 5.5 T, respectively, and the hysteresis loop of the temperature dependence of the effective magnetic moment was determined by magnetic susceptibility measurements. The hysteresis loop is shifted by 0.11 K to lower temperatures on going from 1 T to 5.5 T, indicating a slightly higher stability of the HS state in the stronger magnetic field. The calculated relative shift in the transition temperature is -0.115 K which is in very good agreement with the experiment.

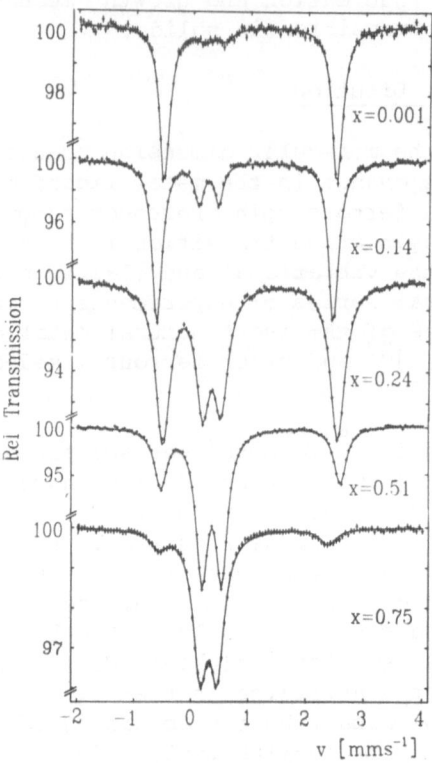

Fig. 7. Mössbauer spectra of $[Fe_xMn_{1-x}(phen)_2(NCS)_2]$ at 5 K and various iron concentrations. The spectra demonstrate that the intensity of the iron(II) HS doublet (outer two lines) increases with decreasing iron concentration.

## 3. $[Fe(2-pic)_3]X_2 \cdot Sol$

Renovitch and Baker have first reported on the existence of a temperature dependent HS $\rightleftharpoons$ LS transition in the $[Fe(2-pic)_3]X_2$ salts (2-pic = 2-aminomethylpyridine = 2-picolylamine; X = $Cl^-$, $Br^-$, $I^-$).[37]

We have examined the spin transition characteristics in some of these compounds in great detail and under various conditions by Mössbauer and magnetic susceptibility measurements. The results obtained from certain "key experiments" such as metal dilution and application of pressure support a recently developed model of elastic interactions suited for the interpretation of spin transition in solids.[38]

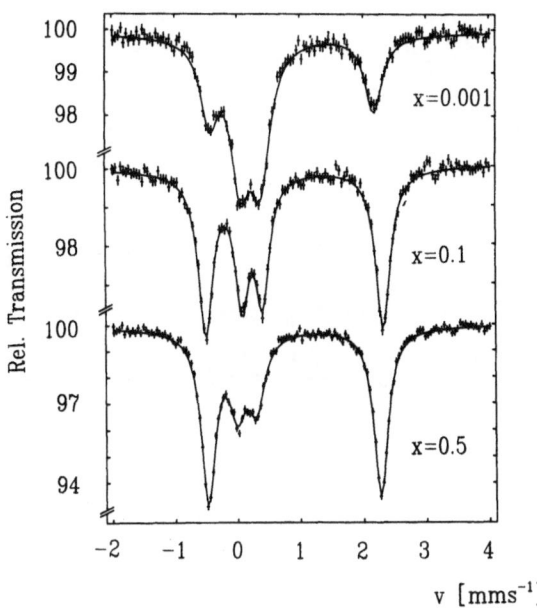

Fig. 8. $^{57}$Fe Mössbauer spectra of $[Fe_xCo_{1-x}(phen)_2(NCS)_2]$ at 260 K and various iron concentrations. The spectra demonstrate that the intensity of the iron(II) LS doublet (inner two lines) increases with decreasing iron concentration.

## 3.1 Unusual Spin Transition Properties of $[Fe(2-pic)_3]Cl_2 \cdot EtOH$

The variable temperature Mössbauer spectra shown in Fig. 9 prove the coexistence of the two spin states, HS and LS in $[Fe(2-pic)_3]Cl_2 \cdot EtOH$ with relaxation times $\tau_{H,L} > \tau_Q$ (reciprocal of quadrupolar precession frequency). The fraction of HS molecules $\gamma_{HS}(T)$ evaluated from the area under the resonance lines and plotted as a function of temperature (s. Fig. 10) exhibits an unusual anomaly in the crossover region which has never been observed for any other spin transition system.[39] It appears that the spin transition takes place in two steps with two transition temperatures, as determined by plotting the first derivative $d\gamma_{HS}(T)/dT$ versus T: $T_c(HS) = 121$ K in the region where $\gamma_{HS} > 0.5$ and thus the lattice properties are governed by the HS molecules; $T_c(LS) = 114$ K in the region where $\gamma_{HS} < 0.5$ and the lattice properties are dominated by the LS molecules. This spin transition anomaly has been reproduced in every detail by magnetic susceptibility measurements in our laboratory.[40] In fact, the two transition curves $\gamma_{HS}(T)$, obtained from Mössbauer and $\chi(T)$ measurements, respectively, are identical within the experimental errors. This, incidentally, proves that the Debye-Waller factors of the HS and LS phases are practically equal, $f_{HS}(T) \approx f_{LS}(T)$, at all temperatures. König and coworkers have analysed the Debye-Waller factors in other spin crossover systems and have always found

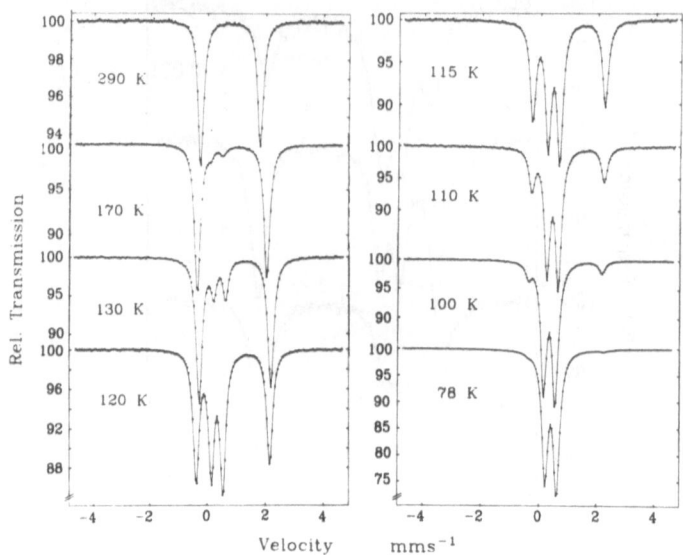

Fig. 9.  $^{57}$Fe Mössbauer spectra of [Fe(2-pic)$_3$]Cl$_2$·EtOH as a function of temperature (source: $^{57}$Co/Rh at 293 K). The outer two lines are the quadrupole doublet of the high-spin (HS) state, the inner two lines are the quadrupole doublet of the low spin (LS) state of iron(II) (from ref. 39).

a somewhat larger value for the LS phase.[41-45]

According to single-crystal X-ray studies at temperatures well above and well below the critical transition region by Mikami et al.[34] the space group for the HS and the LS phase is the same, viz. P2$_1$/c. Thus, a structural phase transition does not seem to accompany the spin transition and may therefore not be considered responsible for the occurrence of this unusual "two-step" spin transition.

In an attempt to explore the origin of this phenomenon, Köppen[46] has considered the bulk modulus K(T)$\sim$$\theta_D^2$(T) within the Debye model. Evaluating $\Theta_D$(T) from the Mössbauer resonance effects $\varepsilon$(T) and inserting into the expression for K(T) he arrives at the results plotted in Fig. 11: the modulus K(T), which is a measure of the compressibility of the lattice, increases nearly linearly with falling temperature (about 20% between 300 and 130 K). In the crossover region, however, where the fraction of HS molecules changes drastically, K(T) decreases to values well below the extrapolated normal function (dashed lines). This lends support to the conclusion that the lattice "softens" somewhat as a consequence of the change in the molecular dimension (r$_{HS}$-r$_{LS}$ $\approx$ 0.18 Å) accompanying the spin transition. This is supported by the observation that the "two-step" spin transition structure disappears gradually by substituting iron for zinc (s.

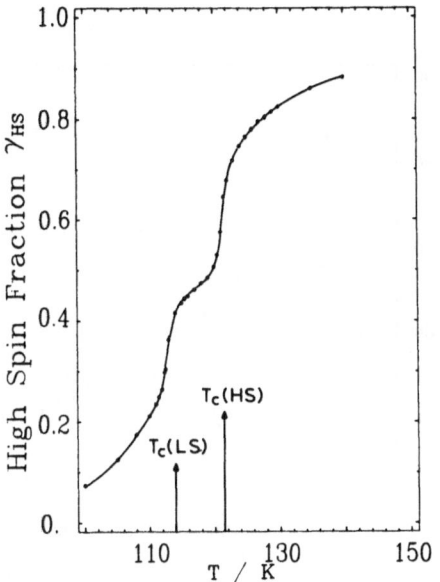

Fig. 10. Temperature dependence of the fraction of HS molecules, $\gamma_{HS}(T)$, evaluated from the area under the resonance lines of the Mössbauer spectra of $[Fe(2\text{-pic})_3]Cl_2 \cdot EtOH$ (from ref. 39).

below). Also, hydrogen bonding seems to play an important role, because the "two-step" transition has not been found in the deuterated system $[Fe(2\text{-pic-}ND_2)_3]Cl_2 \cdot EtOD$ (s. below).

A recent study of the heat capacity $C_p(T)$ of $[Fe(2\text{-pic})_3]Cl_2 \cdot$ $\cdot EtOH$ has again revealed the spin transition occurring in two steps[46] (s. Fig. 12). Köppen has used the measured $C_p(T)$ data to evaluate the effective enthalpy and entropy changes

$$\Delta H_{eff}(T) = \int_{T_O}^{T} C_p \, dT, \quad \Delta S_{eff}(T) = \int_{T_O}^{T} \frac{C_p}{T} \, dT \qquad (2)$$

and it is most interesting to see that these functions show the same "two-step" feature as the spin conversion functions $\gamma_{HS}(T)$ obtained from the Mössbauer and $\chi(T)$ measurements (s. Fig. 13). Therefore, it appears safe to conclude that the spin state conversion is the only process requiring extra heat and entropy in addition to the normal heat capacity of the crystal. Sorai and Seki[17] have analysed the $C_p(T)$ data of the spin crossover systems $[Fe(phen)_2(NCS)_2]$ and $[Fe(phen)_2(NCSe)_2]$. They could show that the major part of the total entropy increase (about 70%), on going from the LS to the HS state, comes from intra- and intermolecular vibrations and the smaller

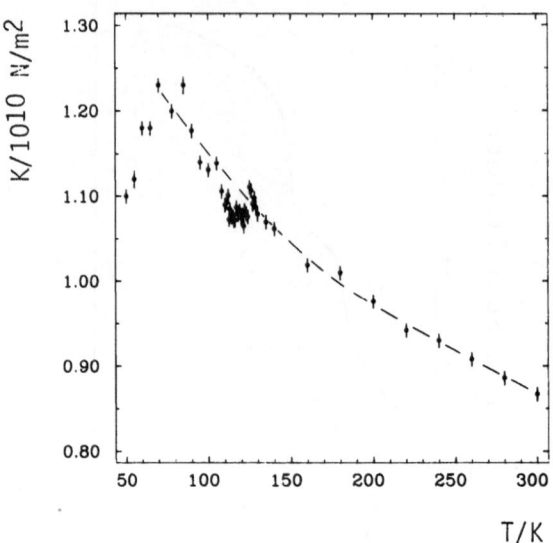

Fig. 11. Temperature dependence of the bulk modulus K(T) of
[Fe(2-pic)$_3$]Cl$_2$·EtOH, derived from a correlation with
the Debye temperature $\Theta_D$, which was evaluated from the
Mössbauer spectra (from ref. 46).

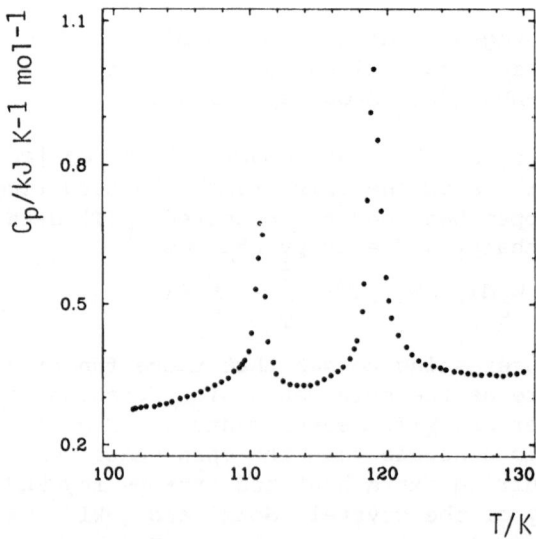

Fig. 12. Heat capacity $C_p$(T) of [Fe(2-pic)$_3$]Cl$_2$·EtOH as a function
of temperature (from ref. 46).

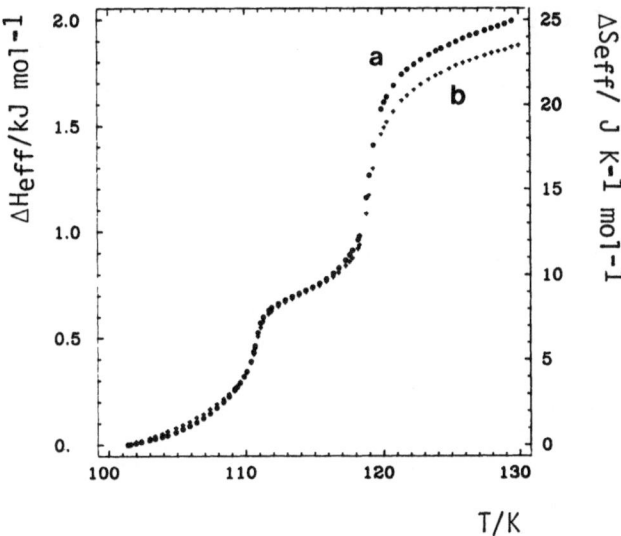

Fig. 13.  Effective enthalpy (a) and entropy (b) changes in the spin transition region of $[Fe(2-pic)_3]Cl_2 \cdot EtOH$ (per mole of the compound, independent of spin state).[46]

contribution from the spin multiplicity change, $\Delta S_{spin} =$ R $[\ln(2S+1)_{HS} - \ln(2S+1)_{LS}] \approx 13$ J $mol^{-1} K^{-1}$. The enormous entropy gain of 50–80 J $mol^{-1}$ $K^{-1}$ observed for spin state conversion process in solids[47,17,48,49] and solution[48,22,9] is the predominant driving force of the spin transition.

Another interesting aspect deduced from the Mössbauer measurements on $[Fe(2-pic)_3]Cl_2 \cdot EtOH$ is demonstrated in Fig. 14: the temperature dependence of the quadrupole splitting of the LS state, QS(LS), which is solely determined by the lattice contribution $(EFG)_{lat}$ to the electric field gradient, follows largely the spin conversion function $\gamma_{HS}(T)$ with two steps near $T_c(LS)$ and $T_c(HS)$. Therefore, it seems likely that the characteristic modes which cause the lattice softening as concluded from the reduction in the bulk modulus K(T) (s. Fig. 11) involve the iron atom and lead to a change in the nearby lattice geometry. The QS(LS) is significantly larger in the HS phase than in the LS phase. QS(HS) shows the usual increase with decreasing temperature characteristic for iron(II) HS complexes, but changes also rather abruptly in the crossover region. This jump to higher values of QS(HS) in the LS phase is qualitatively understood if one takes into account that (i) $(EFG)_{lat}$ is smaller in the LS phase than in the HS phase, and (ii) the valence contribution $(EFG)_{val}$ to the electric field gradient is opposite in sign to $(EFG)_{lat}$, which we have recently proven in a Mössbauer experiment under applied magnetic field.

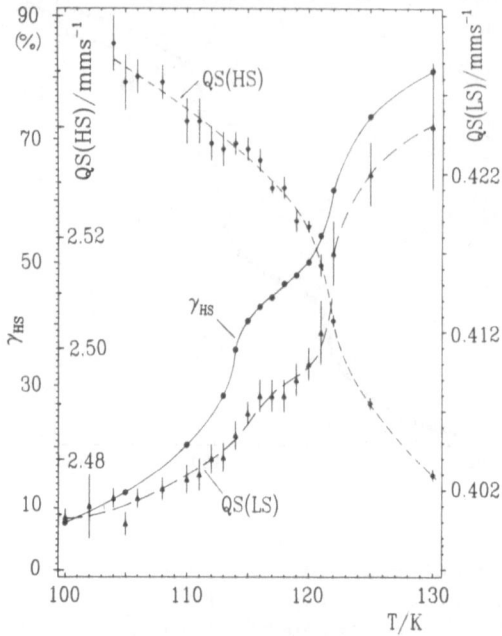

Fig. 14. Temperature dependence of the quadrupole splittings QS(HS) and QS(LS) for the HS and the LS state in $[Fe(2-pic)_3]Cl_2 \cdot EtOH$. Both quantities change rather abruptly in the crossover region, where the HS fraction $\gamma_{HS}(T)$ shows the unusual ("two-step") transition (from ref. 39).

## 3.2 Crystal Structure

Single-crystal structure analysis has been carried out on the solvates of $[Fe(2-pic)_3]Cl_2 \cdot Sol$ with Sol = EtOH,[34,50,51] MeOH,[52,50] and $2H_2O$.[52] The space group is different in the three solvates ($P2_1/c$ for the ethanolate, Pbna for the methanolate, P1 for the dihydrate), but they all have hydrogen bonding bridges

$$-\underset{H}{\overset{|}{N}}H \cdots \underset{HOR}{\overset{|}{X}} \cdots H\underset{H}{\overset{|}{N}}- \qquad (R = H, CH_3, C_2H_5)$$

interconnecting the cationic complexes (meridional in ethanolate and methanolate, facial in dihydrate). The hydrogen bonding network is two-dimensional in the ethanolate and the methanolate, but three-dimensional in the dihydrate. In view of the different nature of R and the different crystal structure it is obvious that the vibrational and elastic properties of the lattice are also different in the three solvates, and it is therefore not surprising that the spin transition characteristics differs also completely with R: the $\gamma_{HS}(T)$ curve of the methanolate is more gradual than that of the ethanolate and its transition temperature is ca. 30 K higher ($T_c \approx 150$ K); the dihydrate shows no spin transition at all and

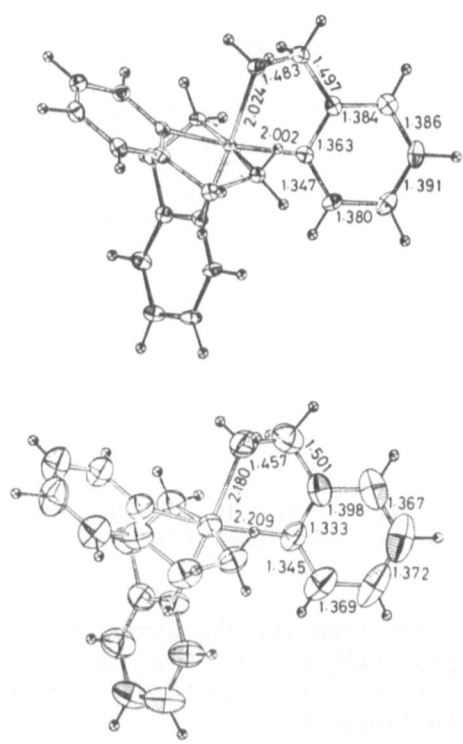

Fig. 15. Crystal structure of the cationic complex in
[Fe(2-pic)$_3$]Cl$_2$·EtOH at 298 K and 90 K (by permission
from ref. 34).

remains in the LS state between 300 and ca. 4 K.[53]

Mikami et al.[34] determined the structure of [Fe(2-pic)$_3$]Cl$_2$·EtOH
at 298 K and 90 K, i.e. well above and well below T$_c$ ≈ 120 K, and
found drastic changes in the Fe-N bond lengths: Fe-N(amino) changes
from 2.180 Å to 2.024 Å, and Fe-N(pyr) from 2.209 Å to 2.002 Å on
going from the HS to the LS phase (s. Fig. 15). The different changes
in the two types of Fe-N bonds is due to π-backbonding in case of
Fe-N(pyr), which is absent in the Fe-N(amino) bond.

It appeared to us that the abrupt change in intramolecular
dimensions of the cationic complexes is a most crucial point in the
cooperative spin transition process in a solid, because it obviously
induces changes in lattice vibrations and elastic properties, which
initiate further spin state changes. This is the background which
stimulated us to develop the "Elastic Interaction and Lattice Expansion (EILE)" model to describe quantitatively the cooperative spin
transition in solids.[38,54] It is, thereby, of fundamental interest
to know what the individual contributions to the total change of

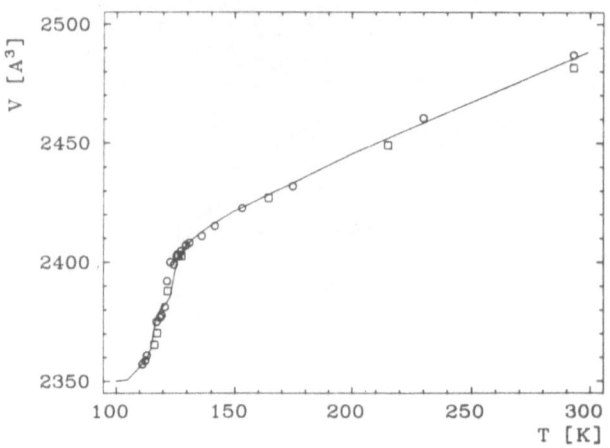

Fig. 16. Temperature dependence of the volume of the monoclinic
unit cell of $[Fe(2-pic)_3]Cl_2 \cdot EtOH$ from single crystal
X-ray measurements on two different single crystals
(O,□) (from ref. 55).

the unit cell volume are from (i) the normal thermal lattice expansion and (ii) the spin state change. Fig. 16 shows the results of a
recent study of $[Fe(2-pic)_3]Cl_2 \cdot EtOH$ in our laboratory.[55] The solid
curve is a fit of the expression

$$V(T) = V_{LS} + \gamma(T) \cdot (V_{HS} - V_{LS}) + c \cdot T \qquad (3)$$

to the measured data. The linear term $c \cdot T$ accounts for the lattice
expansion with increasing temperature. Values of $\gamma_{HS}(T)$ are known
from precise susceptibility measurements.[39] The unit cell volumes
$V_{HS}$ and $V_{LS}$ of the HS and the LS phase, respectively, and c are fit
parameters with the following values:

$$V_{LS} = 2305 \text{ Å}^3, \quad V_{HS} - V_{LS} = 61 \text{ Å}^3,$$
$$c = 0.4 \text{ Å}^3 \text{ K}^{-1}.$$

The essential result is that the volume change due to the thermal
lattice expansion is very much smaller (0.4 $\text{Å}^3 \text{ K}^{-1}$) than that arising
from the spin transition. A similar magnitude of the volume diffe-
rence $V_{HS} - V_{LS}$ was derived from ultrasound absorption measurements
of other spin crossover systems in solution.[23] The volume change is
found to be anisotropic,[56] as can be seen for $[Fe(2-pic)_3]Cl_2 \cdot EtOH$
in Fig. 17: the c-axis (in plane of H-bonding network) decreases
linearly with temperature, but surprisingly shows no additional
change due to the spin transition. The b-axis (in plane of H-bonding
network) also decreases linearly with temperature and reduces sharply
in the spin crossover region. The a-axis ($\angle(a,c)=116°$) shows practically no thermal expansion, but changes sharply with the spin transition. It is most interesting to see that the c-axis, and to a
lesser extent the a-axis, exhibit the same "two-step" anomaly in the

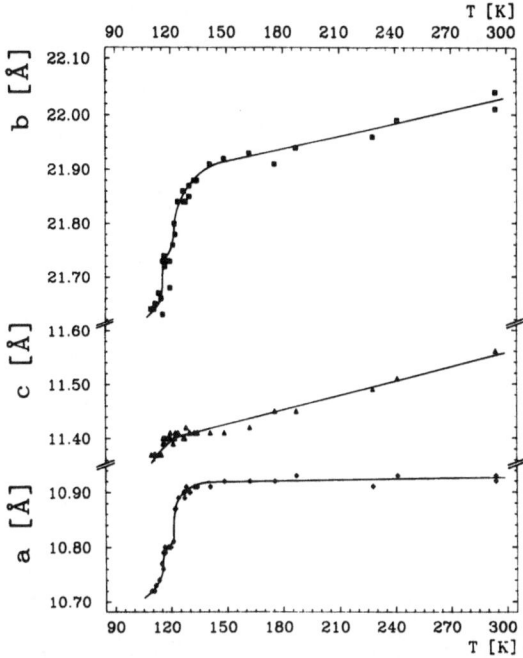

Fig. 17. Thermal expansion of the a,b, and c axes in monoclinic [Fe(2-pic)$_3$]Cl$_2$·EtOH.[56]

crossover region as the spin transition curve $\gamma_{HS}$(T).

## 3.3 Anion Influence

Many examples have become known which show that the nature of the anion alters significantly the spin transition behaviour,[5] even in ionic crystals where the anions are not directly coordinated to the central metal ion. Such an influence is also obvious in the present system, [Fe(2-pic)$_3$]X$_2$·Sol, if one compares the spin transition functions $\gamma_{HS}$(T) of the chloride and the bromide. Fig. 18 shows the most prominent differences in the crossover region of the two salts. Whereas the chloride features the above discussed "two-step" transition with no hysteresis, the transition in the bromide is more complicated with the two steps being much weaker and with a hysteresis in the upper part of the $\gamma_{HS}$(T) function.

The anion effect becomes self-explanatory in a qualitative sense, if one considers the known structure (s. Sect. 3.2). The anion X$^-$ is hydrogen-bonded between the amino groups of two neighboring cationic complexes, -NH$_2$ ... X ... H$_2$N-. This way X$^-$ participates actively in transmitting the "consequences" of a spin state change from one center to another. As the vibrational properties change with the reduced mass of X, it is not surprising that the spin transition be-

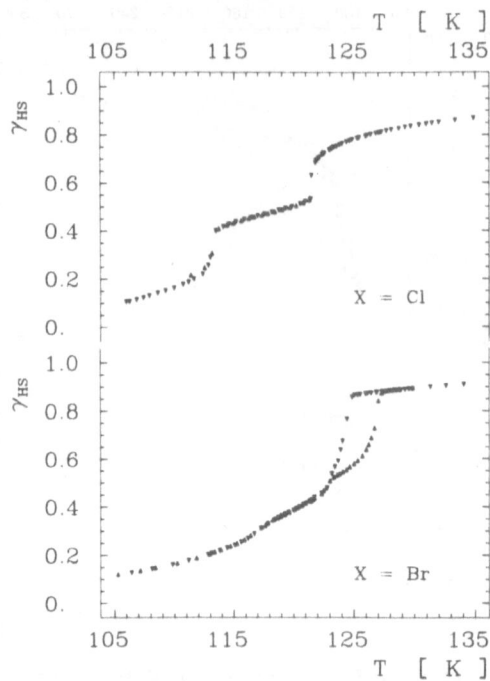

Fig. 18. Temperature dependence of the fraction of HS molecules, $\gamma_{HS}(T)$, derived from magnetic susceptibility measurements on $[Fe(2\text{-pic})_3]X_2 \cdot Sol$ (X = Cl,Br) with falling ($\blacktriangledown$) and rising ($\blacktriangle$) temperature, respectively.[40]

haviour is also different in the two salts. The crystal structures of the two compounds are the same.[34,56] The anion effect, together with the observed pronounced influences of isotopic exchange and iso-morphous metal dilution to be discussed next, supports the suggestion that the spin transition in the solid is a cooperative phenomenon.

### 3.4 Effect of Isotopic Exchange

In a first study of the influence of deuteration on the spin transition behaviour of $[Fe(2\text{-pic})_3]Cl_2 \cdot Sol$ (Sol = $C_2H_3OH$, $CH_3OH$) we have found that the transition curve $\gamma_{HS}(T)$ of both solvates is shifted by ca. 15 K to higher temperatures, if hydrogen is replaced by deuterium in the OH group and - as a chemical necessity - simul-taneously in the $NH_2$ group of the ligand. Both positions are involved in the hydrogen-bonding bridge connecting the cationic complexes, and it is therefore conceivable that deuteration affects the spin transition behaviour. The increase in the relative stability of the

LS state upon deuteration could be qualitatively explained by the difference in the reduced mass within a single harmonic approximation model.[57]

A refined and more extended investigation of the H/D effect in the ethanolate of $[Fe(2-pic)_3]Cl_2$ yielded interesting new results[46] (s. Fig. 19). The $\gamma_{HS}(T)$ curve of the system denoted as "$C_2H_5OD/ND_2$" is not only shifted by ca. 15 K upwards, but also looses completely the "two-step" anomaly in the transition region. In this case the heat capacity measurement yielded a $\lambda$-shaped curve $C_p(T)$ with only one peak (instead of the two peaks of Fig. 12), and the bulk modulus $K(T)$ deduced from a Debye-Waller factor analysis behaves normal in the crossover region. It looks as though the "lattice softening" discussed above for the hydrogenated system disappears in the deuterated "$C_2H_5OD/ND_2$" system. This apparently is not the case in the system denoted as "$C_2D_5OH/NH_2$", where deuteration has only replaced the five H-atoms in the alkyl group (which, as kind of an apendix being attached to the X anion, seems to affect the vibrational properties of the H-bonding bridge to a lesser extent). The "two-step" transition anomaly

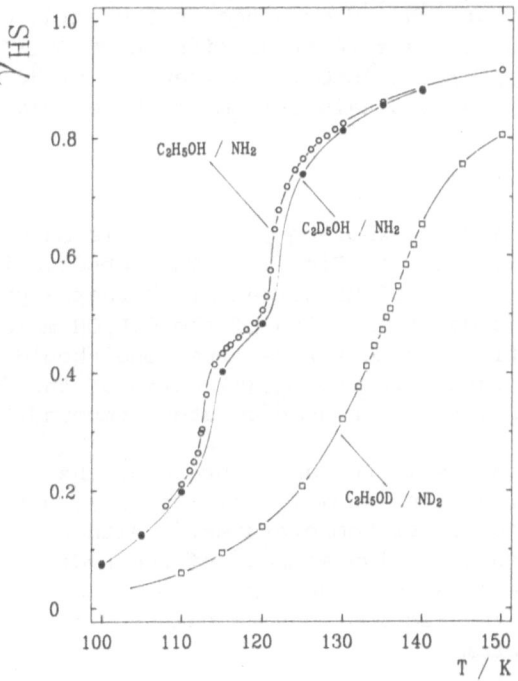

Fig. 19.  Effect of H/D exchange on the spin transition curve $\gamma_{HS}(T)$ of $[Fe(2-pic)_3]Cl_2 \cdot EtOH$ from Mössbauer measurements. The positions of deuteration are indicated in the graph.[46]

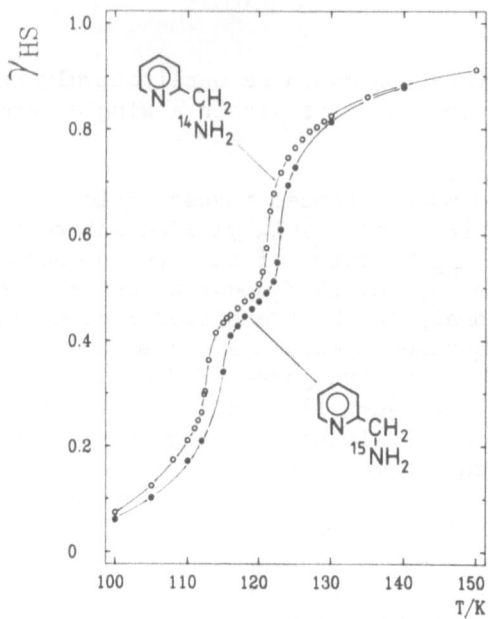

Fig. 20.   Effect of $^{14}N/^{15}N$ exchange in the amino group of the
ligand on the spin transition curve $\gamma_{HS}(T)$ of
$[Fe(2-pic)_3]Cl_2 \cdot EtOH$. The fraction of HS molecules,
$\gamma_{HS}(T)$, were obtained from Mössbauer measurements.[46]


is still there, and the whole $\gamma_{HS}(T)$ curve is only slightly shifted
to higher temperatures (s. Fig. 19). This observation does not
support the suggestion of Mikami et al.[34] that a preceding orientational order-disorder transition of the $C_2H_5OH$ molecule "triggers"
the spin transition. If this were true, one should expect a much
more pronounced change in the $\gamma_{HS}(T)$ curve of the "$C_2D_5OH/NH_2$"
system as compared to the non-deuterated compound.

$^{14}N/^{15}N$ exchange in the amino group of the ligand has also no
effect on the transition anomaly (s. Fig. 20), but shifts the $\gamma_{HS}(T)$
curve slightly to higher temperatures.[46] Although the nitrogen atoms
may be considered as active members of the hydrogen bonding bridge,
the $^{14}N/^{15}N$ effect is relatively small because of the relatively
small change in the reduced mass as compared to H/D exchange in the
"$C_2H_5OD/ND_2$" system.

The effect of H/D exchange has also been studied in systems
with deuterated $-CD_2$ and with deuterated 3,4,5-positions of the pyri-
dinium ring, respectively.[46] The transtition anomaly is preserved
in both cases, and $\gamma_{HS}(T)$ is shifted slightly to lower temperatures
in the former, but to higher temperatures in the latter case. The
overall effect is only marginal, probably because these deuterated

postitions are no longer constituents of the hydrogen bonding network.

It should be emphasized that isotopic exchange is a very subtle modification of the system with no change in the crystal structure to be expected. The only major alteration happens with the reduced mass, which in turn affects the vibrational and elastic properties of the lattice. Thus the observed effects of isotopic exchange on the spin transition should be seen in the light of such changes in elastic properties.

## 3.5  Effect of Metal Dilution

Mössbauer effect measurements of the mixed crystal series $[Fe_xZn_{1-x}(2-pic)_3]Cl_2 \cdot EtOH$, where iron(II) was isotypically substituted by zinc(II) in the concentration range $0.0009 \leqslant x \leqslant 1$,[58,59] have shown that the spin transition is affected in various ways: the transition curve $\gamma_{HS}(T)$ is shifted to lower temperatures, indicating greater relative stability of the HS state, and the slope of $\gamma_{HS}(T)$ at $T_c$ ($\gamma_{HS}=0.5$) becomes more gradual with decreasing iron concentration. Fig. 21 shows these effects for six mixed crystal systems of $[Fe_xZn_{1-x}(2-pic)_3]Cl_2 \cdot EtOH$. This kind of metal dilution effect on the spin transition seems to be general, at least in the picolylamine complexes; it has also been seen in the systems $[Fe_xCo_{1-x}(2-pic)_3]Cl_2 \cdot EtOH$[54] and $[Fe_xZn_{1-x}(2-pic)_3]Cl_2 \cdot CH_3OH$.[60]

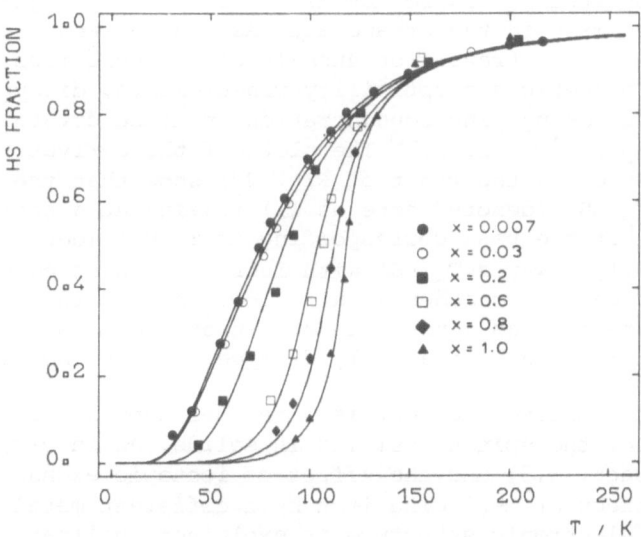

Fig. 21.  Temperature dependence of the HS fraction $\gamma_{HS}(T)$ of mixed crystals of $[Fe_xZn_{1-x}(2-pic)_3]Cl_2 \cdot EtOH$ with variable iron concentrations. The solid curves have been calculated using the EILE model (from ref. 38).

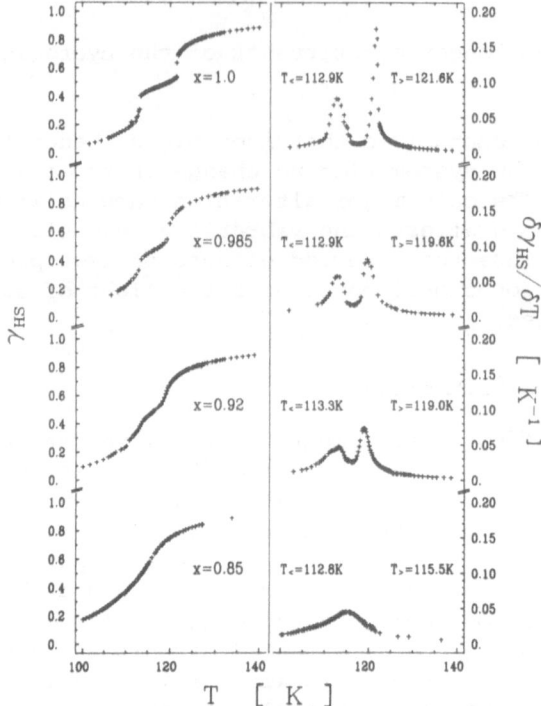

Fig. 22. The effect of metal dilution on the "two-step" spin transition anomaly in [Fe(2-pic)$_3$]Cl$_2$·EtOH (from ref. 40).

The substitution of iron by zinc also affects significantly the
transition anomaly in the ethanolate. As can be seen from Fig. 22,
the "two-step" spin transition anomaly of the pure iron compound,
recorded by magnetic susceptibility measurements, disappears gradually with increasing zinc concentration in mixed crystals of
[Fe$_x$Zn$_{1-x}$(2-pic)$_3$]Cl$_2$·EtOH.[40] The plots of the derivative d$\gamma_{HS}$/dT
vs. temperature (to the right of Fig. 22) show that the peak corresponding to $T_c$(LS) (denoted here as $T_<$) remains at a constant position
near 113 K, but the peak corresponding to $T_c$(HS) (denoted as $T_>$)
moves gradually towards $T_c$(LS) with increasing zinc concentration
until coalescence is reached at $x \approx 0.80$. This would imply that the
"lattice softening" as a possible origin of the "two-step" transition
anomaly (cf. Sect. 3.1) gradually disappears with increasing zinc
concentration.

The metal dilution effect is a further support for the cooperative nature of the spin transition in solids. As in case of the anion
influence (Sect. 3.3) and the effect of isotopic exchange (Sect.
3.4), the effect of replacing iron by a different metal which differs
in size and electronic structure is explained qualitatively on the
grounds of changes in vibrational and elastic properties. These results, together with the fact that $V_{HS} > V_{LS}$ (Sect. 3.2), formed the

basis for an attempt to describe the spin transition in solids quantitatively. The model is called "Elastic Interaction and Lattice Expansion (EILE)" model.[38,54] Some essentials of it will be outlined next.

## 3.6  The "Elastic Interaction and Lattice Expansion (EILE)" Model

Temperature dependent spin transition in liquid solution generally yield $\gamma_{HS}$(T) curves which can be described quantitatively by a Boltzmann distribution law over all vibronic energy levels of the LS and the HS states. Discontinuous changes in the HS fraction $\gamma_{HS}$ with temperature, the existence of hysteresis or residual HS and LS fractions have only been observed in crystals, but not in liquid solution, and are indications for the occurrence of cooperative interactions resulting into extra energy terms to be taken into account.

Several attempts have been undertaken by various research groups to describe quantitatively the spin transition in solids.[61-69] A brief comparative discussion is given in ref. 5. None of these models works fully satisfactorily.

The approach suggested by Slichter and Drickamer[64] starts out with the formulation of the Gibbs free energy per molecule of an ideal solution of HS and LS molecules and includes an extra term $\gamma_{HS}(1-\gamma_{HS}) \Gamma$ to account for interactions between HS and LS molecules:

$$G(T,p,\gamma)/N = \gamma_{HS}G_{HS}(T,p) + (1-\gamma_{HS})G_{LS}(T,p) +$$
$$+ \gamma_{HS}(1- \gamma_{HS}) \Gamma(T,p) - T \cdot S_{mix} \qquad (4)$$

where N is the total number of complex molecules, $\gamma_{HS}$ is the fraction of HS molecules, $G_{HS}$ and $G_{LS}$ are the Gibbs free energies of a HS and LS molecule, respectively, and the mixing entropy is given by $S_{mix} = -K[ \gamma_{HS}\ln \gamma_{HS}+(1-\gamma_{HS})\ln(1-\gamma_{HS})]$. The physical nature of the interaction represented by the parameter $\Gamma$ is not further specified. Nevertheless, the expression for $\gamma_{HS}$(T) derived from the equilibrium condition

$$(\partial G/\partial \gamma_{HS})_{T,p} = 0 \qquad (5)$$

is suited to reproduce many spin transition curves observed in crystals.

H. Spiering in our research group has recently developed a model[38,54] which is based on the observation that the volumes of the HS and LS complex molecules are different, viz. $V_{HS} > V_{LS}$ (s. Sect. 3.2), with the consequence that these molecules fit more or less well into a given lattice. This leads to different elastic energy contributions which have to be included in the free energies of the HS and LS molecules. For LS → HS transitions, the lattice expands linearly with $\gamma_{HS}$ as could be confirmed by variable temperature X-ray studies (s. Sect. 3.2). Furthermore, according to Eshelby (within the approximation of a homogeneous and isotropically elastic medium),[70] the increase of the lattice volume exceeds the one resulting from the spin transition in a single complex molecule by

$$\delta V = (\gamma_o - 1)(V_{HS} - V_{LS}). \tag{6}$$

$\gamma_o$ is the so-called Eshelby factor, which lies in the range $1 \leqslant \gamma_o \leqslant 3$ for solids. This lattice expansion facilitates the transition LS → HS in other complex molecules. This way the complex molecules are coupled with each other via the spin transition induced lattice expansion. This is the basic idea about the mechanism for and the cooperative interactions in spin transitions in solids, and has led us to call this approach the "Elastic Interactions and Lattice Expansion (EILE)" model.

The EILE model has been applied successfully to fit the measured spin transition curves $\gamma_{HS}(T,x)$ in the mixed crystal series $[Fe_xM_{1-x}(2\text{-pic})_3]Cl_2 \cdot Sol$ with M = Zn and Sol = EtOH[38] (s. Fig. 21, where the solid lines were calculated using this model), with M = Co and Sol = EtOH,[54] and with M = Zn,Co and Sol = $CH_3OH$.[60] The average free energy $f(T)$ per complex molecule in the solid solution of an iron concentration x writes:

$$f(T) = x[f_1(T) + f_{El}(T)] + (1-x)f_M(T) \tag{7}$$

where

$$f_1(T) = \gamma_{HS}f_{HS}(T) + (1-\gamma_{HS})f_{LS}(T) - T \cdot S_{mix} \tag{8}$$

represents the average free energy per isolated iron complex molecule in a liquid solution with no elastic interactions. The "elastic" free energy per iron complex arising from the lattice expansion and elastic interactions is given as

$$f_{El}(T) = \Delta_{El}(x)\gamma_{HS} - \Gamma(x)\gamma_{HS}^2. \tag{9}$$

The coefficients $\Delta_{El}(x)$ and $\Gamma(x)$ could be interpreted within the theory of elasticity as

$$\Delta_{El}(x) = \Delta_{El}^o + 2\Gamma(x) \cdot q, \tag{10}$$

where

$$\Delta_{El}^o = \frac{1}{2}K(\gamma_o - 1)(V_{HS} - V_{LS})[(V_{HS} + V_{LS})/V_o - 2 - 2\gamma_o(V_M - V_o)/V_c] \tag{11}$$

is the difference of the elastic self energies of the HS and LS molecules in the isotropic homogeneous medium,

$$\Gamma(x) = x \cdot \Gamma(x=1) = x \cdot \frac{1}{2}K\gamma_o(\gamma_o - 1)(V_{HS} - V_{LS})^2/V_c \tag{12}$$

represents an interaction parameter.
q abbreviates the ratio

$$q = (V_M - V_{LS})/(V_{HS} - V_{LS}). \tag{13}$$

$V_c$ is the average crystal volume of one molecular unit. $V_M$, $V_{HS}$, and $V_{LS}$ are the effective complex volumes of the host compound, and of iron in the HS and LS states, respectively, i.e. approximately the volume of the $MN_6$ core (M = Zn,Fe).[60] $V_0$ is the effective complex volume for which no elastic energy is required if the complex molecule is embedded in the host lattice. K(T) is the bulk modulus, which is slightly temperature dependent, which leads to a change of $\Delta_{El}(x,\Gamma)$, eq.(10), on the order of 10%. This temperature dependence is approximated within the Einstein model.[54] Although K(T) is included in eq. (12) for the interaction parameter $\Gamma(x)$, we have neglected the temperature dependence of $\Gamma(x)$, because the abolute value of $\Gamma(x)$ is small as compared to $\Delta_{El}(x,T)$ (s. below).

Minimizing eq. (7) according to eq. (5) yields an implicit expression for $\gamma_{HS}(T)$. Thereby, the term $(1-x)f_M(T)$ vanishes, since it does not depend on $\gamma_{HS}$. The free energy is expressed in terms of the partition function

$$f = -kT \ln Z. \qquad (14)$$

$Z_{HS}$ and $Z_{LS}$ includes the electronic and vibrational contributions, $Z_k = z_k^e \cdot z_k^V$ (k = HS, LS). $Z_{LS}^e = 1$ for the singlet state $^1A_1$, which serves as the zero reference point. $z_{HS}^e$ is known from the energy level diagram derived from the temperature dependent quadrupole splitting.[71] Each of the vibrational partition functions $z_k^V$ are parameterized by two effective vibrational frequencies, representing the six stretching and nine deformation vibrations of a $MN_6$ core; the wavenumbers in the LS state are higher (350-500 $cm^{-1}$) than in the HS state (200-300 $cm^{-1}$).[72,17] This parameterization has been used to calculate the solid lines of Fig. 21.

It has been found that the energy separation between the lowest vibronic levels of the $^1A_1$ and $^5T_2$ states, which is on the order of 800 $cm^{-1}$ in liquid solution, is decreased by the elastic free energy contribution (eq. (10)) of ca. -700 $cm^{-1}$, which apparently favors massively the HS state. The interaction parameter $\Gamma(x=1)$ turns out to be 130 $cm^{-1}$ for the system [Fe(2-pic)$_3$]Cl$_2$·EtOH.[38] The fit of the parameters to the experimental data of the metal dilution series also confirms the linear concentration dependence of $\Gamma(x)$ as predicted by eq. (12).

## 3.7  Effect of Applied Pressure

The influence of applied pressure on the spin transition in ferrous complexes has already been studied by Drickamer et al.[73] and later on by other research groups.[74,75] Their experiments were all done in the high pressure region ($\gtrsim$ 50 kbar), where drastic spin state conversions in both directions, HS → LS and, surprisingly enough, LS → HS, were observed at constant temperature. The authors explained these findings only qualitatively on the basis of pressure induced changes in the molecular orbital energies of metal-ligand bonds as well as intra-ligand bonds such as to increase or decrease the ligand field strength.

We have studied the effect of low pressure on the spin transition in $[Fe(2\text{-pic})_3]Cl_2 \cdot EtOH^{55}$ with the specific goal to test the EILE model (s. Sect. 3.6). Fig. 23 shows two Mössbauer spectra recorded at 121 K and ambient pressure (1 bar) and hydrostatic pressure of 150 bar, respectively. Obviously, the area fraction of the LS doublet increases by application of pressure, which is expected because of the known volume difference $V_{HS} > V_{LS}$ (s. Sect. 3.2). The spin transition curves $\gamma_{HS}(T)$ measured under various pressures[76] exhibit some interesting trends (s. Fig. 24): the $\gamma_{HS}(T)$ curve is shifted to higher temperatures, implying an increase in relative stability of the LS state, and the "two-step" transition anomaly disappears gradually with increasing pressure. The latter effect again means that the "lattice softening" (s. Sect. 3.1) vanishes under applied pressure.

Employing the EILE model, we have derived from the pressure dependence of $\gamma_{HS}$ a volume change of 55 $Å^3$ per unit cell as a consequence of the spin transition; this is in very good agreement with

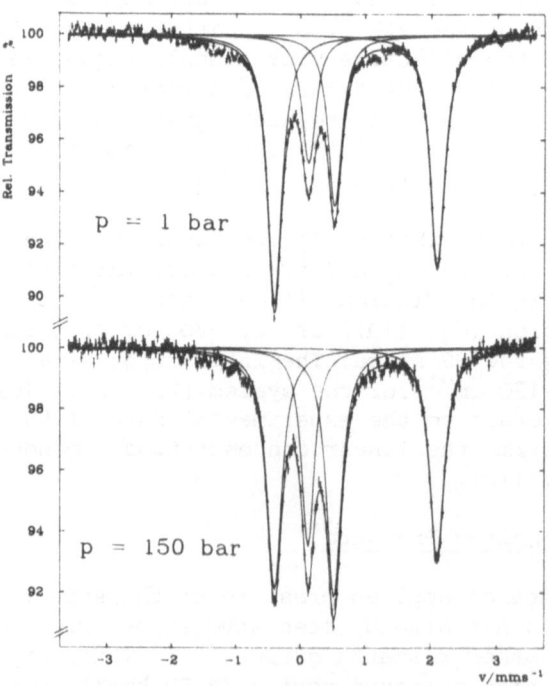

Fig. 23.  Pressure dependence of the Mössbauer spectra of $[Fe(2\text{-pic})_3]Cl_2 \cdot EtOH$ at T = 120.9 K. The area fraction of the LS doublet (inner two lines) increases by application of pressure (p=150 bar, lower spectrum). The upper spectrum is taken at ambient pressure p=1 bar (from ref. 55).

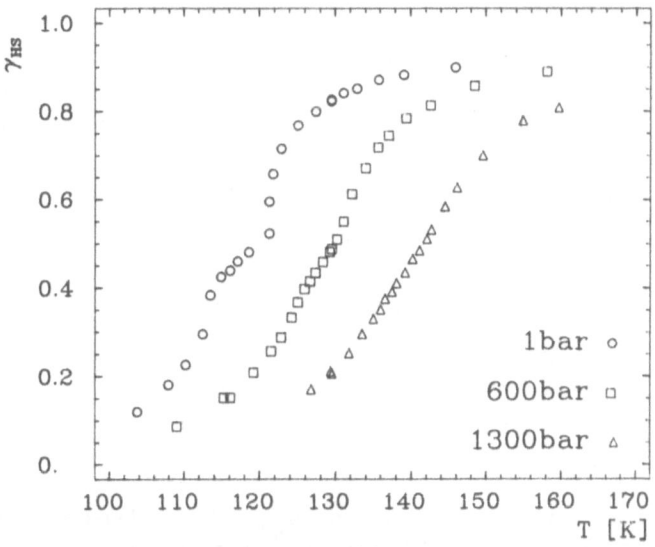

Fig. 24.   Temperature dependence of the HS fraction $\gamma_{HS}$(T), derived
from the Mössbauer spectra of [Fe(2-pic)$_3$]Cl$_2$·EtOH, as a[76]
function of applied pressure.

the value of 61 Å$^3$ obtained from single crystal X-ray measurements.[55]


4.   LIGHT-INDUCED EXCITED SPIN STATE TRAPPING (LIESST)

4.1 First Observation in [Fe(ptz)$_6$](BF$_4$)$_2$

Some members of the class of octahedral complexes [Fe(Rtz)$_6$](BF$_4$)$_2$
with 1-alkyltetrazoles (Rtz) as monofunctional ligands show tempera-
ture dependent HS($^5$T$_2$) $\rightleftharpoons$ LS($^1$A$_1$) spin transition.[77] We have investigated the spin transition characteristics in [Fe(ptz)$_6$](BF$_4$) (R =
i-C$_3$H$_7$) by Mössbauer spectroscopy, magnetic susceptibility, and far
infrared measurements at variable temperatures.[78] The temperature
dependence of the effective magnetic moment $\mu_{eff}$(T) (s. Fig. 25)
shows a sharp drop near 128 K in the cooling direction, but less
abrupt increase near 135 K in the heating mode. The existence of a
hysteresis of ca. 7 K width calls for a (predominantly) first order
nature transition with accompanying structural phase change. The
crystal structure above and below T$_c$ is not yet known. The variable
temperature Mössbauer spectra show the typical quadrupole doublet
with a large and temperature dependent splitting for the iron(II)
HS state, and a resonance singlet for the LS state of iron(II) in
apparently undistorted lattice positions (no quadrupole splitting).[78]

The [Fe(Rtz)$_6$]$^{2+}$ complex possesses remarkable optical properties.
They form crystals which are colorless in the HS state and deeply

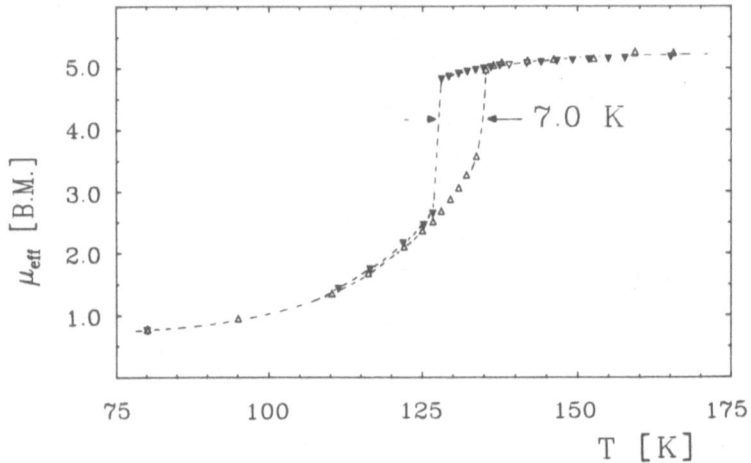

Fig. 25. Effective magnetic moment $\mu_{eff}(T)$ of $[Fe(ptz)_6](BF_4)_2$
as a function of temperature (from ref. 78).

colored (purple) in the LS state. These systems with their energetically high lying charge transfer bands prompted us to investigate in more detail the ligand field optical spectra of the
$[Fe(ptz)_6](BF_4)_2$ complex. In the course of this study we have discovered a surprising effect[79] which is new in spin crossover research
and has, to our knowledge, not been reported in the field of photo-
physics of transition metal compounds: through exciting with white
light in the singlet→singlet absorption bands at sufficiently low
temperatures the excited $^5T_2$(HS) state is quantitatively populated.
Below a critical temperature, which lies near 50 K in case of
$[Fe(ptz)_6](BF_4)_2$, the "trapped" excited HS state possesses a practically infinite lifetime. We call the hitherto unknown phenomenon
"Light Induced Excited Spin State Trapping (LIESST)".

We have identified the existence of the pure HS state produced
by LIESST at T < 50 K with single crystal optical absorption spectroscopy and magnetic susceptibility measurements[80] and as well as by
Mössbauer spectroscopy.[79] Fig. 26 shows a series of spectra of
$[Fe(ptz)_6](BF_4)_2$. (a) is the resonance singlet of the LS state recorded at 15 K before bleaching, (b) is the doublet of the pure HS
state taken at 15 K after bleaching for 1 h. The spectra (c) and (d)
demonstrate that the trapped excited HS state decays gradually back
to the $^1A_1$(LS) ground state upon warming up to 50-55 K. Spectrum
(e), taken after warming up to 97 K, is identical to spectrum (a)
of the pure LS ground state. Finally, if one passes through the
transition temperature of the normal thermally induced spin transition ($T_c \approx$ 135 K) one sees the typical HS doublet again (spectrum f).

The crucial points are the existence of a relaxation path from

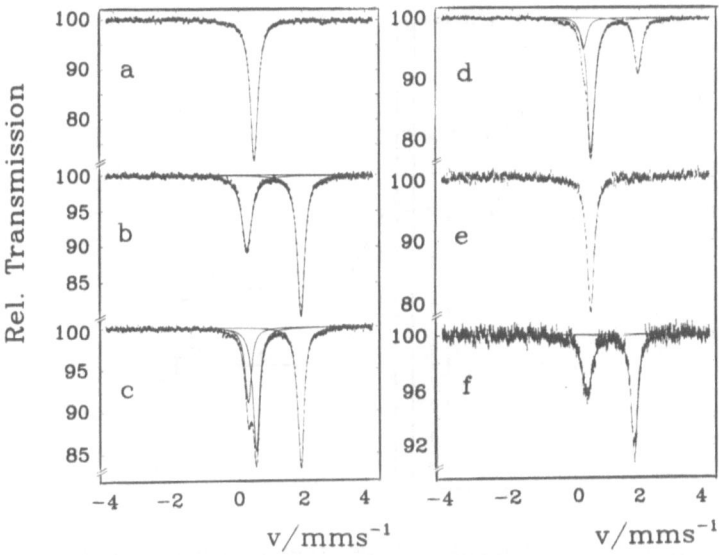

Fig. 26. Mössbauer spectra of $[Fe(ptz)_6](BF_4)_2$. (a) Before bleaching
(measuring temperature $T_M$ = 15 K); (b) after bleaching for
1 h at 15 K ($T_M$ = 15 K); (c) after warming to 50-55 K and
cooling to $T_M$ = 15 K; (d) after second warming to 50-55 K
and cooling to $T_M$ = 15 K; (e) after warming to 97 K ($T_M$ =
97 K); and (f) after warming to 148 K ($T_M$ = 148 K)
(from ref. 79)

the short-lived excited spin singlet states to the $^5T_2$ (HS) state and
the existence of an energy barrier between the $^5T_2$ (HS) and the $^1A_1$ (LS)
potential minima. This is illustrated in the simplified Jablonski
type diagram of Fig. 29. Illuminating the sample with white light
leads to a population of the $^1T_1$, $^1T_2$, and $^1$MLCT states. These states
are very short-lived and can decay rapidly back to the $^1A_1$ ground
state via spin-allowed transitions. As the $^3T_1$, $^3T_2$ ligand field
states are somewhat lower in energy than the $^1T_1$ and $^1T_2$ states, [80]
an alternative decay path, favored by spin orbit coupling, leads to
a population of the spin triplet states (intersystem crossing). Spin-
forbidden transitions also occur when the triplet states decay, initiated again by spin-orbit coupling, either to the $^1A_1$ ground state
or to the excited $^5T_2$ ligand field state. There is no radiative relaxation path between the $^5T_2$ and the $^1A_1$ states. Therefore, the
excited HS($^5T_2$) state remains trapped for practically infinite lifetime provided the temperature is sufficiently low so that the energy
barrier between the HS and the LS potential minima (s. Fig. 2) is not
thermally overcome.

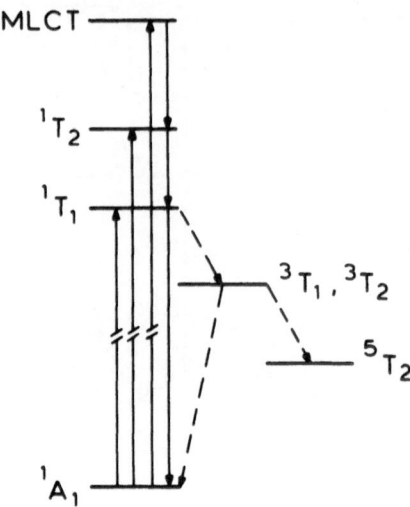

Fig. 27.  Simplified Jablonski type energy level diagram to explain
the observed LIESST phenomenon in ferrous spin crossover
complexes


## 4.2  LIESST in Other Spin Crossover Complexes

With the discovery of LIESST in $[Fe(ptz)_6](BF_4)_2$ the question
arouse whether this is a unique phenomenon occurring only in this
compound, or may also be detected in other systems. As a prerequisite
it appeared to us that the energy difference between the excited HS
state and the LS ground state should be on the order of a few hundred
wavenumbers only as is the case in spin crossover compounds.

Meanwhile we have detected LIESST also in $[Fe(2-pic)_3]Cl_2 \cdot EtOH$
and $[Fe(phen)_2(NCS)_2]$.[81] These compounds are colored in both spin
states and apparently have strong charge transfer bands in the visible
region. Again, Mössbauer spectroscopy proved to be an elegant tool
to identify the pure HS states produced by LIESST (s. Figs. 28 and
29). Spectrum A of Fig. 28 shows the typical LS quadrupole doublet
of $[Fe(2-pic)_3]Cl_2 \cdot EtOH$ at 4.2 K before bleaching. After bleaching
with white light for 1 h the compound has been converted from the
LS to the HS state to more than 80% (s. Fig. 28B). It should be noted
that quantitative LS → HS conversion is possible for a sufficiently
thin absorber. Heating the sample to 30-32 K for a short time causes
the compound to relax gradually back to the LS state as can be seen
from the spectra C, D, and E.

Analogous observations have been made with $[Fe(phen)_2(NCS)_2]$.
Fig. 29 shows the Mössbauer spectra for this compound taken at 6 K

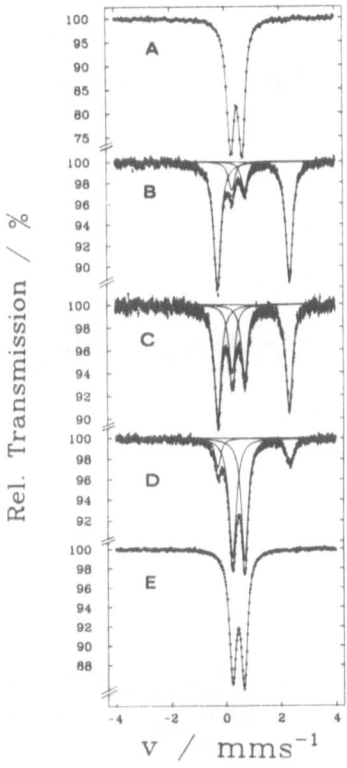

Fig. 28. Mössbauer spectra of $[Fe(2\text{-pic})_3]Cl_2 \cdot EtOH$ recorded at $T_M = 4.2$ K before bleaching (A), after bleaching with white light for 1 h (B), and warming up to 30-32 K for a short time and cooling back to 4.2 K in three cycles (C, D, and E).[81]

before bleaching with white light (A) and after bleaching for 1 h (B), respectively. The critical temperature where the "trapped" excited HS state produced by LIESST begins to depopulate back to the LS state lies around 55 K.


5.    CONCLUSION

     Spin transition, together with valence fluctuation and intra-molecular electron transfer in low-dimensional systems are among the fascinating topics of present research activities in the chemistry and physics of electron dynamics in transition metal compounds.

     Spin transition in transition metal compounds may be induced by chemical and physical means. Ligand molecules may be chosen such as to meet the spin crossover condition $|\Delta - P| \approx k_B T$ (see Fig. 1).

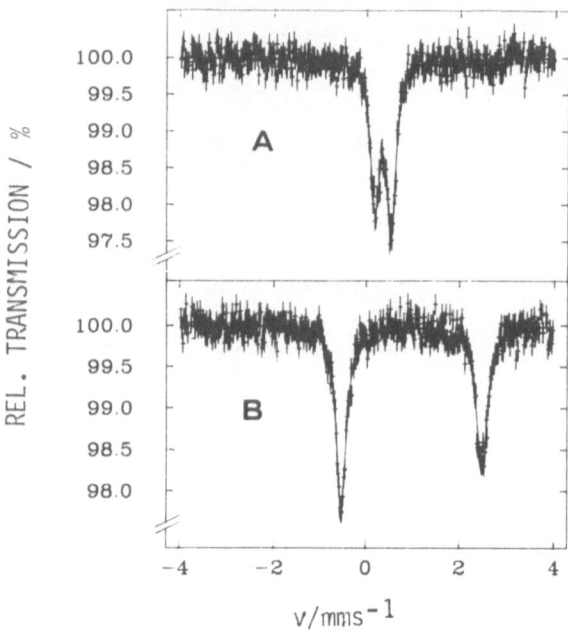

Fig. 29. Mössbauer spectra of $[Fe(phen)_2(NCS)_2]$ recorded at $T_M = 6$ K before bleaching (A) and after bleaching with white light for 1 h (B).[81]

In this case, variation of temperature or pressure may induce a change in spin state. In addition, we have seen that the nature of the anion and crystal solvent molecules, which are not directly co-ordinated to the metal, can affect significantly the spin transition function or even quench the transition totally. The same is true if the spin changing metal is substituted by other metal ions of different size and electronic structure. All these influences are taken as indications for a cooperative nature of the spin transition in crystals, whereby lattice expansion due to $V_{HS} > V_{LS}$ and changes in elastic properties are believed to play a decisive role. Based on these grounds the "Elastic Interaction and Lattice Expansion (EILE)" model has been developed, which allows to describe the spin transition in solids quantitatively.

In the last part of the present account we have reported on a recently observed photophysical effect (LIESST), which demonstrates that quantitative spin state conversion from the stable LS ground state to the excited HS state with practically infinite lifetime at sufficiently low temperatures may also be induced by light. Prospectively, this may eventually be of importance in solar energy conversion and storage.

It may be added that spin transition has also been observed as a consequence of nuclear decay processes in solid complex compounds doped with $^{57}Co$; Mössbauer emission spectroscopy has been the only technique applicable to detect such electronic structure changes on a time scale of $10^{-7}$ to $10^{-9}$ sec. [82]

Future trends in spin transition research will be directed towards better understanding of the mechanism and the nature of the interaction in crystals. A number of theoretical models have been proposed;[61-69] they all have their merits and shortcomings. At present we favor the "Elastic Interaction and Lattice Expansion" model developed in our research group.[38,54] It works satisfactorily in many cases. And yet, it still has its drawbacks, too, because it is still somewhat crude because of lacking data for elastic properties. It is one of our goals in future efforts to fill this gap.

ACKNOWLEDGEMENT

I wish to express my sincere thanks to all my students and co-workers, who have engaged themselves with great enthousiasm in our spin crossover research; their names appear in the list of references.

Particular thanks are extended to Dr. H. Spiering and E. Meissner for fruitful discussions and critical reading of the manuscript, and to Mrs. G. Lehr for typing this article with great care.

I also wish to acknowledge the financial support from the Deutsche Forschungsgemeinschaft, the Fonds der Chemischen Industrie, the Bundesministerium für Forschung und Technologie, the Stiftung Volkswagenwerk, and the University of Mainz.

REFERENCES

1.  L. Cambi, A. Cagnasso, Atti Accad. Naz. Lincei 13, 809 (1931); L. Cambi, L. Szegö, Ber. dtsch. chem. Ges. 64, 259 (1931).
2.  R.H. Martin, A.H. White, Transition Metal Chemistry, 4, 113 (1968).
3.  L. Sacconi, Pure Appl. Chem. 27, 161 (1971).
4.  H.A. Goodwin, Coord. Chem. Rev. 18, 293 (1976).
5.  P. Gütlich, Struct. Bonding 44, 83 (1981).
6.  P.G. Sim, E. Sinn, J. Am. Chem. Soc. 103, 241 (1981).
7.  J.H. Ammeter, L. Zoller, J. Bachmann, E. Baltzer, E. Gamp, R. Bucher, E. Deiss, Helv. Chim. Acta 64, 1063 (1981).
8.  W. Kläui, J. Chem. Soc. Chem. Commun. 700 (1979).
9.  P. Gütlich, B.R. McGarvey, W. Kläui, Inorg. Chem. 19, 3704 (1980).
10. M. Eibschütz, M. Lines, F.J. DiSalvo, Phys. Rev. B15, 103 (1977).
11. S. Ramasesha, T.V. Ramakrishnan, C.N.R. Rao, J. Phys. C. 12, 1307 (1979).
12. H. Imoto, A. Simon, Inorg. Chem. 21, 308 (1982).

13. C.J. Ballhausen, Introduction to Ligand Field Theory, McGraw-Hill, New York 1962.

14. M.F. Tweedle, L.J. Wilson, J. Am. Chem. Soc. 98, 4824 (1976).

15. Y. Maeda, N. Tsutsumi, Y. Takashima, Chem. Phys. Lett. 88, 248 (1982).

16. W.D. Federer, D.N. Hendrickson, private communication.

17. M. Sorai, S. Seki, J. Phys. Chem. Solids 35, 555 (1974).

18. W.A. Baker, H.M. Bobonich, Inorg. Chem. 3, 1184 (1964).

19. E.W. Müller, P. Gütlich, unpublished measurements.

20. M. Cox, J. Darken, B.W. Fitzsimmons, A.W. Smith, L. Larkworthy, K. Rogers, J.C.S. Chem. Comm. 105, 1970.

21. G.W. Flynn, N. Sutin, in: Chemical and Biochemical Applications of Lasers, Vol. 1 (C.B. Moore, ed.), Academic Press New York 1974, p. 309.

22. E.V. Dose, M.A. Hoselton, N. Sutin, M.F. Tweedle, L.J. Wilson, J. Am. Chem. Soc. 100, 1141 (1978)

23. J.K. Beattie, R.A. Binstead, R.J. West, J. Am. Chem. Soc. 100, 3044 (1978).

24. N. Sutin, Acc. Chem. Res. 15, 275 (1982).

25. E. König, K. Madeja, Inorg. Chem. 6, 48 (1967).

26. A.T. Casey, F. Isaac, Aust. J. Chem. 20, 2765 (1967).

27. P. Ganguli, P. Gütlich, E.W. Müller, W. Irler, J.C.S. Dalton 441 (1981).

28. E.W. Müller, H. Spiering, P. Gütlich, Chem. Phys. Lett. 93, 567 (1982).

29. E.W. Müller, H. Spiering, P. Gütlich, J. Chem. Phys. 79, 1439 (1983).

30. M.S. Haddad, W.D. Federer, M.W. Lynch, D.N. Hendrickson, J. Am. Chem. Soc. 102, 1468 (1980); Inorg. Chem. 20, 123 (1981); ibid. p. 131.

31. E.W. Müller, Dissertation, University of Mainz, Chemistry Dept. 1982.

32. C.N.R. Rao, K.J. Rao, "Phase Transitions in Solids", McGraw-Hill, New York 1978.

33. K. Madeja, W. Wilke, and S. Schmidt, Z. anorg. Chem. 346, 306 (1966).

34. M. Mikami, M. Konno, Y. Saito, Chem. Phys. Lett. 63, 566 (1979).

35. R.D. Shannon, C.T. Prewitt, Acta Crystallogr. B26, 1076 (1970).

36. Y. Qi, E.W. Müller, H. Spiering, P. Gütlich, Chem. Phys. Lett. 101, 503 (1983).

37. G.A. Renovitch, W.A. Baker Jr., J. Am. Chem. Soc. 89, 6377 (1967).

38. H. Spiering, E. Meissner, H. Köppen, E.W. Müller, P. Gütlich, Chem. Phys. 68, 65 (1982).

39. H. Köppen, E.W. Müller, C.P. Köhler, H. Spiering, E. Meissner, P. Gütlich, Chem. Phys. Lett. 91, 348 (1982).

40. C.P. Köhler, in partial fulfilment of the doctoral thesis, University of Mainz, Chemistry Dept., 1984.

41. E. König, Ber. Bunsengesellsch. phys. Chem. 76, 975 (1972).

42. E. König, G. Ritter, H.A. Goodwin, Chem. Phys. 1, 17 (1973).

43. E. König, G. Ritter, H.A. Goodwin, Chem. Phys. 5, 211 (1974).
44. E. König, G. Ritter, W. Irler, H.A. Goodwin, B. Kanellakopulos, J. Phys. Chem. Solids 39, 521 (1978).
45. E. König, G. Ritter, W. Irler, S.M. Nelson, Inorg. Chim. Acta 37, 169 (1979).
46. H. Köppen, in partial fulfilment of his doctoral thesis, University of Mainz, Chemistry Dept., 1984.
47. D.M.L. Goodgame, A.A.S.C. Machado, Inorg. Chem. 6, 2031 (1969).
48. M.A. Hoselton, L.J. Wilson, R.S. Drago, J. Am. Chem. Soc. 97, 1722 (1975).
49. P. Gütlich, H. Köppen, R. Link, H.G. Steinhäuser, J. Chem. Phys. 70, 3977 (1979).
50. B.A. Katz, C.E. Strouse, J. Am. Chem. Soc. 101, 6214 (1979).
51. A.M. Greenaway, C.J. O'Connor, A. Schrock, E. Sinn, Inorg. Chem. 18, 2692 (1979).
52. A.M. Greenaway, E. Sinn, J. Am. Chem. Soc. 100, 8080 (1978).
53. M. Sorai, J. Ensling, K.M. Hasselbach, P. Gütlich, Chem. Phys. 20, 951 (1977).
54. I. Sanner, E. Meissner, H. Köppen, H. Spiering, P. Gütlich, Chem. Phys. (1984), in press.
55. E. Meissner, H. Köppen, H. Spiering, P. Gütlich, Chem. Phys. Lett. 95, 163 (1983).
56. L. Wiehl, C.P. Köhler, P. Gütlich, to be published.
57. P. Gütlich, H. Köppen, H.G. Steinhäuser, Chem. Phys. Lett. 74, 475 (1980).
58. M. Sorai, J. Ensling, P. Gütlich, Chem. Phys. 18, 199 (1976).
59. P. Gütlich, R. Link, H.G. Steinhäuser, Inorg. Chem. 17, 2509 (1978).
60. P. Adler, H. Spiering, P. Gütlich, to be published.
61. D.B. Chesnut, J. Chem. Phys. 40, 405 (1964).
62. J. Wajnflasz, Phys. stat. sol. 40, 537 (1970).
63. R.A. Bari, J. Sivardiére, Phys. Rev. B5, 4466 (1972).
64. C.P. Slichter, H.G. Drickamer, J. Chem. Phys. 56, 2142 (1972).
65. R. Zimmermann, E. König, J. Phys. Chem. Solids 38, 779 (1977).
66. T. Kambara, J. Chem. Phys. 70, 4199 (1979).
67. T. Kambara, J. Phys. Soc. Japan 49, 1806 (1980).
68. S. Ramasesha, T.V. Ramakrishnan, C.N.R. Rao, J. Phys. C (Sol. State Phys.) 12, 1307 (1979).
69. S. Ohnishi, S. Sugano, J. Phys. C (Sol. State Phys.) 14, 39 (1981).
70. J.D. Eshelby, Sol. State Phys. 3, 79 (1956).
71. R. Zimmermann, H. Spiering, Phys. Stat. Sol. B67, 487 (1975).
72. J.H. Takemoto, B. Hutchinson, Inorg. Chem. 12, 705 (1973).
73. H.G. Drickamer, C.W. Frank: Electronic Transitions and the High Pressure Chemistry and Physics of Solids, Chapman and Hall, London 1973.
74. J.R. Ferraro, J. Takemoto, Appl. Spectroscopy 28, 66 (1974).

75. G.J. Long, L.W. Becker, B.B. Hutchinson, in: "Mössbauer Spectroscopy and its Chemical Applications" (J.G.Stevens, G.K. Shenoy, eds.) Adv. Chem. Series, Am. Chem. Soc., Washington, D.C. 1981, p. 453.

76. E. Meissner, in partial fulfilment of the doctoral thesis, University of Mainz, Physics Department, 1984.

77. P.L. Franke, J.G. Haasnoot, A.P. Zuur, Inorg. Chim. Acta 59, 5 (1982).

78. E.W. Müller, J. Ensling, H. Spiering, P. Gütlich, Inorg. Chem. 22, 2074 (1983).

79. S. Decurtins, P. Gütlich, C.P. Köhler, H. Spiering, Chem. Phys. Lett. 105, 1 (1984).

80. S. Decurtins, P. Gütlich, A. Hauser, K.M. Hasselbach, H. Spiering, to be published in Inorg. Chem.

81. S. Decurtins, P. Gütlich, C.P. Köhler, H. Spiering, J. Am. Chem. Soc., submitted.

82. R. Grimm, P. Gütlich, E. Kankeleit, R. Link, J. Chem. Phys. 67, 5491 (1977).

ZERO AND HIGH FIELD MÖSSBAUER SPECTROSCOPY STUDIES OF THE MAGNETIC

ORDERING BEHAVIOR OF ONE, TWO AND THREE DIMENSIONAL SYSTEMS

William Michael Reiff

Department of Chemistry
Northeastern University
Boston, Massachusetts 02115    USA

ABSTRACT

The use of zero and high field iron-57 Mössbauer spectroscopy in the study of magnetic ordering phenomena is reviewed. Highlights of cooperative magnetic behavior are presented for 3D salt, 2D layer, and 1D chain examples. Correlations with the results of single crystal structure determinations as well as susceptibility measurements are emphasized.

## I. INTRODUCTION

In many contexts the nucleus is essentially independent of the chemical state of its associated (electronic) atom owing to the normally large differences in nuclear and chemical energies. However, there are a number of weak nuclear-chemical environment interactions such that the line widths of the typical Mössbauer gamma rays are <u>smaller</u> than the interactions allowing their study by Mössbauer spectroscopy.(1,2) Two of these interactions: (a) the quadrupole splitting resulting from the interaction of a nuclear quadrupole moment with an electric field gradient and (b) the magnetic hyperfine splitting resulting from the nuclear Zeeman effect of the large magnetic fields originating from the paramagnetism of an atom's own electrons (orbital and spin moments) are <u>inherently directional</u>. This leads to a variety of magneto-structural correlations based on the zero and high field Mössbauer spectroscopy of cooperatively ordered systems.

It is very fortunate that iron-57 exhibits one of the readily observed Mössbauer effects. The range of spin states for the common oxidation states of iron is extensive: $S = 0$, $S = 1$, and $S = 2$ for iron II; $S = 1/2$, $S = 3/2$ and $S = 5/2$ for iron III. In addition, there are now a number of well documented $S = 1$ iron IV (3) systems and $S = 1/2$ iron I compounds.(4,5) Structural isomorphism of iron compounds to other first transition series metal ion analogues (especially those of Co, Mn, Ni) is often the rule. Thus, sound inferences concerning the magnetic and electronic properties of compounds of such ions are feasible either through simple comparison or doping studies. Finally, to this investigator's experience, there is no type of magnetic phenomena, save perhaps spin-Peirls transition, that is not emphasized in some iron compound. These include behavior such as: (1) rapidly relaxing Curie and Curie-Weiss paramagnetism; (2) slowly relaxing paramagnetism; (3) spontaneous three dimensional cooperative ordering, typified in ferromagnetism, antiferromagnetism, or ferrimagnetism; (4) magnetic low dimensionality in one dimensional chain and two dimensional layer structures; (5) temperature dependent magnetic anisotropy-Morin transition; (6) intramolecular anti- and ferromagnetism in dimeric and higher polynuclear clusters. In applied fields one can observe: (1) spin-flop and (2) metamagnetic transitions as well as (3) field induced ordering and (4) field induced slow paramagnetic relaxation. In what follows, we give illustrative examples of many of the preceding magnetic phenomena as studied by iron-57 Mössbauer spectroscopy, the details of basic nuclear gamma resonance theory as well. Other applications of the Mössbauer effect are given in a number of books and review articles.(6,7,8,9,10,11).

For many problems in magnetism, Mössbauer spectroscopy is an excellent "stand alone" zero field technique. However, it is especially powerful when its application is correlated with the results of classical susceptibility, magnetic heat capacity, neutron diffraction, and electron spin resonance measurements. This aspect will be also emphasized where possible. For low dimensionality magnetic materials, it should be emphasized that using Mössbauer spectroscopy, one usually observes only the ultimate 3D ordering owing to interchain or interlayer interactions in 1D or 2D magnets, respectively. The resolution of 3D from lower dimensionality magnetic exchange correlation effects in the same measurement has been achieved through studies of the temperature dependence of magnetic heat capacity $C_M$. Here one generally observes the characteristic sharp $\lambda$ anomaly corresponding to 3D order and loss of "spin entropy". At somewhat higher temperatures a broad maximum in $C_M$ vs. T is evidence of a low dimensionality (1D or 2D) effect. The other technique readily capable of resolving the preceding three and lower dimensionality effects is of course neutron diffraction. Finally, classical susceptibility measurements ($\chi$ vs. T) on low dimensionality compounds compliment Mössbauer studies in that they emphasize the low dimensionality behavior as opposed to 3D. Thus, for instance, broad maxima in $\chi$ vs. T are observed corresponding to 1D or 2D antiferromagnetism while a sharp Néel 3D maximum in $\chi$ vs. T is often not resolved. Before considering various examples of magnetic ordering behavior, we discuss some fundamental aspects of nuclear Zeeman splitting in zero and non-zero external fields.


## II. HYPERFINE INTERACTIONS

### (A) Effective magnetic fields

In general, the effective magnetic field $H_{eff}$ at a Mössbauer nucleus is related to the applied field ($H_o$) and internal hyperfine field ($H_n$) by

$$H_{eff} = H_o + H_n \tag{1}$$

We are interested in spectra of quadrupole and non-quadrupole split rapidly relaxing paramagnetic systems as well as the cases of slow relaxation and pure Zeeman interaction which are considered first. The following situations arise where $\Delta E$ is the quadrupole splitting.

(1) $\Delta E = 0$, $H_n \neq 0$. In this instance the characteristic six-line pure Zeeman pattern (Figure 1(a), (b)) of a magnetically ordered system is observed. A similar spectrum is also observed for a slowly relaxing paramagnet in the limit of infinitely long relaxation time between the components of its Kramers´ doublets. The spectra correspond to the normally allowed transitions ($\Delta M_I = 0, \pm 1$) among the magnetically split $I = 1/2$ ground and $I = 3/2$ excited states with magnetic field relaxation time long relative to the nuclear Larmor precession frequency.

(2) $\Delta E = 0$, $H_n = 0$, $H_0 \neq 0$ and parallel to the $\gamma$-ray. This situation corresponds to an effective magnetic field equal to the applied, as in the case of a diamagnet or a rapidly relaxing paramagnet whose metal ions are in sites of cubic symmetry. The result is the spectrum of Figure 1(c). One sees that transitions 2 and 5 vanish owing to their $\sin^2\theta$ dependence. The angle $\theta$ between the direction of $\gamma$-ray propagation and $H_0$ is zero for a longitudinally (axial) applied field. The energy separation of transitions 1-6 is $3\alpha + \beta$ and is, of course, proportional to the applied field. The

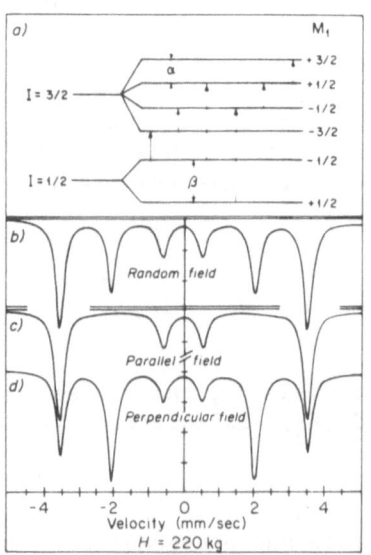

Figure 1. Pure Zeeman Splitting

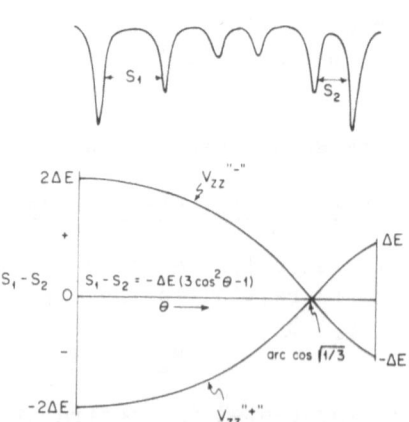

Figure 2. Combined Zeeman Splitting and Quadrupole Interaction.

quantities $\alpha = g_{3/2}\beta_N H$ and $\beta = g_{1/2}\beta_N H$ where $g_{3/2}$ and $g_{1/2}$ are the excited and ground state gyromagnetic ratios and $\beta_N$ is the nuclear magneton. For iron-57, the ratio $\beta/\alpha = g_{1/2}/g_{3/2}$ has been determined using Mössbauer spectroscopy (12) and has the value -1.715. The ground state <u>gyromagnetic</u> ratio has been measured independently using NMR (13) and is 0.1828. Thus with the preceding values of $\alpha, \beta$ etc. and a known applied field, the total splitting $3\alpha + \beta$ is determined. In the case of the Mössbauer calibrant $\alpha$-iron; the splitting is 10.626 mm/sec, corresponding to $H_n$ = 330 kGauss. These numbers correspond to the saturated ferromagnetic state at ambient temperature. Single-line ($\Delta E = 0$) diamagnetic absorbers such as $K_4Fe(CN)_6$ and TiFe (14) alloy also exhibit spectra such as that illustrated in Figure 1(c) in large external (<u>axial</u>) fields. Hence the perturbed Mössbauer spectra of such a system in conjunction with the internal field of $\alpha$-iron can be used for calibration of superconductive magnet.

(3)  $\Delta E = 0$, $H_n = 0$, $H_o \neq 0$ and perpendicular to the $\gamma$-ray. The comments for this case are similar to those for case (2), except that transitions 2 and 5 are intensified relative to 1, 3, 4, and 6 which are all expected to be weaker from their $1+\cos^2\theta$ dependence (Figure 1(d)).

(4)  $\Delta E \neq 0$, $H_n = 0$, $H_o \neq 0$, $\Delta E \gg H_o$ and $H_{eff} = H_o$. This corresponds to the perturbed spectrum of a <u>quadrupole split dia-magnet</u> (as in Figure 3 for ferrocene) or rapidly relaxing para-magnet at room temperature and where the quadrupole splitting of the nuclear levels is assumed to be a larger perturbation than nuclear Zeeman splitting. Under these circumstances the Curie susceptibility is either zero or small while spin-lattice relaxation times are also small. Hence an externally applied magnetic field causes negligible magnetization and $H_{eff} \approx H_o$. The approximate value of $H_{eff}$ is determined from the observed triplet splitting ($\Delta_t$) for a spectrum such as shown in Figure 3 using the relation

$$H_{eff} \approx (\Delta_t/10.626)((3\alpha + \beta)/(2\alpha + \beta))330 \qquad (2)$$

where $3\alpha + \beta$ and $2\alpha + \beta$ are the theoretical doublet and triplet splittings respectively. Since $\beta = 1.715\alpha$, equation (2) simplifies to

$$H_{eff} \approx (\Delta_t/10.626)(4.715/3.715)330 = 39.29 \, \Delta_t \qquad (3)$$

Some brief comment about the relative magnitudes of the doublet and triplet splittings is in order. The theoretical values for these quantities differ. However, for the perturbed spectrum of an isotropic powder, the doublet to triplet splitting ratio is in practice nearly unity. This is due to the "band" nature of a powder spectrum with the greatest broadening usually occuring for the $\pi$ transition. Thus the <u>apparent</u> doublet and triplet splittings are nearly equal.

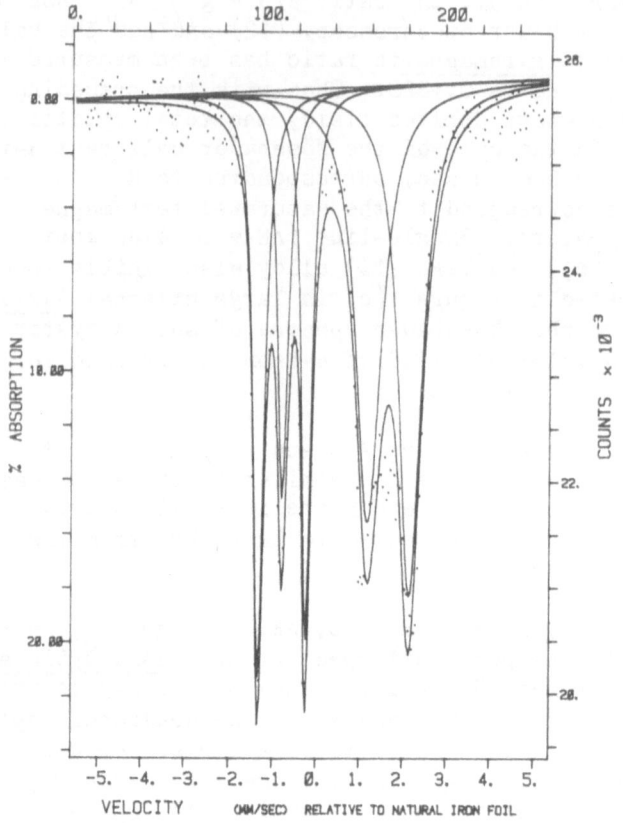

Figure 3. Ferrocene, $H_o$ = 46.5 kG, T = 4.2K

(5) $\Delta E \neq 0$, $H_o \neq 0$, $\Delta E > H_o$, $H_{eff} > H_o$ or $H_{eff} < H_o$. Paramagnetic systems at low temperature may exhibit the foregoing behavior. Spin-lattice relaxation times are now longer and Curie suscepti-bility high. Thus a relatively small external field results in large magnetization. The effective field as measured by the triplet splitting can be larger (or smaller) than the applied field depending on the sign and magnitude of the internal hyperfine field $H_n$. The various contributions to $H_n$ are considered subsequently and can often be expressed as a Brillouin function of $H_o$, magnetic

moment and temperature. A spectrum like that of Figure 3 may still be observed or the triplet replaced by a resolved quartet. Finally, significant magnetic anisotropy and crystallite orientation in the applied field may be evident from a triplet (quartet) to doublet splitting ratio deviating from unity.

(B)  Combined Nuclear Zeeman and Quadrupole Splitting, $H_n \gg \Delta E$, $\underline{H_o = 0}$

The spectrum in this case is like that of Figure 1(a,b) except that the center of the inner four transitions is shifted toward lower or higher energy relative to the center of transitions 1 and 6 because of the non-zero quadrupole. This is shown as an asymmetric pattern in Figure 2. For axial symmetry this shift may be related to the angle $\theta$ between $H_n$ and the principal axis of $V_{zz}$ by the relation

$$S_1 - S_2 = -\Delta E(3 \cos^2\theta - 1) \qquad (4)$$

where $S_1$ is the separation of transitions 1 and 2 and $S_2$ that of transitions 5 and 6, and $\Delta E$ is the quadrupole splitting determined from the paramagnetic phase. Thus for pure Zeeman splitting, as in Figures 1(a)-(d), $S_1-S_2 = 0$ and a symmetric pattern is evident. A graph of $S_1-S_2$ versus $\theta$ along with an exemplary Mössbauer spectrum is given in Figure 2. In this figure, it is assumed that the Zeeman splitting is a significantly greater perturbation than the quadrupole interaction. It is seen that for the $S_1-S_2 < \Delta E$, two values of $\theta$ corresponding to opposite signs for $V_{zz}$ are possible and the angle $\theta$ cannot be determined from the six-line combined interaction spectrum alone. It is of course precisely the angle $\theta$ that is desired for a magneto-structural correlation. An independent determination of the sign of $V_{zz}$, e.g. from the magnetically perturbed spectrum at a temperature where the material is simple rapidly relaxing paramagnetic or a study of the $\pi/\sigma$ intensity in a single crystal, enables unique determination of the angle $\theta$ from the Zeeman-quadrupole spectrum. From Equation 4 it is apparent that a symmetric spectrum ($S_1-S_2=0$) also results when arc $\cos\theta = (1/3)^{1/2}$ ($\sim 54°$). Thus the observation of a symmetric hyperfine split spectrum cannot be taken as conclusive evidence of the absence of a quadrupole interaction. In general, spectra such as shown in Figure 2 are exhibited by three dimensionally ordered materials at $T < T_{critical}$ or paramagnets in the limit of slow relaxation and where the metal ions are in sites of lower than cubic symmetry. In both instances there is almost always a large internal hyperfine field in zero field applied. The only exception to this is the rare case in which the various contributions to $H_n$ happen to cancel e.g. the antiferromagnetically ordered state of $FeCl_2$. The various contributions to $H_n$ are now considered.

## (C)  Contributions to the internal hyperfine field ($H_n$)

The large internal hyperfine fields exerted on the nuclei of atoms in exchanged coupled, cooperatively ordered or slowly relaxing systems ultimately arise from the paramagnetism of the atom's electrons and can be divided into three components (6,7):

(1)  $H_S \alpha \{|\Psi_\uparrow(0)|^2 - |\Psi_\downarrow(0)|^2\}$ (the Fermi contact term);

(2)  $H_L \alpha \langle r^{-3}\rangle \langle L_z\rangle$ or $H_L \alpha \langle r^{-3}\rangle (g-2)\langle S_z\rangle$ (the orbital moment contribution); and

(3)  $H_D \alpha \langle r^{-3}\rangle \langle 3\cos^2\theta - 1\rangle \langle S_z\rangle$ or alternatively $H_D \alpha\ V_{zz}/e\langle S_z\rangle$ (the dipolar interaction).

Thus, $H_{eff} = H_o + H_n = H_o + H_S + H_L + H_D$. It is seen that (1) is the result of an imbalance of "s" electron $\alpha$ and $\beta$ spin density at the nucleus. This imbalance results from the polarization caused by the differential interaction of "s"($\alpha$) and "s"($\beta$) with unpaired valence shell "d" and "f" electrons. On the other hand, (2) is directly related to the magnitude of orbital angular momemtum (L) for a particular unpaired valence shell electron of radius r. It is clear that (3) is operative only for sites of less than cubic symmetry and is the result of a through space interaction of the valence shell spin angular momentum with that of the nucleus. $H_L$ and $H_D$ can either oppose or add to $H_S$. For the case of iron, $H_S$ is ∿11T/unpaired electron (IT = 10 kiloGauss). Thus for high spin iron III, $^6A$ ground term, in cubic symmetry where L, $H_L$, and $H_D$ are all ∿0, limiting low temperature values of $H_n$ ranging between 40 to ∿60T can be observed. For high spin iron II, somewhat smaller values of $H_n$ (to ∿35T) are observed owing to a combination of the smaller value of ⟨S⟩, namely two, and the possibility of an orbital contribution, $H_L$ that generally opposes $H_S$. Reduction of $H_n$ for any oxidation state is also related to variable covalency and delocalization effects depending on the ligands. This can lead directly to delocalization of metal ion "S" spin density and an increase of $\langle r^3\rangle$ for the valence shell electrons. These are effects that result in a decrease of any of $H_S$, $H_L$ or $H_D$ in view of their respective equations. An additional effect leading to substantially reduced values of $H_n$ is the so-called "zero point spin reduction" present in certain low dimensional magnetic systems (15). An example of this will be discussed subsequently.

## III. SINGLE ION ZERO FIELD SPLITTING AND SLOW PARAMAGNETIC RELAXATION VERSUS COOPERATIVE THREE DIMENSIONAL ORDER

Slow paramagnetic relaxation-hyperfine splitting is a dynamic single ion effect resulting in part from the zero field splitting of the ground (electronic) spin manifold. This is shown in Figure 4 for the spin sextet of high spin iron III. Since this is a

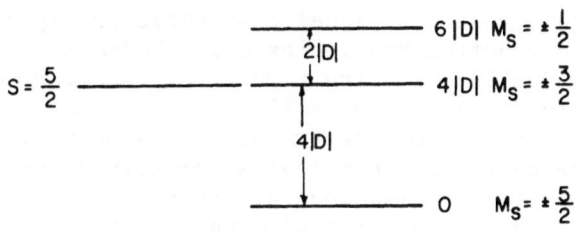

HIGH SPIN FERRIC
D   Negative

Figure 4.   Zero Field Splitting for High Spin $Fe^{3+}$.

Kramers ion, the ground state of the zfs ion must be the doubly
degenerate $M_S$ = ±1/2 (D>0) or $M_S$ = ±5/2 (D<0) in zero applied
magnetic field.   Relaxation in the former is rapid while that in
the latter as well as $M_S$ = ±3/2 is slow.   We will consider only the
case high spin iron III with D large and negative and at low temperatures. In this situation the dominant relaxation mechanism is
spin-spin relaxation (interatom exchange of $S_z$ values) as opposed
to spin-lattice that is more important to L ≠ 0 ions such as high
spin $Fe^{2+}$.   If the metal ions are closely situated so as to allow
for rapid interatom spin flips via direct dipolar interactions (but

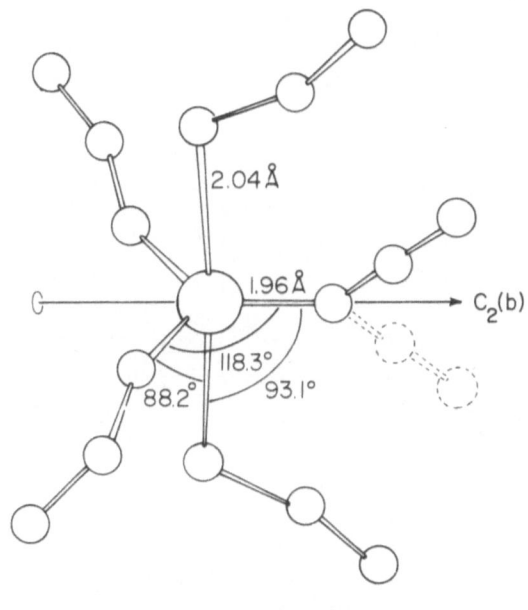

$[(\phi)_4 As]_2 Fe(N_3)_5$

Figure 5.   Local Coordination (16) of the $Fe(N_3)_5^{2-}$ Anion.

not close enough for direct magnetic exchange or superexchange) there will be no spectral broadening even though a "slowly" relaxing ($M_S = \pm 5/2$) doublet is being populated. However, dilution of the metal ions to distances typically $\geq 7.5$ Å leads to longer spin-spin relaxation times whose reciprocal corresponds to a dynamic, temperature dependent frequency that eventually becomes comparable to the Larmor precession frequency of the nuclear moment (17). This typically leads to the <u>gradual development</u> of a non-zero time average value of $H_n$ and <u>gradual</u> Zeeman splitting of the Mössbauer spectrum as the $M_S^n = \pm 5/2$ doublet is progressively populated. One observes a <u>single</u> six line pattern in the limit of low T since the $M_S = \pm 3/2$ at 4D is little populated. An example (18) of this is shown in Figure 6 for $[\Phi_4 As]_2 [Fe(N_3)_5]$ (Figure 5) whose ferric ions are already <u>highly diluted</u> (smallest Fe-Fe distance $\sim 10$ Å) in the <u>undiluted compound</u>. The limiting value of $H_n$ is well defined for each doublet and changes relatively little with temperature over the course of the hyperfine splitting process. More complicated situations such as slow relaxation for high spin $Fe^{2+}$, small D values and thus participation of more than one Kramers doublet (e.g. Fe(acetylacetonate)$_3$, D $\sim 0.02$ cm$^{-1}$, or the now famous case of hydrated ferric ammonium sulfate), are not considered here. Finally, one can have slow relaxation in combination with quadrupole splitting. In this case, knowledge of the angle between the fluctuating $H_n$ and the principal axis of $V_{zz}$ is necessary for full

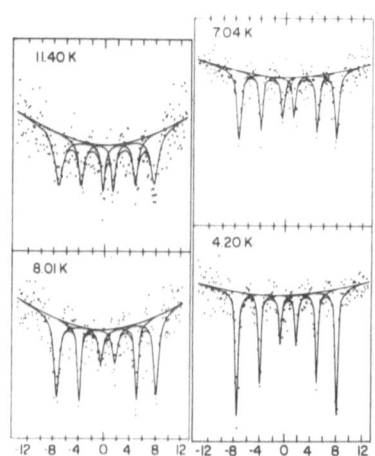

Figure 6. Slow Paramagnetic Relaxation (18) for $Fe(N_3)_5^{2-}$.

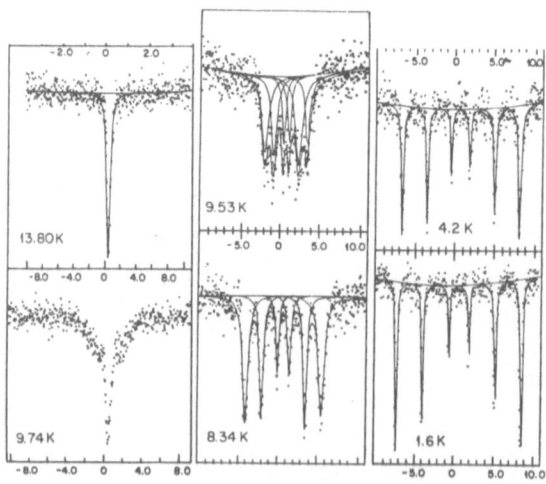

Figure 7. Cooperative 3D-Ordering of Co (1,2-propanediamine)$_3$FeCl$_6$ (21, 22).

interpretation of experimental spectra.  Stochastic models for such relaxation have been published.(19)  A rather complete treatment of relaxation effects in Mössbauer spectra is given by Wickman and Wertheim (20) in Goldanskii and Herber.

The limiting (zero field) low temperature spectra corresponding to three dimensional magnetic ordering processes (antiferromagnetism, ferro- and ferrimagnetism) are identical in appearance to those for slow relaxation.  The difference is that $H_n$ at the individual metal ion sites now corresponds to and is generally thought of as collinear with a spontaneous magnetization developing in the bulk sample.  The latter originates from exchange interactions (either direct e.g. $\alpha$-Fe, or super-exchange e.g. $\alpha$-Fe$_2$O$_3$) between the metal ions that become comparable to or greater than the thermal spin randomization effects as the temperature is decreased and a molecular exchange field develops.  In general and in contrast to slow relaxation, the hyperfine splitting process occurs "suddenly" over a small temperature interval reflecting the cooperative (usually second order) phase transformation of magnetic ordering.  In addition the internal hyperfine field is very temperature dependent in the vicinity of $T_{critical}$ and only levels off as magnetic saturation is reached at low temperatures, often as $T \rightarrow 0^\circ K$.  All of these features are seen in Figure 7 for [Co(1,2-propane-diamine)$_3$][FeCl$_6$] (S = 0 Co$^{3+}$ cation, S = 5/2 Fe$^{3+}$ anion).  The magnetic exchange in this system is embodied in close contacts between centers of delocalized metal ion spin density, namely the chlorine ligands of the paramagnetic anions.  Since the quadrupole splitting is zero for the high symmetry [FeCl$_6^{3-}$] anion, this is an example of pure Zeeman splitting.  The type of magnetic order, antiferromagnetic in this case, is determined from complimentary temperature dependence of susceptibility and field dependence isothermal magnetization studies.  The type of magnetic order can also be inferred from Mössbauer spectra determined in externally applied magnetic fields as discussed subsequently.

The Néel temperature as measured via extrapolation of $H_n \rightarrow 0$ is 9.4K in reasonable agreement with susceptibility results.  Actually, at values for critical temperatures as determined by more precise methods such as classical measurements of the temperature dependence of magnetic heat capacity, Mössbauer spectra sometimes exhibit a substantial non-zero value of $H_n$.  This problem is not dealt with here.  The dimensionality of magnetic order (1, 2, or 3) is related to the so-called "critical exponent" which can be determined from fits to functions such as:

$$H_n(T)/H_n(0^\circ K) \;=\; F(T_c - T)^\beta/T_c \qquad (5)$$

in temperature ranges as close as feasible to $T_{critical}$ ($T_c$) where $\beta$ is the critical exponent.  Aspects of critical exponent theory relevant to Mössbauer spectroscopy studies have been discussed by

Shenoy.(23)  It should be mentioned that hyperfine splitting and ordering may not be sharp but can actually be spread over larger temperature intervals as exemplified in finely divided superparamagnetic materials or highly defect structures.  Rather characteristic Mössbauer spectra are observed in these cases.  Finally, no hyperfine splitting may occur even though other techniques give clear evidence of magnetic order.  This can occur when the various contributions to $H_n$, namely $H_S$, $H_L$, $H_D$ fortuitously cancel e.g. as the case of anhydrous ferrous chloride for which magnetic susceptibility studies indicate $T_{Neel}$ = 23K.  Excellent, detailed examples of the Mössbauer spectroscopy study of <u>cooperative 3D magnetic order</u> for the cases of pure Zeeman and combined Zeeman-quadrupole interaction are $FeF_3$ and $FeF_2$, respectively.(24,25)

## IV.  INTENSITIES - SINGLE CRYSTAL STUDIES

So far all of the discussion has referred to the Mössbauer spectra of isotropic, polycrystalline powders or powders whose paramagnetic moments are polarized to the direction of an applied field, $H_o$.  We now focus on oriented single crystals in zero and non-zero magnetic fields.  For convenience throughout the rest of this chapter, should a field be applied, it will be longitudinal (axial) i.e. parallel to the direction of gamma ray propigation, $\vec{E}_\gamma$.  The orientation will then refer to either the angle ($\theta$) between the principal axis of $V_{zz}$ and $\vec{E}_\gamma$ for a quadrupole split material or the angle between the <u>easy axis or plane of magnetization</u> and $\vec{E}_\gamma$ for a three dimensionally ordered material.  Hopefully the foregoing axes will be collinear with a convenient laboratory crystallographic axis of a favorable unit cell, e.g. orthorhombic or tetragonal.  As with other spectroscopies, the utility of the Mössbauer effect vis ã vis determination of local electronic and extended magnetic structure is significantly enhanced when applied to single crystal samples and this will be seen for several examples (<u>vide infra</u>).  In addition one can also determine <u>polarized</u> gamma ray Mössbauer spectra for single crystal samples through the use of a polarized, magnetically ordered gamma ray source, e.g. a magnetic $\alpha$-Fe foil matrix onto which $Co^{57}$ activity has been electroplated and then subsequently annealed to the interior.  (The I = 5/2 precursor nuclear spin level ultimately leading to the 14.4 kev I = 3/2 Mössbauer excited gamma ray level of iron-57 is populated by the electron capture decay of $Co^{57}$: $Co^{57} + e^- \rightarrow Fe^{57}$.  In any event, the subject of polarized spectra is beyond the scope of the present article and is only referenced here.(26)  For now we summarize the <u>basic angular dependency</u> factors for pure quadrupole and Zeeman split spectra.  The results tabulated refer only to I = 3/2 $\rightarrow$ I = 1/2 gamma ray transitions and come from the more general theory of coupling of two angular momenta states by various types of electro-magentic radiation.

## Magnetic Dipole (M1) Transitions

| Transition | $\Delta M_I$ | Angular Dependence | $\theta = 0$ | $\theta = 90^o$ |
|---|---|---|---|---|
| $\lvert 1/2, \; 1/2 \rangle \rightarrow \lvert 3/2, \; 3/2 \rangle$ | +1 | $3 + 3\cos^2\theta$ | 6 | 3 |
| $\lvert 1/2, \; 1/2 \rangle \rightarrow \lvert 3/2, \; 1/2 \rangle$ | 0 | $4 \sin^2\theta$ | 0 | 4 |
| $\lvert 1/2, \; 1/2 \rangle \rightarrow \lvert 3/2, \; -1/2 \rangle$ | -1 | $1 + \cos^2\theta$ | 2 | 1 |
| $\lvert 1/2, \; -1/2 \rangle \rightarrow \lvert 3/2, \; 1/2 \rangle$ | +1 | $1 + \cos^2\theta$ | 2 | 1 |
| $\lvert 1/2, \; -1/2 \rangle \rightarrow \lvert 3/2, \; -1/2 \rangle$ | 0 | $4 \sin^2\theta$ | 0 | 4 |
| $\lvert 1/2, \; -1/2 \rangle \rightarrow \lvert 3/2, \; -3/2 \rangle$ | -1 | $3 + 3\cos^2\theta$ | 6 | 3 |

## Electric Quadrupole (E2) Transitions

| Transition | | Angular Dependence | $\theta = 0$ | $\theta = 90^o$ |
|---|---|---|---|---|
| $(\sigma)\lvert 1/2, \; \pm 1/2 \rangle \rightarrow \lvert 3/2, \; \pm 1/2 \rangle$ | - | $2 + 3 \sin^2\theta$ | 2 | 5 |
| $(\pi)\lvert 1/2, \; \pm 1/2 \rangle \rightarrow \lvert 3/2, \; \pm 1/2 \rangle$ | - | $3 + 3 \cos^2\theta$ | 6 | 3 |

From the table for E2 transitions, one sees that for $\vec{E}_\gamma \lvert\lvert V_{zz}$, $I(\sigma)/I(\pi) = 0.3$. The less intense $\sigma$ transition occurs at lower Doppler velocity, i.e. $E_\sigma < E_\pi$ when $V_{zz}$ is positive. One considers the integral quotient,

$$I_\pi / I_\sigma = \int_o^\pi (3 + 3\cos^2\theta)d\theta / \int_o^\pi (2 + 3\sin^2\theta)d\theta$$

for polycrystalline (powder) samples. This ratio is unity for such samples in the absence of texture effects or anisotropy of the recoil free fraction (Karyagin-Goldanski effect)(27). Similarly from the M1 results for the polycrystalline powder form of a 3D-ordered ferro or anti-ferromagnet and thus a random orientation of magnetic domains in zero applied field, one expects the familiar 3:2:1:1:2:3 pattern of integrated intensity as in Figure 1B for thin, untextured absorbers. Otherwise for $\vec{E}_\gamma$ parallel or perpendicular to an easy axis (or plane) of magnetization (H again equal to zero) one sees 3:0:1:1:0:3 (Figure 1c) or 3:4:1:1:4:3 (Figure 1d) respectively. The $\Delta M_I = \pm 2$ transitions are normally forbidden or of low intensity and are not considered further herein. The application of these results in obtaining useful magneto-structural correlations for a number of specific examples is considered subsequently.

# V. HIGH FIELD MÖSSBAUER SPECTRA OF ORDERED SYSTEMS AND THE DETERMINATION OF THREE DIMENSIONAL GROUND STATE

(i) <u>Antiferromagnets</u>. We shall now show how the determination of Mössbauer spectra in external fields can aid in distinguishing type of magnetic order present, i.e. the nature of the 3D-ordered ground state, even for polycrystalline samples. Probably the simplest type of antiferromagnet is the two sublattice <u>collinear</u> Néel magnet in which the magnetization and internal hyperfine field of one sublattice $H_n(\alpha)$ is antiparallel to and exactly opposes that of the other $H_n(\beta)$. For a random <u>polycrystalline powder</u> form of such a uniaxial <u>antiferromagnet</u> at $T < T_N$, one expects little if any

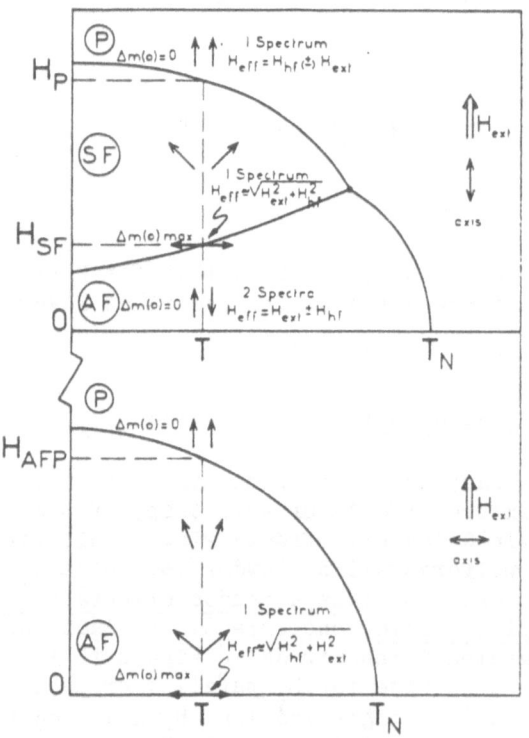

Figure 8. Phase Diagrams for a Uniaxial Antiferromagnet (28).

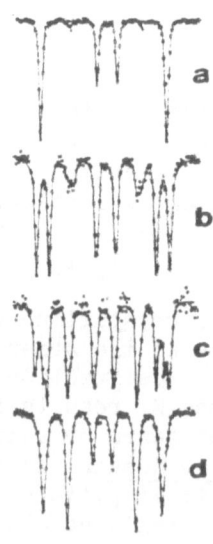

Figure 9. Spectra (29) of single crystal $\alpha$-$Fe_2O_3$ at 4.2K with (a) $H_0 = 0$; b) $H_0 = 65$ kOe; c) $H_0 = 66.5$ kOe; d) $H_0 = 70$ kOe.

effect from an applied field ($H_o$) on its Mössbauer spectrum for moderate applied fields. That is, an approximate 3:2:1:1:2:3 intensity pattern is maintained and the average effective Zeeman-splitting ($H_{eff}$) is unchanged, where $H_{eff} = H_o + H_n$. One simply observes broadening owing to the distribution of angles of sublattice magnetizations with respect to $H_o$. The preceding is the case provided $H_o$ is less than $H_{sf}$, the so-called spin-flop field at which a first order phase transformation corresponding to the flopping of the spins to a direction normal to $H_o$ occurs. The appropriate phase diagrams are shown in Figure 8 (28). Note that for the single crystal form of the uniaxial (collinear Néel) antiferromagnet, $\alpha$-$Fe_2O_3$, in zero field or in a longitudinal field such that $H_o < H_{sf}$ and applied along the easy axis, the $\Delta M_I = 0$ transitions have zero intensity, i.e. a 3:0:1:1:0:3 pattern is observed (Figure 9a). $H_o$ is parallel to the opposing sublattice magnetic hyperfine fields, $H_n(\alpha)$, and $H_n(\beta)$. Thus the highest and lowest velocity transitions of the spectrum are normally observed to be split into symmetric doublets corresponding to the addition of $H_o$ to one of the antiferromagnetic sublattices, say $H_n(\alpha)$ and subtraction of $H_o$ from the other, $H_n(\beta)$ (Figure 9b). For $H_o > H_{sf}$, the foregoing splitting vanishes and the intensity pattern changes to 3:4:1:1:4:3 in accord with the angular components of the selection rules for $\theta = 90^\circ$ (Figure 9c, d for single crystal $\alpha$-$Fe_2O_3$). Note that only one (six line) spectrum is now observed since at the stage of Figure 9d, $H_o > H_{sf}$. We are "well into" the spin-flop phase where $H_o$ is perpendicular to both $H_n(\alpha)$ and $H_n(\beta)$, and thus these sublattices are no longer inequivalent. From 9(b) it is evident that spin flopping has already occurred in some portions of the single crystal. In addition, Figure 9b shows that the transitions just above and below zero velocity have broadened considerably relative to 9a ($H_o = 0$) but are not observed to split into doublets. Clearly the energies of the inner $\Delta M_I = \pm 1$ transitions are not as sensitive to $H_o$ as the outer $\Delta M_I = \pm 1$ transitions. The continuous second order antiferro to "paramagnetic" phase transition of a uniaxial antiferromagnet (bottom of Figure 9) can also be observed when $H_o \perp$ the easy axis. It is clear that the entire $H_o$ vs. T phase diagram of an ordered antiferromagnet can be mapped using Mössbauer spectroscopy. This has been done for the single crystal forms of metamagnetics such as $FeCl_2 \cdot 2\,H_2O$, $FeCl_2$ and more complex systems such as spiral spin structures, e.g. $FeCl_3$ (30).

Before leaving this section on single crystal antiferromagnets, we briefly consider a zero field transition for $\alpha$-$Fe_2O_3$ in which, owing to a small but temperature dependent magnetic anisotropy, the easy axis of magnetization changes orientation relative to $\vec{E}_\gamma$. There is a spin reorientation effect within the A F ordered phase (31). This phenomenon can be seen with somewhat more labor for a polycrystalline sample in terms of a change in the sign of the differential quadrupole shift ($S_1$-$S_2$), vide supra. The spin reorientation is clearly more spectacular and readily observed as a single crystal of $\alpha$-Fe$_2$O$_3$ ($\vec{E}_\gamma$ || C axis) is cooled through the Morin transition temperature ($T_M \simeq 260K$, $T_N = 950K$). For $T_N > T > T_M$, the spins are $\perp$ to the C axis and hence $\vec{E}_\gamma$ thus leading to the approximate 3:4:1:1:4:3 spectrum at the bottom of Figure 10. For $T < T_M$ the spins reorientate || to C and $\vec{E}_\gamma$ and hence the 3:0:1:1:0:3 spectrum at the top of Figure 10 results. Similar "Morin-like" behavior has recently (32) been conjectured from zero field Mössbauer spectroscopy studies of polycrystalline FePO$_4$. Both $\alpha$-Fe$_2$O$_3$ and FePO$_4$ contain high-spin Fe$^{3+}$ with its essentially spherically symmetric $^6$A ground term. It is clear that the small single ion magnetic anisotropy normally expected for this term is one of the requirements necessary for ready observation of Morin-like phenomena or relatively <u>low-field</u> spin flop behavior, for that matter.

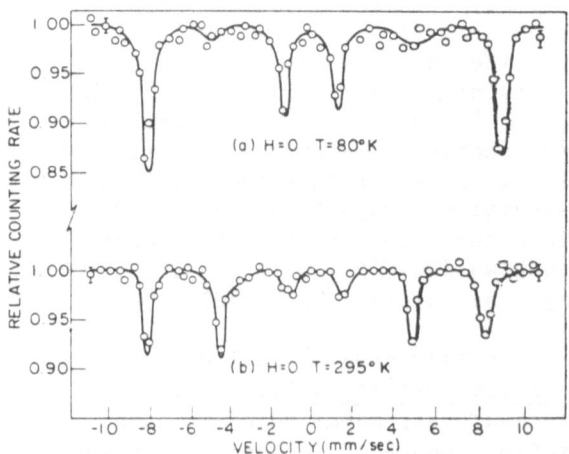

Figure 10. Mossbauer spectra (29,30) of single crystal $\alpha$-Fe$_2$O$_3$ with $\gamma$ || c - axis, above and below $T_m$.

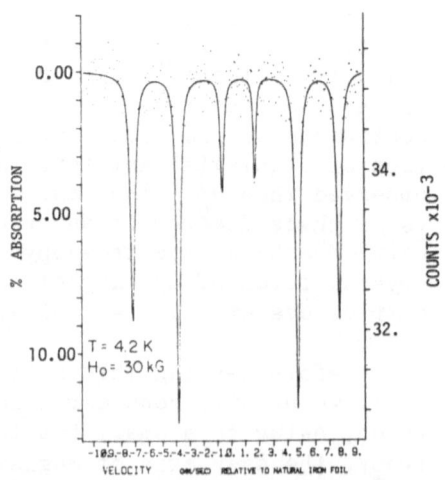

Figure 11. Mössbauer spectrum (22) of Co(1,2-propane-dia-mine)$_3$ FeCl$_6$, $H_0 = 0$; $H_0 = 30$ kG.

The importance of the strength of exchange interaction and degree of magnetic anisotropy are exemplified in the observation of a low-field spin-flop transition for the <u>polycrystalline</u> form of the previously mentioned $[Co(1,2\text{-propanediamine})_3][FeCl_6]$. Susceptibility, magnetization and zero field Mössbauer spectra indicate that this complex is an essentially <u>cubic anti-ferromagnet</u>, $T_N \simeq$ 9.4K. The intensity pattern observed for the powder at $H_o$ = 30 kG is close to 3:4:1:1:4:3, strongly suggesting that the sample microcrystals are polarized by the applied field and that essentially all of the spins have in fact "flopped" when $H_o$ has reached 30kG (Figure 11)(22). The preferential orientation perpendicular $H_o$ with increasing $H_o$ is expected from the fact that for a uniaxial antiferromagnet at $T < T_N$, $\chi_\perp > \chi_{||}$. For comparison, a similar spin-flop transition is observed at $H_o \backsim$ 27 kG for $FePO_4$ ($T_N$ = 25K). This value of $H_{sf}$ comes from detailed analysis of the intensity variation of the Mössbauer spectra of a polycrystalline sample in longitudinal fields (33). There is no evidence of spin-flop behavior in the field dependence of the magnetization for $Co(1,2\text{-propane-diamine})_3[FeCl_6]$ up to $H_o \backsim$ 14 kG. Thus, one concludes that <u>the</u> spin-flop field is somewhere between $\backsim$14 and 30 kG, and that a "sharp" spin-flop transition should be observed in this field range for an appropriately oriented single crystal sample. The spin-flop field ($H_{sf}$) can be approximated by $H_{sf}$ = (2 $H_a H_{ex}/1-[\chi_{||}/\chi_\perp])^{1/2}$ where $H_a$ is the so-called anisotropy field and $H_{ex}$ the exchange field. At $T < < T_N$, $\chi_{||}/\chi_\perp \to 0$ and thus $H_{sf} \simeq$ (2 $H_a H_{ex})^{1/2}$. Hence, one might expect a low value of $H_{sf}$ for $[Co(1,2\text{-propane-diamine})_3][FeCl_6]$ whose exchange is weak and whose ion's local and packing symmetry are high.

(ii) <u>Ferromagnets</u>. Three dimensional ferromagnets are decidedly rarer than antiferromagnets. Nevertheless, they are fundamentally less complex than the latter in that they can be generally viewed as <u>single</u> lattice systems with all spins parallel. There is typically a very large susceptibility even at temperatures slightly above $T_{Curie}$. Thus, <u>unlike antiferromagnets</u>, relatively small values of $H_{ur}$ lead to substantial magnetization and effective internal fields at temperatures just above $T_{Curie}$. An example of this phenomena is shown in Figure 12 (34) for the linear chain ferromagnet, $[(CH_3)_3NH]FeCl_3 \cdot 2H_2O$ (FeTAC), $T_{Curie}$ = 3.10K, whose structure (the iron and cobalt analogs are isomorphous) is shown (35) in Figure 13. In zero field at 3.15K, the spectrum consists of a broad doublet, while an applied field of only 3.06 kG results in essentially fully resolved magnetic hyperfine splitting. An external field of only 5 kG leads to a large hyperfine splitting and an effective field of some 60 kG in the case of the zig-zag chain ferromagnetic polymer ($T_{Curie}$ = 5.0K), $[Fe(bipyridine)Cl2]_\infty$ (36). Isotropic ferromagnets also exhibit rather characteristic high field Mössbauer spectra below $T_{Curie}$. Relatively small axial fields lead to ready polarization for polycrystalline powders, foils, or unoriented single crystals and domain alignment <u>parallel</u>

to H$_0$ and $\tilde{E}_\gamma$. Thus one observes 3:0:1:1:0:3 spectral patterns reminiscent of Figures 1c, 9a, or 10(top). However, in contradistinction to an antiferromagnet with H$_0$ || the easy axis, one never sees more than four transitions since there is only <u>one magnetic lattice</u>, namely the field oriented and now extensive domain of parallel spins.

(iii) <u>Ferrimagnets</u>. Finally we consider the simplest case for a ferrimagnet, a collinear Néel type. A good example is the inverted ferrimagnetic spinel NiFe$_2$O$_4$ (37). Since it is inverted, there must be ferric ions on both A(T$_D$) and B(O$_h$) sites. A collinear spinel of this type will exhibit features characteristic of an antiferromagnet, i.e. two opposing sublattices and thus more than four transitions in longitudinal applied fields. On the other hand, due to the fact that by the definition of a ferrimagnet the sublattices do not completely cancel, there is a large spontaneous magnetization and <u>uncompensated moment</u> in zero field for the ferrimagnetic state. Thus Mössbauer spectra of random polycrystalline powders will exhibit large and <u>facile polarization-domain alignment</u> effects reminiscent of ferromagnetism and achievement of quasi

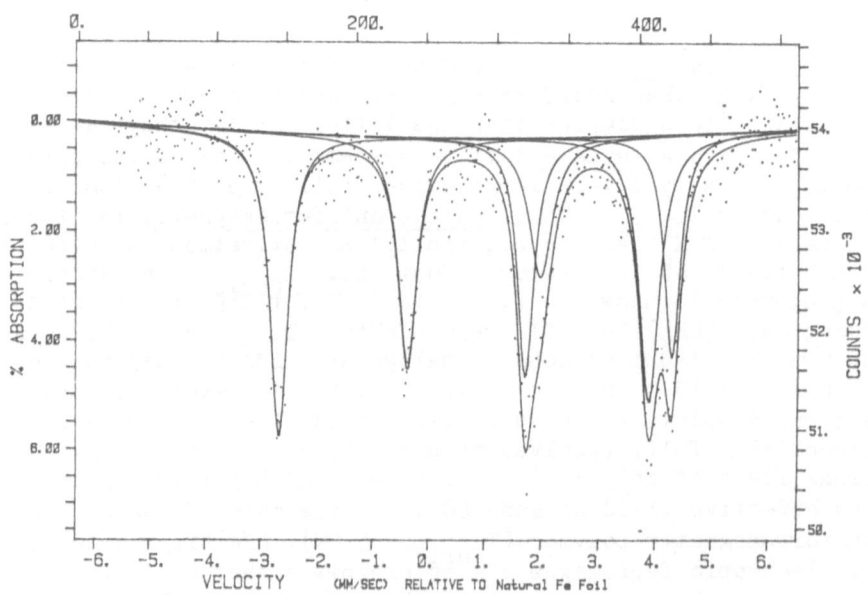

Figure 12.  Spectrum of FeTAC in a Transverse Field of 3.06 kG at 3.15K (34).

3:0:1:1:0:3 intensity patterns in moderate applied fields. All of these features are illustrated in Figure 14 (37). In these powder spectra of $NiFe_2O_4$, the inequivalent sublattices are obvious for even zero applied field, rather definitive evidence for ferrimagnetic nature of the ground state.

## VI. SOME RECENT APPLICATIONS

### (A) Identification of Magnetically Interesting Isomeric Ordered Materials

We now discuss some examples of the application of Mössbauer spectroscopy to the study of the magnetic ordering behavior of a number of novel, new 3D and lower dimensionality systems. These eminate from the laboratory of the author and are as yet largely unpublished, or of only very recent publication.

The chemical literature for the anhydrous ferric chloride-mono α-diimine system is confused. Many purported structures are not well characterized or supported with the available physical data. Part of the problem arises from the high lability of high spin ferric centers and thus simple recrystallization-purification processes may result in a radical change of structure. In addition, there are many possible isomers consistent with the simplest empirical formula, $FeLCl_3$, where L is an α-diimine such as the familiar 2,2´-bipyridine or 1,10-ortho-phenanthroline. In this concluding

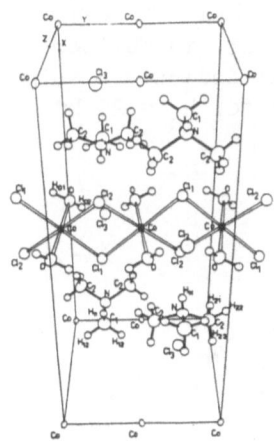

Figure 13. Chain Polymer Structure of Co(TAC) (35).

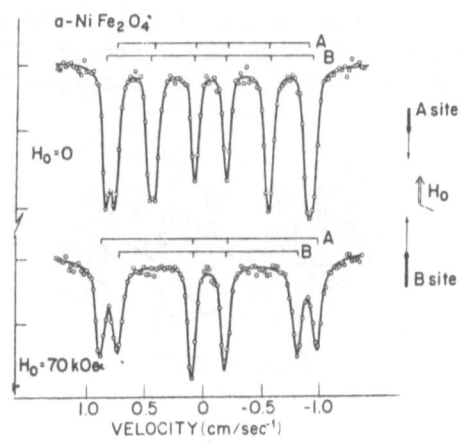

Figure 14. Zero and high field spectra (37) of the spinel α-$NiFe_2O_4$.

section, we show the power of Mössbauer spectroscopy in elucidating the magnetic properties of two isomers in this system. There are: (a) the complex bimetallic salt cis[Fe(2,2´-bipyridine)$_2$Cl$_2$][FeCl$_4$] (S = 5/2 cis-dichloro octahedral cation, S = 5/2 tetrahedral anion) whose structure and ionic packing have been recently determined (38) and (b) [Fe(2,2´-bipyridine)Cl$_2^+$]$_\infty$Cl$_\infty^-$ (S = 5/2 cation) an infinite cationic zigzag chain polymer. The structure of isomer (b) (Figure 15) is simply an extension of that found for the zig chain ferromagnetic polymer, [Fe(2,2´-bipyridine)Cl$_2$]$_\infty$, that contains high spin iron II. Isomer (b) clearly contains one kind of FeIII while (a) contains two. Without laboring the reader with all of the details, the Mössbauer spectra of (a) show gradual slow paramagnetic relaxation-hyperfine splitting for the complex cation over the range 12K to 4.2K giving seven transitions (Figure 16) while at ᷍3.2K the system further undergoes ordering as a ferrimagnet whose saturation spectrum at 1.8K exhibits the expected twelve transitions. Coincidental with the hyperfine splitting at 3.2K is a sharp rise in magnetic moment (μ) typical of a ferrimagnet and for which the paramagnetic Curie temperature, θ, is negative (-7K). The inequivalent lattices of the ferrimagnet are apparently the cationic and anionic sublattices, respectively. The chain polymer, isomer (b), (Figure 15) has ionic chloride between the chains, and exhibits a sharp transition to a 3D antiferromagnetically ordered ground state, $T_N$ = 3.0K in agreement with susceptibility data. Some pertinent zero field spectra are shown in Figure 17 where the very small value of $H_n$ (only 310 kG versus an expected ᷍500 kG for S = 5/2 ferric) correlates well with a high degree of zero point spin reduction, a hallmark of 1D (chain) low dimensionality. Note that only one six line hyperfine pattern is evident consistent with the presence of only one kind of iron site. The AF ground state of [Fe(2,2´-bipyridine)Cl$_2^+$]$_\infty$Cl$_\infty^-$ is further confirmed by high field spectra (Figure 18) that indicate a spin flop transition is near complete for the powder for $H_o$ = 25 kG.

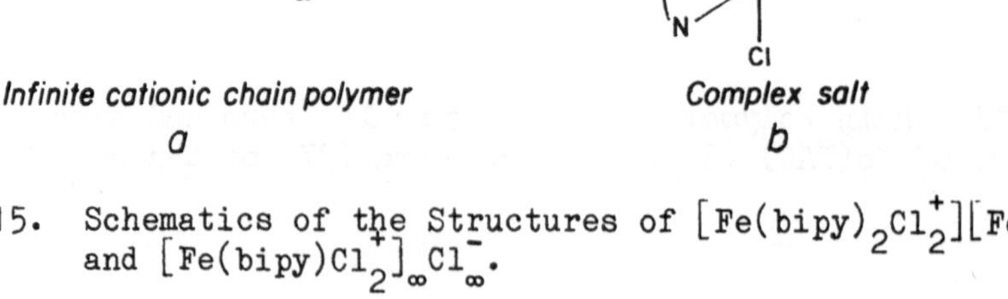

[Fe(Bipy)Cl$_2^+$]$_\infty$ Cl$_\infty^-$

FeCl$_4^-$

Infinite cationic chain polymer

Complex salt

a

b

Figure 15. Schematics of the Structures of [Fe(bipy)$_2$Cl$_2^+$][FeCl$_4^-$] and [Fe(bipy)Cl$_2^+$]$_\infty$Cl$_\infty^-$.

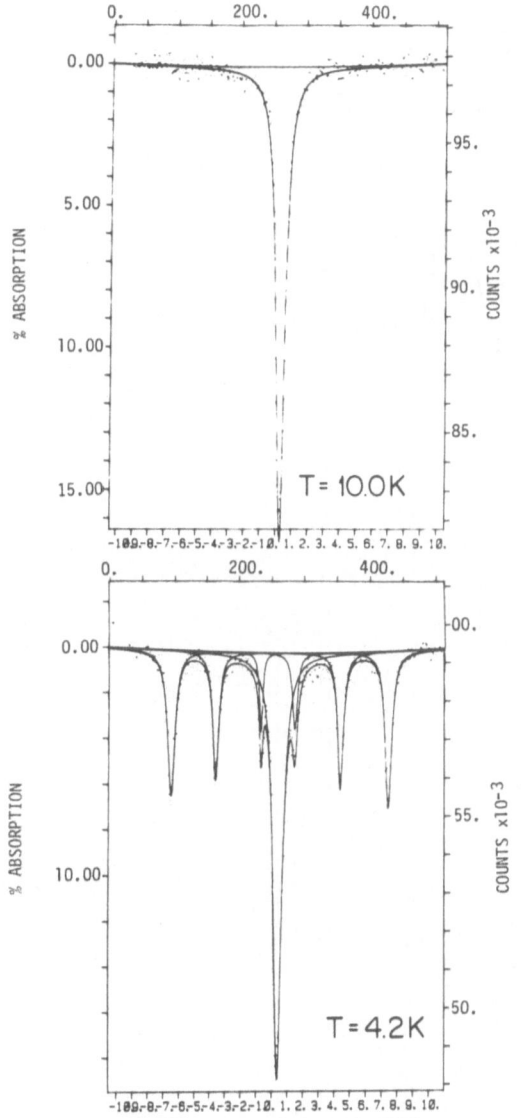

Figure 16. Some zero field spectra of [Fe(bipy)$_2$Cl$_2^+$][FeCl$_4^-$].

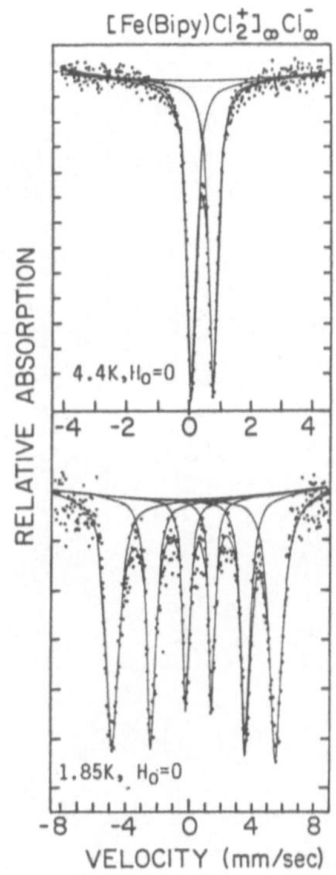

Figure 17. Some zero field spectra for [Fe(bipy)Cl$_2^+$]$_\infty$Cl$_\infty^-$.

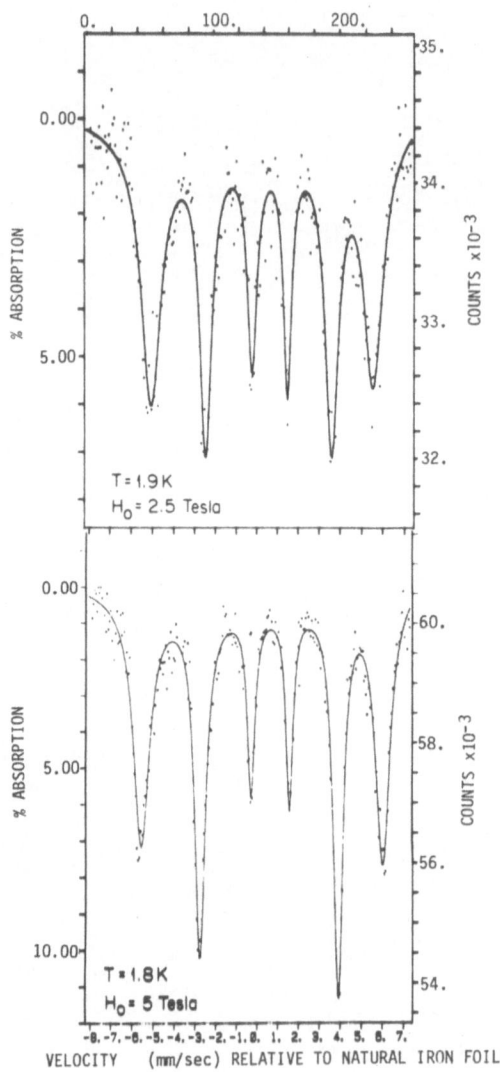

Figure 18. Some high-field spectra for $[Fe(bipy)Cl_2^+]_\infty Cl_\infty^-$.

## (B) 3D Ordering of Complex Bimetallic Salts

There are a number of experiments and systems in this area that will hold our interest for the next several years. We are particularly excited about our recent findings for ferricinium tetrachloroferrate, $[Fe(CP)_2^+][FeCl_4^-]$ and emphasize these here. We were led to this system by our study of the foregoing $[Fe(bipy)_2 Cl_2][FeCl_4^-]$, an S = 5/2, S = 5/2 system. The ferricinium salt is an S = 1/2, S = 5/2 bimetallic system whose single crystal x-ray structure (39) corresponds to isolated layers of the S = 1/2 cations and S = 5/2 anions. A drawing of this based on the published atomic coordinates is shown in Figure 19.

The effective magnetic moment of $[Fe(CP)_2^+][FeCl_4^-]$ per metal atom is 4.41 $\mu_B$ as determined from least squares fits of $\chi_M^{-1}$ vs. T to a Curie-Weiss law (θ = -5.8K, C = 2.44 emu/mole) over the range 50 to 300K. This value is close to the theoretically predicted spin-only moment per metal atom (4.35 $\mu_B$) for a lattice of independent S = 1/2 and S = 5/2 spins in 1:1 ratio. Below ∿26K, one finds the moment per metal atom to decrease dramatically from 4.36 $\mu_B$ to a low of 1.94 $\mu_B$ at 1.66K. The molar susceptibility ($\chi_M$) exhibits a sharp maximum at ∿3.8K below which it becomes strongly field dependent. Essentially coincidental with the preceding magnetic susceptibility transition is (zero field) magnetic hyperfine splitting of the Mössbauer spectrum such that a fully resolved twelve transition spectral pattern is observed resulting from Zeeman splitting of the spin doublet and sextet sublattices, Figure 20. We interpret these preliminary results in terms of a novel complex antiferromagnetic ordering of the material. It may be ferrimagnetic; we are not sure yet. High field Mössbauer spectroscopy and additional magnetization studies should shed light on this problem. In any event, the potential of this system is considerable. It is soluble in organic solvents such as $CH_2Cl_2$ and crystallizes in an orthorhombic space group (Pna2$_1$), suggesting reasonably easy single crystal studies (39). The central question is: what is the nature and mechanism of magnetic exchange in systems such as this? There are no directly bonded inter-cation, inter-anion, or intercation-anion super exchange pathways. We hope to approach these problems as we attempt the modulation of the magnetic behavior of this system by frozen glass studies for iron-57 enriched preparations. This may also be effectuated by preparation of the 1,1′-dimethyl and the decamethyl ferricinium analogs. Further, it should be possible to prepare the isomorphous cobalticinium variant using Co-57, i.e., $[Co^{57}(CP)_2^+][FeCl_4^-]$ (S = 0, S = 5/2) for emission Mössbauer spectroscopy studies. The diamagnetic $Co(CP)_2^+$ can hopefully function as a probe for transferred hyperfine interaction from the paramagnetic $FeCl_4^-$ sublattice in the ordered state. This is, provided the cation of the latter salt can withstand the after effects of nuclear decay. The study of $[Co(CP)_2^+][FeCl_4^-]$ will allow one to investigate magnetic

$[Fe CP_2^+] [FeCl_4^-]$

*Isolated Layers of $S = \frac{1}{2}$ Cations and $S = \frac{5}{2}$ Anions*

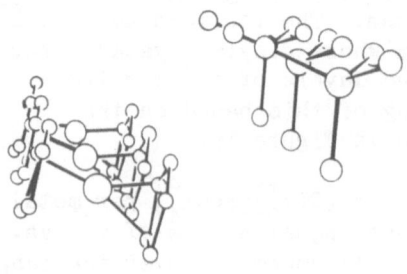

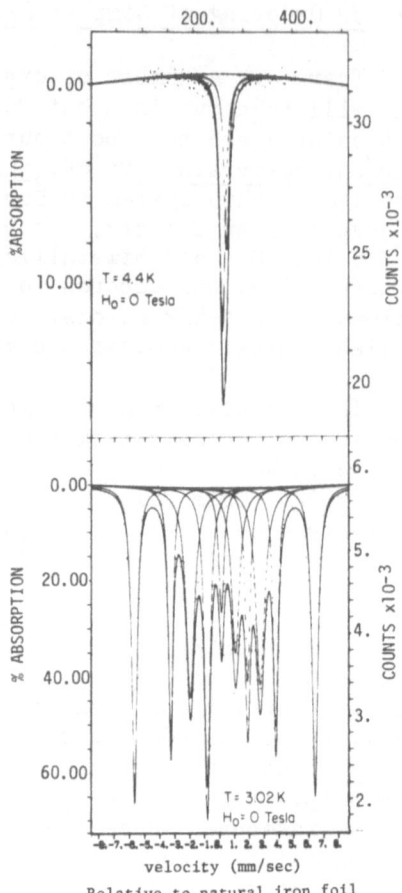

Figure 19. Layer structure (39) of $[Fe(CP)_2^+][FeCl_4^-]$, alternating layers of cations and anions in ortho-rhombic symmetry.

Figure 20. Zero field Mössbauer spectra of $[Fe(CP)_2^+]$ $[FeCl_4^-]$ at 4.40K (top) and 3.02K (bottom).

exchange confined primarily within the anion lattice. Finally, there are no complicating quadrupole splitting effects in the interpretation of its Zeeman split Mössbauer spectra since the EFG tensor at each site is ~0. This is a unique system whose detailed study is clearly worth the effort.

(C)   A 2D Magnet Based on $FeIIIClMoO_4$ Layers

The high spin FeIII compound, $FeIIIClMoO_4$, is a new (40) tetragonal 2D-AF system whose intralayer exchange owing to a tetragonally symmetric net of long (~7A) O-Mo-O pathways is remarkable ($T_{xmax}$ (2D) = 104K, J/kE = -4.1K). This value of J is comparable to that of $[CH_3NH_3]_2MnCl_4$, a 2D-AF having short single atom bridges within the layer. The structure of the new molybdate system is shown in Figures 21 and 22 while some Mössbauer spectra showing its 3D-AF order and powder susceptibility data illustrating the strong 2D effect are given in Figures <u>23, 24, and 25</u>, respectively.  The susceptibility data below 50K to as low as 1.5K show no other magnetic phase transitions.  Furthermore, separate 3D effects in $\chi_M$ coincidental with $T_{N\acute{e}el}$ via Mössbauer are not evident.  The <u>interlayer exchange</u> is the result of a non-bonded FeIII---Cl (3.015 Å) interaction where the axial FeIII---Cl bond (2.209 Å) vector of one layer "completes the octahedron" of the five coordinate iron III atom in the next layer above.  Even though this is a nonbonded interaction and the interlayer Fe-Fe distance is 5.23 Å along the C (tetragonal) axis, the interlayer exchange is quite strong, $T_{N\acute{e}el}$ = 69.2K.  The ambient temperature moment is 4.9 $m_\beta$.  This is a full Bohr magneton below the spin only ( 35) value expected for spin sextet iron III and clear evidence of very strong intralayer exchange extending to high temperatures.  The results suggest that this system is an almost textbook example for the study of isotropic (Heisenberg) 2D-AF behavior.  The local electronic and essential 2D magnetic structures of this system are further confirmed by single crystal Mössbauer spectra for $T > T_{N\acute{e}el}$

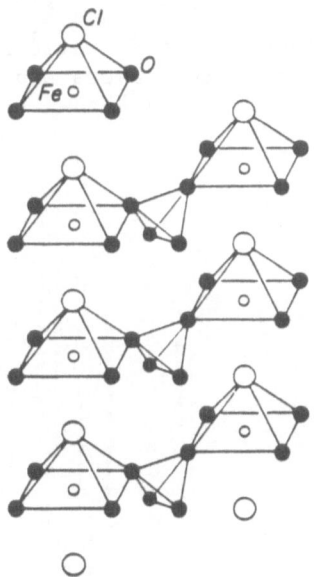

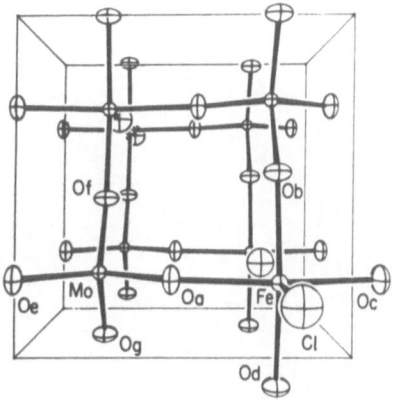

Figure 21. Structure (40) of FeClMoO$_4$ showing layers stacked along the tetragonal (C) axis.

Figure 22. Structure (40) of FeClMoO$_4$ showing how each chlorine atom nestles into the pocket formed by the four FeO$_4$Cl oxygen atoms of a neighboring molecule.

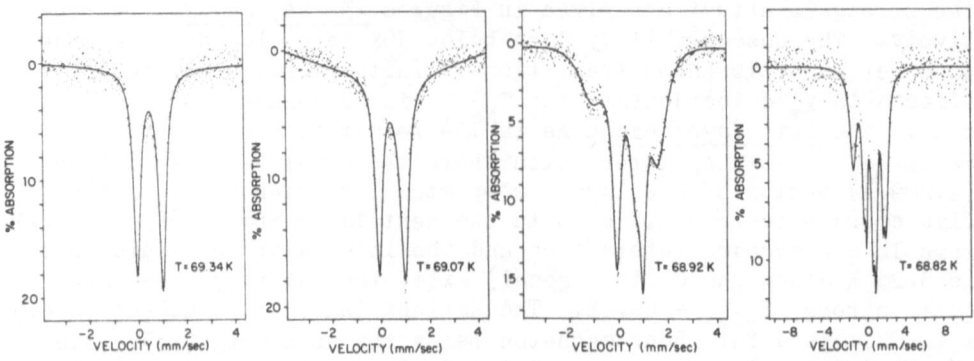

Figure 23.   Some zero field Mössbauer spectra (40) of FeClMoO$_4$ in
the range 69.34K to 68.82K, emphasizing 3D ordering.


and $T < T_{Neel}$. For $T > T_{Neel}$, spectra taken with $\vec{E} \parallel C$ show that
$V_{zz}$ is positive. This is precisely the result expected for a
regular, five coordinate, square pyramidal $[FeO_4Cl]$ chromophore
using simple point charge calculations of the ligand contributions
to the electric field gradient tensor. For $T < T_{Neel}$ and $\vec{E}_Y$ again
$\parallel C$, a 3:4:1:1:4:3 hyperfine split Mössbauer spectral pattern is
observed. This indicates magnetization within the layers and
perpendicular to C.

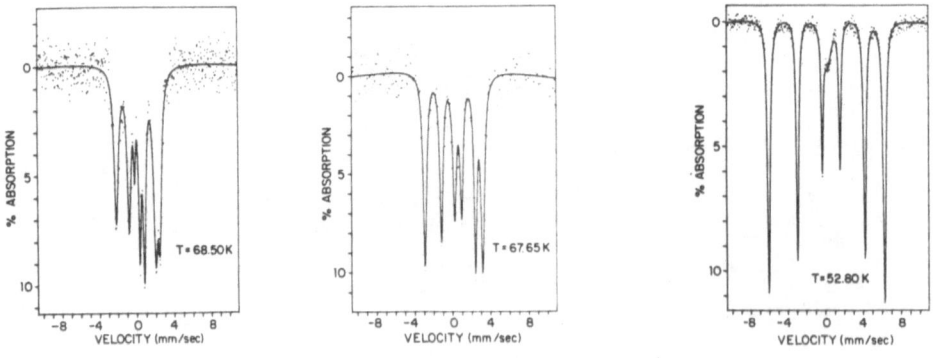

Figure 24.   Some zero field Mössbauer spectra (40) of FeClMoO$_4$ in
the range 68.50K to 52.80K.

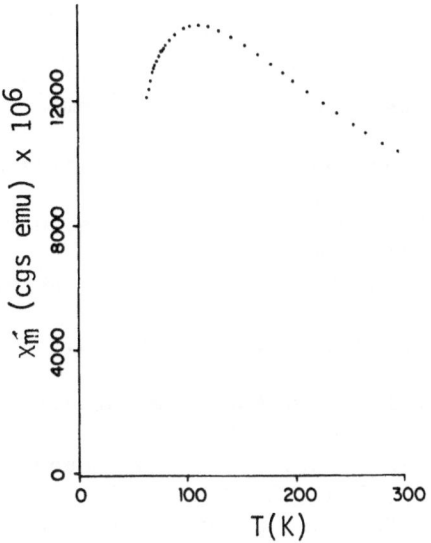

Figure 25. The molar susceptibility (40) $\chi_M^{'}$ vs. T for $FeClMoO_4$ emphasizing a 2D maximum in $\chi_M^{'}$ at $\sim$105K.

CONCLUSION AND ACKNOWLEDGEMENT

There are a number of other important, exciting new areas in which the Mössbauer effect has direct application in the present context, e.g. Mössbauer spectroscopy observation of the high field enhancement and diminution of critical ordering temperatures for low dimensionality systems (41) or observations of solitary excitations (42) (solitons) through careful linewidth measurements just above and below $T_{N\acute{e}el}$. For lack of space and time, we can only reference these recent developments herein. In any event, it is hoped that this chapter has shown the reader the power of Mössbauer spectroscopy as an analytical technique in discovering and understanding magnetic ordering phenomena.

The author is grateful for the continued support of the National Science Foundation, Division of Materials Research, Solid State Chemistry Program, especially under recent grants NSF-DMR 8016441 and 8313710. He also thanks the organizer (Rolfe Herber) for his invitation to present a lecture and prepare this chapter. Finally, he acknowledges fruitful collaborative interactions and discussions with Dr. Charlie Torardi (DuPont Central Research), Dr. Chris Landee (Clark University) and, of course, his own graduate students and postdoctorals at Northeastern University.

REFERENCES

1.  Goldanskii, V.I., and Herber, R.H. (Eds.), Chemical Applica-
    tions of Mössbauer Spectroscopy, Academic Press, New York,
    1968.
2.  Wertheim, G.K., Mössbauer Effect: Principles and Applica-
    tions, Academic Press, New York, 1964.
3.  Reiff, W.M., Dockum, B., and Ooosterhuis, W.T., J. Chem.
    Phys. 67, (1977) p. 3537.
4.  Oosterhuis, W.T., and Lang, G., J. Chem. Phys., 50 (1969)
    p. 4381.
5.  Johnson, C.E., Richards, R., Hill, H.A.O., J. Chem. Phys. 50
    (1969) p. 2594.
6.  Gibb, T.C., "Principles of Mössbauer Spectroscopy," Chapman
    and Hall, Ltd., London, 1976.
7.  Greenwood, N.N., and Gibb, T.C., "Mössbauer Spectroscopy,"
    Chapman and Hall, Ltd., London, 1971.
8.  Bancroft, G.M., "Mössbauer Spectroscopy - An Introduction for
    Inorganic Chemists and Geochemists," McGraw-Hill, London,
    1973.
9.  Gruverman, I.J., Ed., "Mössbauer Effect Methodology," Volumes
    1-10, Plenum Press, 1965-1975.
10. Bancroft, G.M., and Platt, R.H., Adv. Inorg. Chem. and
    Radiochem., 15, (1972) p. 59.
11. Reiff, W.M., Coord. Chem. Rev. 10, (1973) p. 37.
12. Preston, R.S., Hanna, S.S., and Heberle, J., Phys. Rev., 128
    (1962) p. 2207.
13. Ludwig, G.W., and Woodbury, H.H., Phys. Rev., 117 (1960)
    p. 1286.
14. Swartzendruber, L.J., and Bennett, L.H., J. Res. Nat. Bur.
    Stand., Sect. A, 74 (1970) p. 691.
15. Johnson, C.E., "Proceedings of the International Conference
    on the Applications of the Mössbauer Effect," Jaipur, India
    (1981) p. 72.
16. Drummond, J., and Wood, J.S., Chem. Commun. (1969) p. 1373.
17. Wignall, J.W.G., J. Chem. Phys., 44 (1966) p. 2462.
18. Reiff, W.M., Wong, H., Tuiroc, M. and Eisman, G., J. de
    Physique, Colloque C2, 40, C2-234 (1979).
19. Blume, M., and Tjon, J.A., Phys. Rev., 165 (1968) p. 446.
20. Wickman, H.H. and Wertheim, G.K., "Spin Relaxation in Solids
    and After Effects of Nuclear Transformations" in Chemical
    Applications of Mössbauer Spectroscopy, Goldanskii, V.I.,
    and Herber, R.H. (Eds.), Academic Press, New York, 1968.
21. Scoville, A. N., Witten, E., and Reiff, W.M., Inorg. Paper
    No. 22, National ACS Meeting, Houston, Texas, March 1980.
22. Scoville, A. N., Lazar, K., Reiff, W. M., and Landee, C.,
    Inorg. Chem. 22, 3514 (1983).
23. Shenoy, G.K., in "Perspectives in Mössbauer Spectroscopy",
    Cohen, S., and Pasternack, M., Eds., Plenum Press, New
    York, (1973) p. 141.

24. Wertheim, G.K., Guggenheim, H.J., and Buchanan, D.N.E., Phys. Rev., 169, (1968) p. 465.

25. Wertheim, G.K., and Buchanan, D.N.E., Phys. Rev., 161, (1967) p. 478.

26. Perlow, G.J., Hanna, S.S., Hamermesh, M., Littlejohn, C., Vincent, D.H., Preston, R.S., and Heberle, S., Phys. Rev. Letters, 4, (1960) p. 74.

27. Karyagin, S.V., Dokl. Phys. Chem., 148 (1964) p. 110.

28. Chappert, J., J. de Physique, Colloque C 6, C6-71 (1974) p. 35.

29. Blum, N., Freeman, A.J., Shaner, J.W., and Grodzins, L., J. Appl. Phys., 36, (1965) p. 1169.

30. Frankel, R.B., "Mössbauer Effect Methodology," 9, (1974) p. 151 and references therein.

31. Morin, F.J., Phys. Rev. 78, (1950) p. 819.

32. Battle, P.D., Cheetham, A.K., Gleitzer, L., Harrison, W.T.A., Long, G.J., and Longworth, G., J. Phys. C. Solid State Phys. 15, (1982) L919.

33. Beckmann, V., Bruckner, W., Fuchs, W., Ritter, G., and Wegener, H., Phys. Stat. Solidi, 29, (1968) p. 781.

34. Reiff, W. M. and Landee, C., unpublished results.

35. Losee, D. B., McElearney, J. N., Shankle, G. E., Carlin, R. L., Cresswell, P. J. and Robinson, W. T., "An Anisotropic Low-Dimensional Ising System, $[(CH_3)_3NH]CoCl_3 \cdot 2H_2O$: Its Structure and Canted Antiferromagnetic Behavior," "Physical Review B., Volume 8, No. 5, September 1973.

36. Reiff, W.M., Dockum, B., Torardi, C., Frankel, R.B., and Weber, M.A., in "Extended Interactions Between Transition Metal Ions in Transition Metal Complexes," A.C.S. Symposium Series No. 5, A.C.S. Washington, April 1974.

37. Chappert, J. and Frankel, R.B., Phys. Rev. Letters, 15, (1967) p. 70.

38. Witten, E., Reiff, W.M., Foxman, B., and Sullivan, B., Inorganic Paper No. 133, National ACS Meeting, New York, NY, August 1981.

39. Paulus, F. F., and Schäfer, L., J. Organometallic Chem. 144, 205 (1978).

40. Torardi, C. C., Calabrese, J. C., Lazar, K., and Reiff, W. M., J. Solid State Chem. 51, 376 (1984).

41. Cooper, D.M., Dickson, D.P.E., and Johnson, C.E., Hyperfine Interactions, 10, (1981) p. 783.

42. Thiel, R.C., de Graaf, H., and de Jongh, L.J., Phys. Rev. Letters, 47, (1981) p. 1415.

# A MÖSSBAUER EFFECT AND MAGNETIC STUDY OF $Fe_2(SO_4)_3$ AND $Fe_2(MoO_4)_3$, TWO L-TYPE FERRIMAGNETS

Gary J. Long
Department of Chemistry
University of Missouri-Rolla
Rolla, MO 65401

## INTRODUCTION

In his 1948 classification[1,2] of magnetic interactions Néel proposed that a material which was basically antiferromagnetic could show ferrimagnetic behavior, perhaps over a limited temperature range, if the moments of the magnetic ions on the antiferromagnetically coupled sublattices were slightly different. He classified this type of magnetic behavior as L-type ferrimagnetism. The different moments on the magnetic ions might arise from inequivalent exchange coupling pathways in the different sublattices. This difference could occur if the magnetic ions, which may be in quite similar chemical environments, are on crystallographically distinct lattice sites. The different moments on the antiferromagnetically coupled sublattices would lead to ferrimagnetism with a spontaneous magnetization and a large susceptibility. If the chemical environments of the crystallographically distinct magnetic ions are quite similar, then the moments should saturate at low temperatures to similar values. At these low temperatures the material would exhibit normal antiferromagnetic behavior. Hence the magnetic susceptibilty would exhibit a field dependent peak at the temperature with the maximum difference in the magnetic moments, and would fall to a low value at low temperature. This type of behavior is shown schematically in Figure 1.

The Mössbauer effect is an ideal technique to detect L-type ferrimagnetic behavior because the internal hyperfine field observed for a specific iron site is very sensitive to the moment on that site. Indeed the Mössbauer effect, when combined with magnetic susceptibility and neutron diffraction studies, can

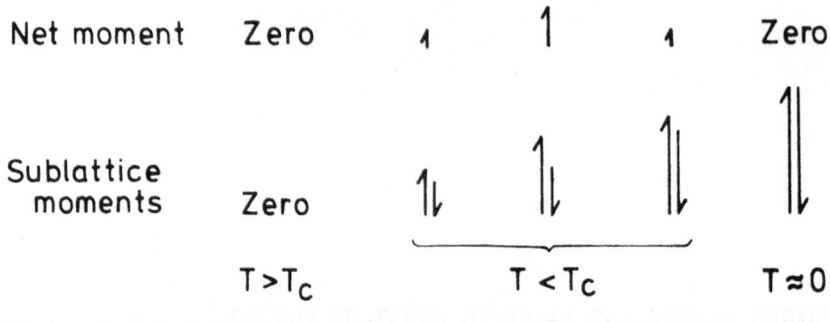

Fig. 1   A schematic representation of the sublattice moments as a
         function of temperature in an L-type ferrimagnet.

provide a detailed picture of the magnetic interaction in an L-
type ferrimagnet.   By a combination of these techniques, it has
been shown that anhydrous iron(III) sulfate[3] and molybdate[4,5]
represent the first examples of L-type ferrimagnets.

## CRYSTAL AND MAGNETIC STRUCTURES

The room temperature crystal structure[6,7] of anhydrous
monoclinic iron(III) sulfate, $Fe_2(SO_4)_3$, reveals an infinite
network of iron-oxygen-sulfur bonds in which the oxygen atoms
coordinate the iron atoms in a nearly   pure octahedral
coordination geometry.   These oxygen atoms also coordinate the
sulfur atoms in tetrahedral geometry in such a way that the
octahedra and tetrahedra share corners.   The resulting structure
is shown in Figure 2, which indicates that the iron atoms are
present on two crystallographically distinct sites in this
monoclinic, $P2_1/n$ (b-unique), structure.   A Rietveld line profile
analysis[3] of powder neutron diffraction data (obtained on the D1A
powder diffractometer at the Institut Laue - Langevin, Grenoble,
France) has revealed the same structure at 4.2K and gives average
iron-oxygen bond distances of 1.992Å and 1.976Å and average
oxygen-iron-oxygen bond angles of 90.1° and 90.0° for the Fe(1)
and Fe(2) sites respectively.   This indicates that there are no
crystallographic phase changes between room temperature and 4.2K
and further that the two crystallographically distinct iron atoms
are present in very similar chemical environments, with close to
octahedral symmetry.

A line profile analysis of the magnetic neutron scattering at
4.2K   has   revealed   that   monoclinic   $Fe_2(SO_4)_3$   is   basically

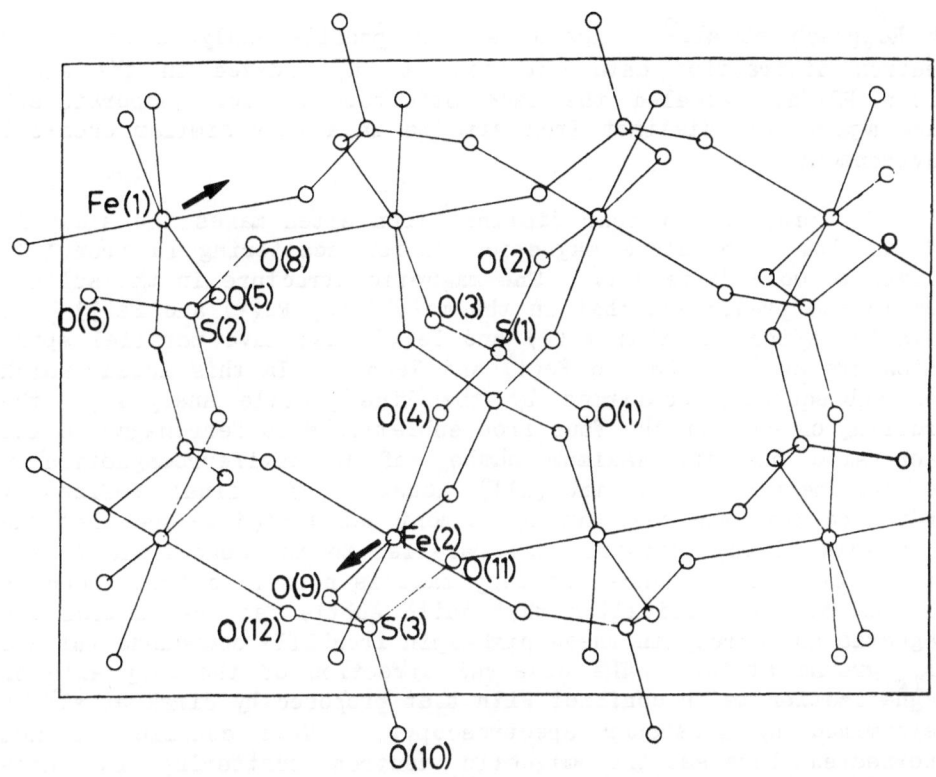

Fig. 2  A projection of the atomic and magnetic structure of
        iron(III) sulfate at 4.2K normal to the crystallographic
        ac plane.

antiferromagnetic with all the Fe(1) ions on one sublattice
antiferromagnetically coupled to the Fe(2) ions with the easy axis
of magnetization normal to the monoclinic b-axis.    This magnetic
structure is represented by the arrows in Figure 2.    The 4.2K
value of the magnetic moment was $4.52(8)\,\mu_B$ for each site.    The
neutron diffraction data was not accurate enough to distinguish
any inequivalence in the moments on the different iron sites, and
in the final analysis, the moments on the two sites were
constrained to be equal.[3]


    The molecular structure of iron(III) molybdate, $Fe_2(MoO_4)_3$,
is very similar to that of the sulfate, but with a longer iron-to-
molybdenum pathway.    Unfortunately its crystal structure is more
complex and has either four crystallographically distinct
iron(III) sites if the structure is in the centerosymmetric space
group $P2_1/a$ (b-unique) found by Chen[8] or eight distinct iron(III)
sites as found in the noncentrosymmetric space group $P2_1$ reported

by Rapposch et al.[9]   Again, a line profile analysis of powder neutron diffraction data obtained at 2K, indexed in the space group $P2_1/a$, revealed the same structure at low temperature.[4] Once again each distinct iron site is in a very similar chemical environment.

The presence of four distinct iron sites makes the analysis of the low temperature magnetic neutron scattering in iron(III) molybdate more difficult.   The magnetic structure in the sulfate led to the prediction that in the molybdate, Fe(1) and Fe(3) have parallel spins and that Fe(2) and Fe(4) also have parallel spins which are antiparallel to Fe(1) and Fe(3).   In this model, which was subsequently confirmed by the line profile analysis,  the ordering on each of the four iron sublattices is ferromagnetic and each site has the maximum number of 12 antiferromagnetically coupled near neighbor iron(III) ions.   The final refinement indicated that the best average moment was $4.34(3)\,\mu_B$ and that the easy axis of magnetization was parallel to the monoclinic b-axis of the crystal.   Hence the easy axis is normal to that found in the sulfate, an indication that quite subtle factors control the magnetic anisotropy in these high-spin iron(III) compounds  with a $^6A_{1g}$ ground state.   The observed direction of the easy axis of magnetization is in conflict with that proposed by Jirak et al[5] as determined by Mössbauer spectroscopy.   This conflict is not unexpected because the magnetic neutron scattering is quite sensitive to the direction of the easy axis, whereas the Mössbauer effect hyperfine parameters are rather insensitive to this direction because of the very small magnitude of the quadrupole interaction at the nearly cubic site symmetry found for the iron(III) ions (see below).

MÖSSBAUER-EFFECT SPECTRAL RESULTS

The Mössbauer-effect spectra obtained for iron(III) sulfate between 300K and 28.8K are illustrated in Figure 3.   At room temperature the spectrum consists of two unresolved doublets, one arising from each iron site; the average isomer shift of 0.49mm/s and quadrupole interaction of 0.29mm/s are very typical of high-spin octahedral iron(III).   This assignment agrees with that of Haven and Noftle[10] but conflicts with that of Bristoli et al,[11] who assign each observed line to a specific iron site in the iron(III) sulfate.   This assignment, however, seems unreasonable in view of the similarity of the bonding for the two iron sites as revealed in the X-ray and neutron diffraction studies.   It is also in conflict with the low temperature data discussed below.

The Mössbauer spectra for iron(III) molybdate are very similar to that of the sulfate between 300K and 11.9K.   However, the room temperature isomer shift of 0.42mm/s  and quadrupole

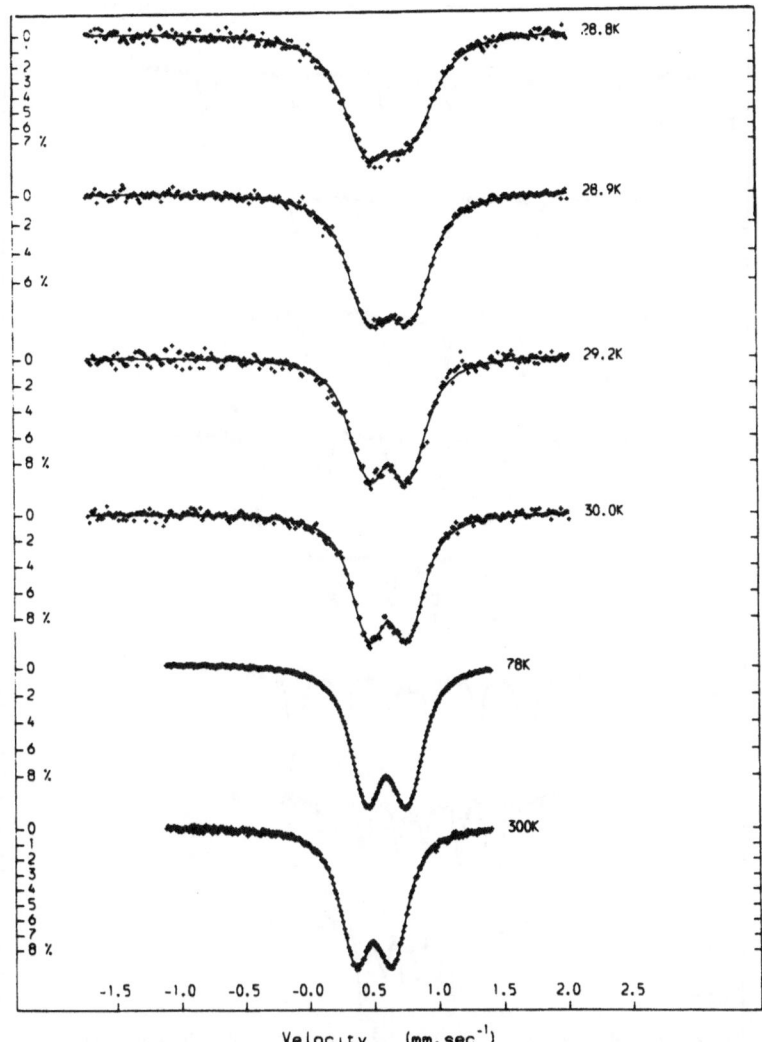

Fig. 3   The Mössbauer-effect spectra of paramagnetic iron(III)
         sulfate obtained at 28.8K and higher temperatures.


interaction of 0.18mm/s are, as expected, smaller than that
observed in the sulfate.   From these results we can conclude that
both the sulfate and molybdate contain rapidly relaxing high-spin
iron(III) ions at 28.8K and 11.9K and above respectively.

    At temperatures of 28.6K and below, monoclinic iron(III)
sulfate exhibits long range magnetic order as illustrated in
Figure 4.   The most interesting feature in these spectra is the
development of two different internal hyperfine fields as the
temperature is lowered.   This is quite apparent from the two

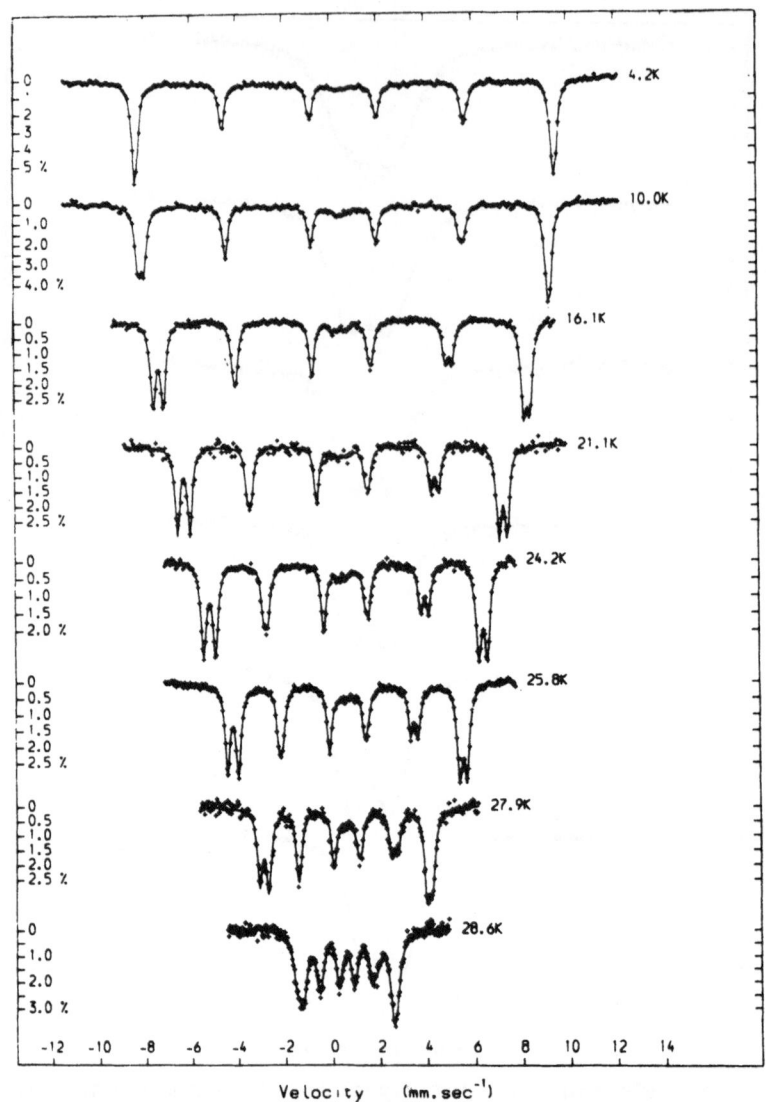

Fig. 4    The Mössbauer-effect specra of ordered iron(III) sulfate
         obtained at 28.6K and lower temperatures.

partially resolved sextets observed between 27.9 and 16.1K.    At
temperatures below ca. 10K the two hyperfine fields again approach
the same value.    A summary of the resulting spectral parameters
is presented in Table I (see reference 3 for more details) and the
temperature dependence of the hyperfine fields for the two sites
is shown in Figure 5.    Once again the hyperfine parameters are
typical  of high-spin iron(III)  and the internal hyperfine field

Table I.  Mossbauer Effect Parameters for monoclinic anhydrous Iron (III) Sulfate–
Antiferromagnetic Phase[a]

| T, °K | Fe(1) Sublattice | | | | | | | Fe(2) Sublattice | | | | | | | Absolute Area[e] | $\chi^2$ |
|---|---|---|---|---|---|---|---|---|---|---|---|---|---|---|---|---|
| | $\delta$ | H[b] | QS | $\Gamma_{\frac{1}{2}}$ | $\Delta\Gamma$[c] | Area | I[d] | $\delta$ | H[b] | QS | $\Gamma_{\frac{1}{2}}$ | $\Delta\Gamma$[c] | Area | I[d] | | |
| 4.2 | 0.55 | 550 | -0.04 | 0.25 | 0.01 | 49 | 1.92 | 0.61 | 556 | 0.03 | 0.23 | 0.04 | 46 | 0.78 | 10.0 | 1.3 |
| 10.0 | 0.56 | 532 | -0.04 | 0.26 | 0.01 | 47 | 1.39 | 0.57 | 543 | 0.04 | 0.25 | 0.02 | 48 | 1.33 | 9.1 | 1.3 |
| 16.1 | 0.56 | 473 | -0.04 | 0.25 | 0.01 | 42 | 1.33 | 0.56 | 495 | 0.05 | 0.27 | 0.01 | 52 | 1.49 | 8.4 | 1.6 |
| 21.1 | 0.58 | 406 | -0.05 | 0.23 | 0.02 | 48 | 1.51 | 0.57 | 433 | 0.05 | 0.23 | 0.02 | 47 | 1.53 | 8.4 | 1.2 |
| 24.2 | 0.58 | 346 | -0.04 | 0.25 | 0.02 | 48 | 1.50 | 0.58 | 373 | 0.04 | 0.21 | 0.06 | 47 | 1.40 | 8.3 | 1.5 |
| 25.8 | 0.66 | 290 | -0.04 | 0.22 | 0.05 | 47 | 1.45 | 0.65 | 314 | 0.04 | 0.24 | 0.02 | 48 | 1.61 | 8.4 | 1.3 |
| 27.9 | 0.59 | 207 | -0.04 | 0.23 | 0.04 | 51 | 1.59 | 0.60 | 226 | 0.04 | 0.22 | 0.05 | 43 | 1.46 | 8.1 | 1.2 |
| 28.6 | 0.59 | 118 | -0.05 | 0.35 | 0.03 | 50 | 1.84 | 0.61 | 130 | 0.04 | 0.38 | 0.05 | 44 | 1.47 | 9.4 | 1.0 |

[a]Data in mm/s relative to natural $\alpha$-iron foil.  [b]Internal magnetic hyperfine field in KOe.  [c]The incremental linewidth increase for the outermost magnetic lines.  [d]Intensity ratio of line 2 (or 5) to line 3 (or 4).  [e]Absolute area in counts·mm/s.

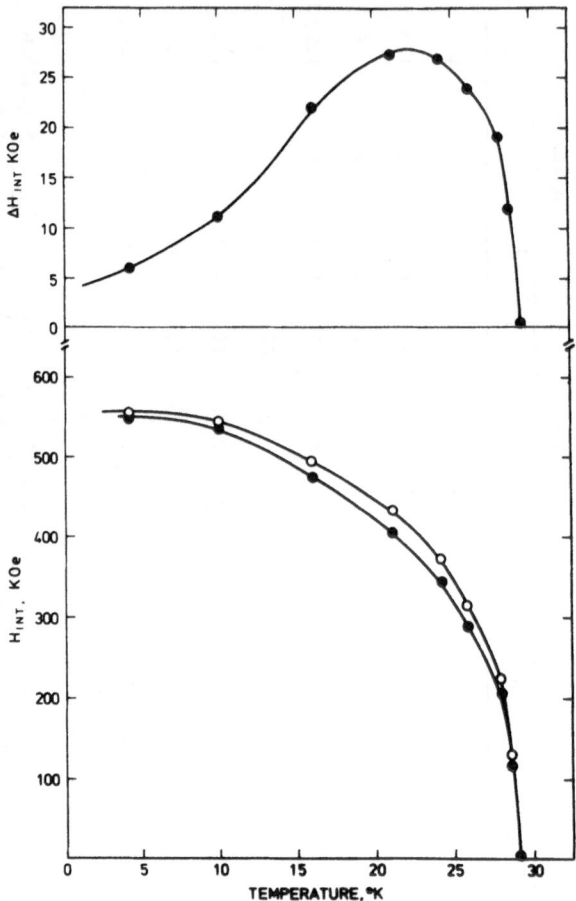

Fig. 5 A plot of the internal hyperfine field on each sublattice in iron(III) sulfate and the difference between these fields, Δ sulfate and function of temperature.

saturates at a value of ca. 550kOe as expected.[12] It is apparent that the two different iron(III) sites in iron(III) sulfate follow different magnetization curves and that the maximum difference in the moments on the two sites occurs at ca. 23K as illustrated at the top of Figure 5.

The Mössbauer spectra of iron(III) molybdate are, as expected, more complex than those of the sulfate because of the presence of the four crystallographically distinct iron sites. Fortunately, the four resulting sextets are partially resolved. As is illustrated in Figure 6, iron(III) molybdate shows the first indication of ordering at 11.72K. Over the small temperature range of 11.72 11.59K there is evidence for the presence of

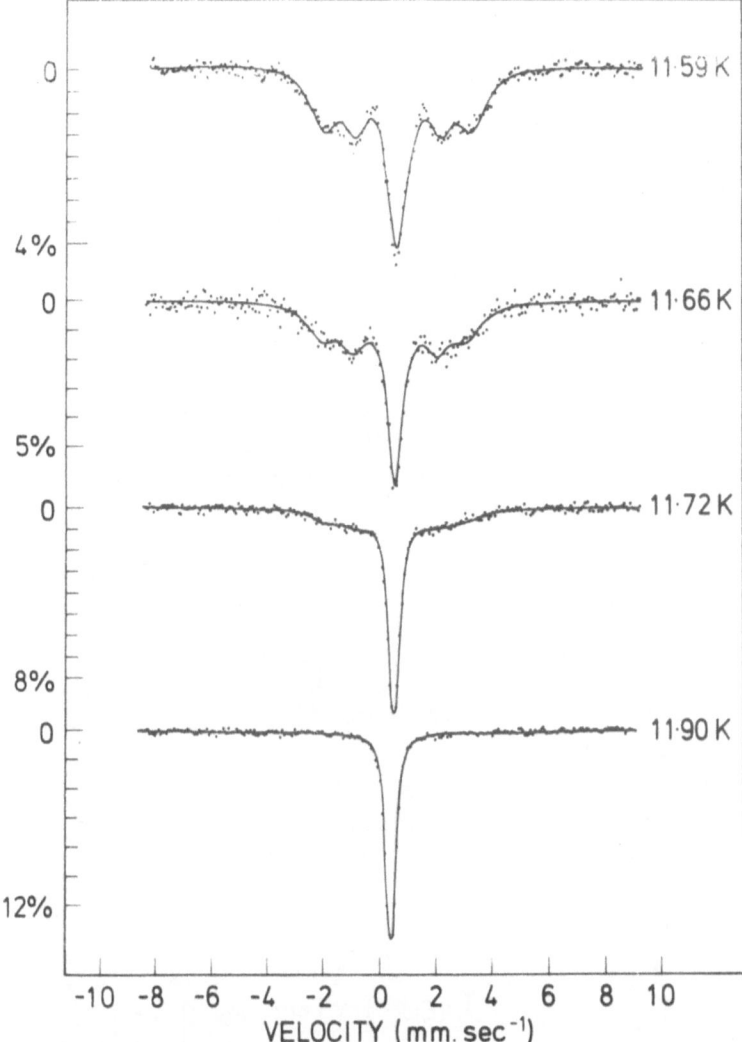

Fig. 6 The Mössbauer-effect spectra of iron(III) molybdate
obtained between 11.9 and 11.59K.

both ordered and paramagnetic components in the spectra. This
probably results from the presence of different particle sizes in
this fine powder material. Below 11.2K, as illustrated in Figure
7, the presence of partially resolved sextets with differing
hyperfine fields is apparent; the area ratio of the different
components is three to one. The hyperfine fields are resolved

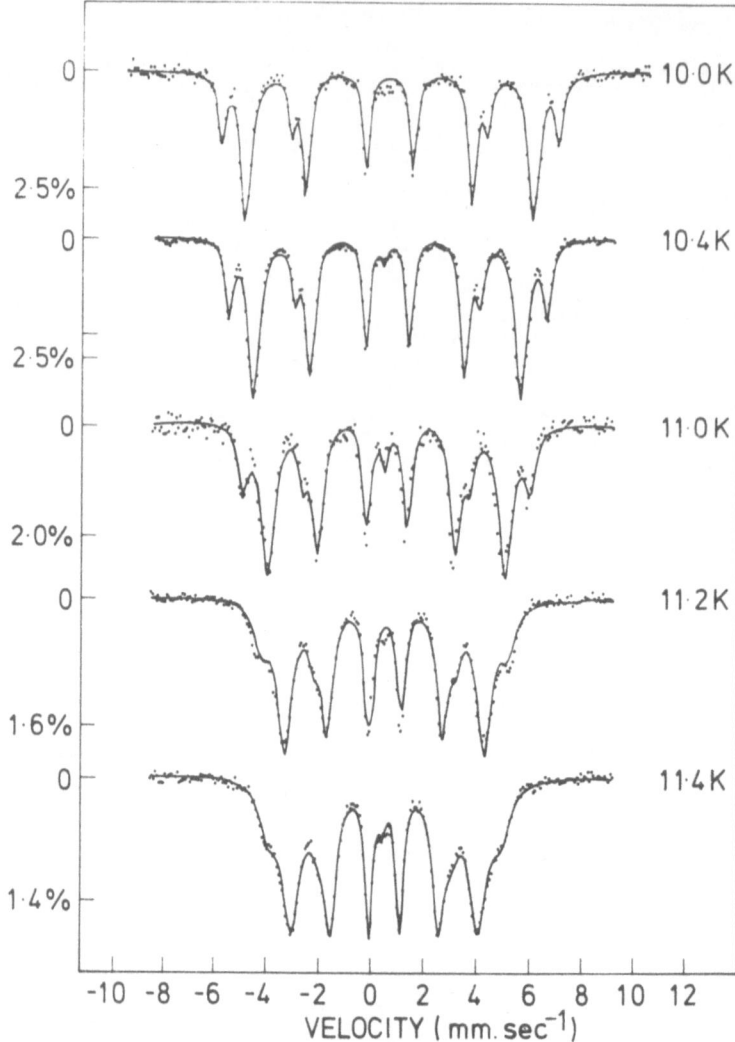

Fig. 7 The Mössbauer-effect spectra of iron(III) molybdate obtained between 11.4 and 10.0K.

down to ca. 3.5K but are virtually equivalent in the spectrum observed at 1.3K as shown in Figure 8. In this case the analysis is more complex, but it is possible to use computer least-squares minimization techniques to estimate the fields at each site. The resulting fields are shown in Figure 9A and illustrate that

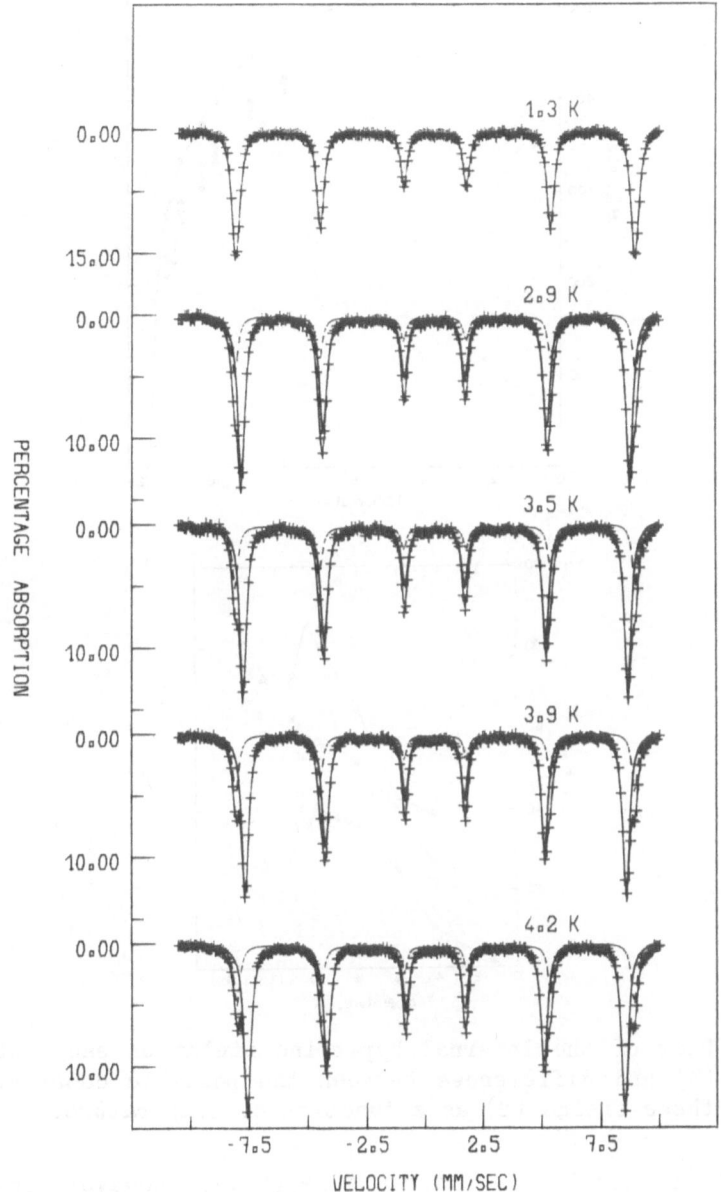

Fig. 8 The Mössbauer-effect spectra of iron(III) molybdate obtained between 4.2 and 1.3K.

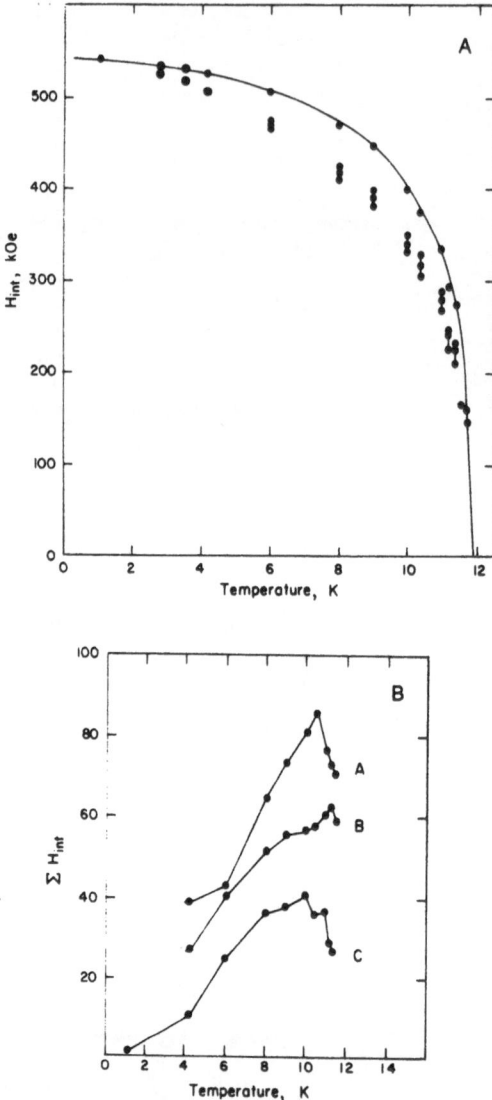

Fig. 9   Plot of the internal hyperfine fields on each sublattice
(A) and differences between the possible combinations of
these fields (B) as a function of temperature.

$Fe_2(MoO_4)_3$ is similar in nature to $Fe_2(SO_4)_3$. However, $Fe_2(MoO_4)_3$
is more complex because of the four different sites present and
the coupling of the internal hyperfine fields could occur in
various combinations.   We believe that the coupling, $H_c + H_d - H_a$
$- H_b$, represented by curve C in Figure 9B is most reasonable
because the net internal hyperfine field, and hence moment, drops
to ca. zero at very low temperature a behavior which is consistent
with the magnetic  properties discussed below.   Unfortunately  it

is not possible in this case to associate a specific magnetic sextet in the Mössbauer spectrum with a specific crystallographic lattice site. It would be useful if this were possible because this would help in trying to understand why one of these lattice sites has a higher moment than the other sites. This no doubt occurs because of differences in the exchange coupling pathways for the different sites, but the quality of the neutron and X-ray diffraction data is not sufficient to permit the identification of the site or pathway. More details are presented in reference 4 and an alternative approach is discussed in reference 5.

## MAGNETIC SUSCEPTIBILITY STUDIES

If the above Mössbauer-effect studies are an indication that $Fe_2(SO_4)_3$ and $Fe_2(MoO_4)_3$ are L-type ferrimagnets, then the magnetic susceptibility must show a temperature range over which the susceptibility is field dependent and much higher than would be expected for typical antiferromagnetic behavior. At high temperature the material should show paramagnetic behavior with a roughly linear inverse susceptibilty and a negative Curie-Weiss temperature behavior expected for a material with strong antiferromagnetic coupling between sublattices. At very low temperatures, the susceptibilty should fall to a small value as the sublattice moments saturate to roughly the same moment. Thus at high and low temperature the field dependence of the magnetization should extrapolate to zero at zero applied field whereas in the ferrimagnetic region the magnetization should extrapolate to a non-zero value at zero field, corresponding to the expected spontaneous magnetization. This is exactly the behavior found for both $Fe_2(SO_4)_3$ and $Fe_2(MoO_4)_3$.

A plot of the molar magnetic susceptibilty $\chi_M$ and inverse molar magnetic susceptibilty for monoclinic anhydrous iron sulfate is shown in Figure 10. The inverse susceptibility data above 60K extrapolates to a Curie-Weiss value of -82.0K and the resulting effective moment of ca. $5.9\mu_B$ corresponds to that expected of a high-spin iron(III) compound. Below about 33K the susceptibility increases dramatically and shows a strong field dependence and a spontaneous magnetization. The maximum susceptibility occurs at ca. 25K and then drops sharply to a small value at 4.2K.[5] The behavior for $Fe_2(MoO_4)_3$ is quite similar and shows a field dependent susceptibilty with a peak at ca. 9K.

The magnetic studies thus indicate that both $Fe_2(SO_4)_3$ and $Fe_2(MoO_4)_3$ exhibit ferrimagnetic behavior with a maximum in the magnetic moment at a temperature which corresponds to the maximum difference in the internal hyperfine field as measured in zero applied field by the Mössbauer effect. This correspondence is exactly that expected for L-type ferrimagnetism.

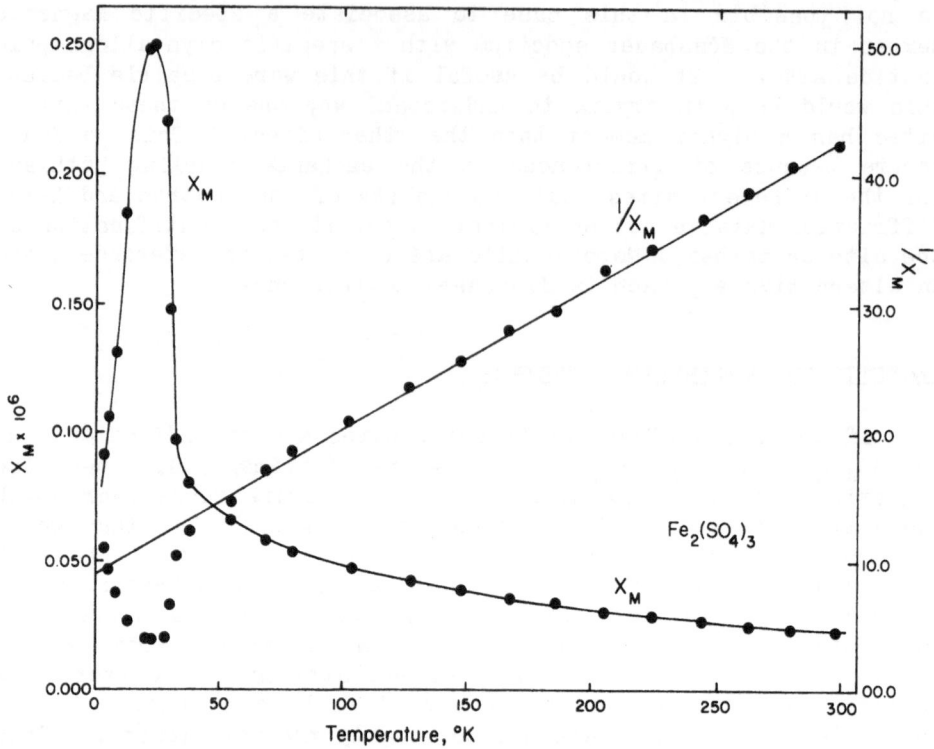

Fig. 10 A plot of the molar magnetic susceptibility and the inverse susceptibility for anhydrome monoclinic iron(III) sulfate as a function of temperature.

## MAGNETIC ORDERING MODEL

The rather unusual magnetic properties of $Fe_2(SO_4)_3$ and $Fe_2(MoO_4)_3$ call for a discussion as to why these materials should exhibit L-type ferrimagnetism. Specifically we need to understand why the magnetization on the two magnetic sublattices is different at intermediate temperatures. Because of the very similar chemical environment of the different iron(III) ions in each of these compounds, it is reasonable that the saturation moment observed on the different sublattices at temperatures significantly below the ordering temperature should be very similar. At temperatures close to the ordering temperature the moment will to some extent depend upon the ferrimagnetic exchange coupling constant for a given sublattice. Different coupling constants would then produce different antiferromagnetically coupled moments with a net ferrimagnetic component.

In the following discussion we consider in detail the structure of monoclinic iron(III) sulfate. Similar, although more involved arguments apply for the iron(III) molybdate. In $Fe_2(SO_4)_3$ each iron(III) ion is coordinated to six oxygen atoms which are a part of six different sulfate groups (see Figure 2). Thus each iron atom is connected to ten surrounding iron atoms by 18 bridging Fe-O-S-O-Fe pathways. The nature of these linkages is illustrated in Figure 11 which is a representation of the unit cell of $Fe_2(SO_4)_3$ shown in Figure 2. The various intersublattice antiferromagnetic exchange pathways (with an exchange coupling constant $J_{ab}$,) are represented by solid lines; the intra-sublattice ferrimagnetic exchange pathways (with exchange constants $J_{aa}$ and $J_{bb}$ are represented by broken lines. The number of exchange pathways between a specific iron atom and a given neighbor varies from 1 to 4; the specific values are given in Figure 11. Specifically, an Fe(1) site is connected to three other Fe(1) atoms via two $J_{aa}$ pathways each for a total of six. In addition the Fe(1) atom is connected to seven different Fe(2) atoms via a total of 12 different $J_{ab}$ exchange pathways. Indeed, four exchange pathways connect any specific pair of Fe(1) and Fe(2) atoms. Because of the similarity of the various superexchange pathways, we believe that all the Fe...Fe interactions are intrinsically antiferromagnetic, but that the resultant antiferromagnetic coupling between the crystallographically distinct Fe(1) and Fe(2) stems from the relative number of exchange pathways shown in Figure 11.

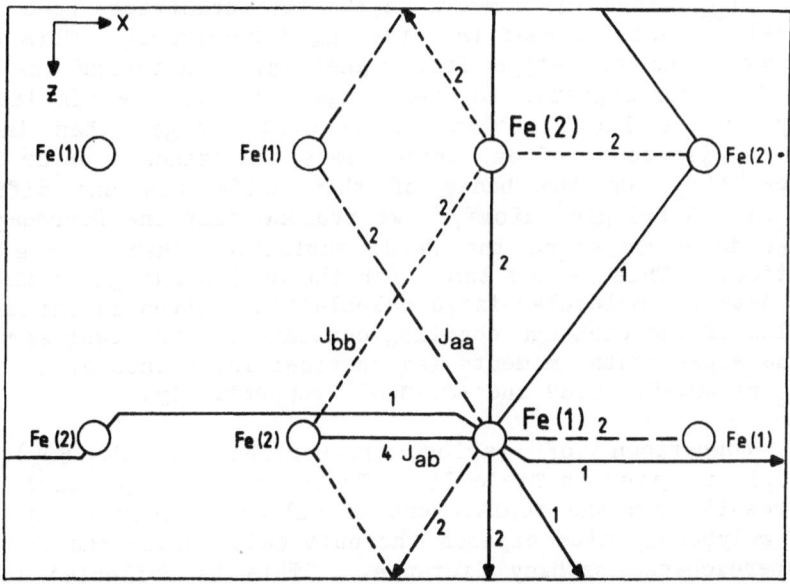

Fig. 11 A projection of the iron(III) sulfate iron atomic coordinates normal to the crystallographic ac plane and showing the various superexchange plathways.

Table II.  Comparison of $Fe_2(SO_4)_3$ and $Fe_2(MoO_4)_3$

| Property | $Fe_2(SO_4)_3$ | $Fe_2(MoO_4)_3$ |
|---|---|---|
| Unit cell volume, $Å^3$ | 1578(1) | 2140(1) |
| Superexchange distance, Å | 6.92(1) | 7.41(1) |
| Curie-Weiss temperature, K | -82(2) | -56(2) |
| Magnetic ordering temperature, K | 28.8(1) | 11.5(1) |
| Saturation $H_{int}$, kOe | 556(2) | 529(3) |
| Average $\delta$ at 296K, mm/s | 0.58(1) | 0.52(1) |
| Magnetic moment, $\mu_\beta$ | 4.52(8) | 4.34(3) |
| Covalency sum, % | 6.1(1.4) | 9.2(1.2) |

If the crystallographically distinct iron atoms are antiferromagnetically coupled, as is clear from the neutron diffraction results, then the interactions between crystallographically identical iron atoms must be ferromagnetic in nature. In other words, $J_{ab} < 0$ and $J_{aa}$ and $J_{bb} > 0$. However, if $J_{aa} > J_{bb}$, then the moments on the two sublattices need not be identical, especially near the ordering temperature. This would, of course, lead to L-type ferrimagnetism. Although, as noted above, the ferromagnetic superexchange pathways are similar, the pathway on the Fe(1) sublattice is 0.04Å longer than that of Fe(2), a difference of ca. three times the standard error in the distance.[6,7] On the basis of this difference and different angles at the bridging atoms,[3] we propose that the ferromagnetic exchange is stronger on the Fe(2) sublattice than on the Fe(1) sublattice. This is the basis for the assignment given in Table I. A detailed molecular field calculation[13] gives an estimate of the value of the exchange coupling constants. The best agreement with the experimental moments was obtained for values of $J_{ab}$, $J_{aa}$, and $J_{bb}$ of -0.55, +0.62 and +0.43$cm^{-1}$ respectively.

A comparison of various properties of $Fe_2(SO_4)_3$ and $Fe_2(MoO_4)_3$ is given in Table II. There are two major differences which result from the replacement of sulfur by molybdenum. The larger molybdenum atom expands the unit cell volume and increases the superexchange pathway distance. This is reflected in both the reduction of the Curie-Weiss temperature and the significant decrease in the ordering temperature. It is also partly responsible for the reduced saturation hyperfine field at the iron site. These results also support the higher degree of covalency expected in iron(III) molybdate as compared with iron(III) sulfate. This is observed both in the lower value of the saturation hyperfine field[14] and the average isomer shift in the molybdate. The smaller isomer shift implies a greater s-electron density at the iron nucleus in the molybdate, suggesting that ligand to metal 4s-orbital donation is an important covalent interaction. This increased covalency is also observed in the higher covalency sum[15] in the neutron diffraction study of the molybdate.

## ACKNOWLEDGEMENTS

The author wishes to thank Drs. P. D. Battle, D. Beveridge, A. K. Cheetham, J. W. Culvahouse, P. Day, and G. Longworth, all of whom were actively involved in this work over the past several years. I would also like to thank my various colleagues at Oxford University, AERE Harwell and the University of Liverpool who have made my various visits so pleasant, and Ms. S. R. Owen for help in the preparation of this typescript. Financial support from NATO, NSF, and PRF is greatly appreciated.

## REFERENCES

1. Néel, L. *Ann. Phys. (Paris)* 1948, *3*, 137.
2. Goodenough, J. B. "Magnetism and the Chemical Bond" Wiley-Interscience, New York, 1963.
3. Long, G. J.; Longworth, G.; Battle, P. D.; Cheetham, A. K.; Thundathil, R. V; Beveridge, D. *Inorg. Chem.* 1979, *18*, 624.
4. Battle, P. D; Cheetham, A. K.; Long, G. J.; Longworth, G. *Inorg. Chem.* 1982, *21*, 4223.
5. Jirak, Z.; Salmon, R.; Fournes, L.; Menil, F.; Hagenmuller, P. *Inorg. Chem.* 1982, *21*, 4329.
6. Moore, P. B.; Araki, T. *Neues Jahrb. Mineral. Abh.* 1974, *121*, 208.
7. Christidis, P. C.; Rentzeperis, P. J. *Z. Kristallogr., Kristallgeom., Kristallphys., Kristallchem.* 1975, *141*, 233.
8. Chen, H. *Mater. Res. Bull.* 1979, *14*, 1583.
9. Rapposch, M. H.; Anderson, J. B.; Kostiner, E. *Inorg. Chem.* 1980, *19*, 3531.
10. Haven, Y.; Noftle, R. E. *J. Chem. Phys.* 1977, *67*, 2825.
11. Bristoli, A.; Viccaro, P. J.; Kunrath, J. I.; Brandao, D. E. *Inorg. Nucl. Chem. Lett.* 1975, *11*, 253.
12. Johnson, C. E. in "Hyperfine Interactions in Excited Nuclei," Goldring, G.; Kalish, R.; Eds; Gordon and Breach, New York, 1971, p. 803.
13. Culvahouse, J. W. *J. Magn. Magn. Mater.* 1980, *21*, 133.
14. Sawatzky, G. A.; van der Woude, F. J. *J. Phys., Colloque (Orsay, France)* 1974, *34* (C-6), 47.
15. Tofield, B. C.; Fender, B. E. F. *J. Phys. Chem. Solids* 1970, *31*, 2741.

# BIOMINERALIZATION OF $Fe_3O_4$ IN BACTERIA

R. B. Frankel and G.C. Papaefthymiou

Francis Bitter National Magnet Laboratory
Massachusetts Institute of Technology
Cambridge, MA   02139

## INTRODUCTION

Iron is a constituent of the active sites of many important proteins including hemoglobin and myoglobin, the cytochromes, iron-sulfur proteins, nitrogenase and others.  Because of the importance of the iron-containing proteins to the physiology of a wide variety of organisms, their physical and electronic structures have been extensively studied.  Since the iron atom or atoms often play a central role in the function of the protein, Mössbauer spectroscopy is one of the most important tools in these studies.

In most cases iron is incorporated as an isolated atom or part of a small cluster of atoms at specific sites and is ligated by oxygen, nitrogen, or sulfur atoms of the polypeptide chain.  In these systems the iron atoms are paramagnetic or diamagnetic depending on whether the number of 3d electrons is odd or even.  An important exception is the iron storage protein ferritin which can accommodate up to 4000 atoms per molecule.  The iron is deposited as a hydrous iron oxide mineral in a spherical cavity in the protein.  Exchange interactions between the iron atoms are strong and the mineral is magnetically ordered at low temperature.[1]

In addition to ferritin, and the related iron-storage material hemosiderin, it is now known that organisms including bacteria and higher plants and animals can produce other iron deposits of varying crystallinity.[2]  These deposits can occur extracellularly or intracellularly, and include the minerals ferrihydrite, geothite lepidocrocite, and magnetite.  All these minerals are magnetically ordered at low temperature and magnetite is magnetically ordered even at room temperature.  The processes by which organisms deposit

iron minerals are interesting because they involve biological inter-
vention in essentially inorganic processes. In discussing biomineralization of calcium and silica as well as iron deposits, Lowenstam[2,3] has distinguished "biological induced" mineralization from
"organic matrix-mediated" mineralization. In the former process,
cellular export of metabolic end products leads to precipitation of
metal ions in the environment. In the latter process, the minerals
are deposited in a preformed organic matrix produced by the organism.
Biologically induced mineralization results in mineral forms of
varying crystallinity with crystal structures and habits similar to
those produced by inorganic processes. On the other hand, organic
matrix-mediated mineralization typically results in crystals with
definite morphologies, narrow size ranges and often, definite
orientations in the matrix. Both types of processes might contribute to mineralization in certain cases.

An example of a biologically induced iron mineralization process might be FeS precipitation in marine sediments, resulting
from $S^=$ ions produced in the metabolism of sulfate reducing bacteria
such as <u>Desulfovibrio</u>. In contrast, magnetotactic bacteria,[4] such
as <u>Aquaspirillum magnetotacticum</u>, produce uniformly sized and shaped
crystals of $Fe_3O_4$ in an intracellular sheath. The morphologies of
the particles are apparently species specific, indicating a matrix-
mediated precipitation process.[5] A possible example of a mixed
mineralization process concerns the so-called iron bacteria, such
as <u>Leptothrix</u>, which precipitate hydrous iron oxides extracellularly.[6]
The precipitate is x-ray amorphous but the precipitation process
could involve polysaccharides on the surface of the cells.

MÖSSBAUER SPECTROSCOPY OF FERRITIN

The iron core of the protein ferritin is the most extensively
studied iron biomineral.[7] Ferritins occur widely in the living
world, from bacteria to man. Mammalian ferritin consists of a
spherical protein shell with $\sim$ 12 nm outer diameter encasing a 7
nm hydrous iron oxide core which is associated with phosphate.
The apoprotein shell consists of 24 identical protein subunits each
of molecular weight 18,500. The number of iron atoms in a molecule
can vary from zero to approximately 4000. While ferritin is easily
crystallized, there is no unique orientation of the hydrous iron
oxide cores with respect to the apoferritin shells. Hence, a precise x-ray determination of the crystal structure of the core material has not been possible. However, it has been suggested that
the hydrous iron oxide cores of ferritin consist essentially of the
mineral ferrihydrite, with six fold oxygen coordinated ferric iron
and hexagonal close packing of the oxygen atoms.[8,9]

Iron in ferritins from several sources have been studied by Mössbauer spectroscopy.[1,10-12] The high temperature spectrum (T > 60 K) typically consists of a broadened quadrupole doublet with isomer shift and quadrupole splitting characteristic of $Fe^{+3}$. At helium temperatures (T ∿ 4.2 K) the spectrum is magnetically split with a 490 kOe field at the nucleus. The lines are broad and some authors have used a distribution of hyperfine fields to fit the spectra. As the temperature is increased from 4.2 K, the magnetically split spectrum gradually decreases in intensity and the quadrupole doublet increases in intensity. The temperature range over which the magnetically split spectrum and quadrupole doublet coexist depends on the source of the ferritin (mammalian, plant or bacterial in origin) and the degree of iron loading of the ferritin molecules.

The temperature dependence of the Mössbauer spectrum is due to superparamagnetic behavior in the small particles of the ferritin core.[1] In this phenomenon, the iron atoms are antiferromagnetically coupled at low temperature. The sublattice magnetizations lie along particular crystallographic orientations in the crystal, the easy magnetic axes. At finite temperature there is a certain probability that the sublattice magnetizations will undergo a transition to an energetically equivalent easy axis. The sublattice relaxation time τ is an exponential function of the magnetic anisotropy K, the volumne V of the particle, and the temperature:

$$\tau = \tau_o . \exp \ [KV/kT] \tag{1}$$

where $\tau_o$ is a constant and k is Boltzmann's constant. When the relaxation time in a particle is of the order of or faster than the Larmor precession time of the 14.4 keV excited state, the hyperfine field will be wiped out and the spectrum will consist of the quadrupole doublet. For particles of a given volume, this condition will occur at a definite temperature called the blocking temperature. If there is a distribution of particle volumes in the sample, there will be a distribution of blocking temperatures, and the magnetic spectrum and the quadrupole doublet corresonding to the larger particles and the smaller particles, respectively, can coexist. Determination of the relative intensities of the two subspectra as a function of temperature is a means of measuring the distribution of particle volumes. For example, Williams et al.[12] found that ferritins reconstituted from apoferritin and iron under different conditions (e.g., presence or absence of phosphate) had different distributions of particle volumes.

The spectra of ferritin from the fungus Phycomyes[11] and of an unusual bacterial ferritin with associated heme groups from Azotobacter[13] are similar to mammalian ferritin except that the superparamagnetic behavior is observed over lower temperature ranges. If the core compositions in those ferritins are similar to that in the horse spleen ferritin, we can assume that the decreases in the blocking temperatures reflect smaller particle sizes in the plant and bacterial ferritins.

An iron-rich storage material also referred to as bacterial ferritin has been found in E. coli and other prokaryotes.[14] However, the Mössbauer spectra for these materials are distinctly different than the ferritin from Azotobacter. The Mössbauer spectrum for T > 4 K of the ferritin from E. coli is a quadrupole doublet with parameters characteristic of high spin $Fe^{3+}$. A six line magnetic hyperfine spectrum with an effective magnetic field at the nucleus of 430 kOe is observed at T < 1 K. Above 1 K the lines broaden and the splitting decreases with increasing T, and collapses into the quadrupole doublet at about 3.5 K. Between 1.2 and 3.5 K the doublet and sextet are superposed, indicating a spread of magnetic transition temperatures. This indicates lower energy magnetic interactions between iron atoms than in other ferritins, perhaps reflecting less dense packing of the iron atoms or less crystallinity in the core.

Cohen et al.[15] have discovered an interesting dynamic effect in the Mössbauer spectra of crystals of horse spleen ferritin molecules when the crystals are warmed through the freezing point of water ($\sim$ 265 K). Below 265 K the spectrum consists of the quadrupole doublet referred to above. Above 265 K, the spectrum consists of a narrow line quadrupole doublet superposed on a broad spectrum of width $\sim$ 4 cm/s. This effect has been interpreted in terms of bounded or localized diffusive motions of the ferritin molecules after interstitial water in the crystal has melted. Analysis has been made in terms of discrete transitions between a number of fixed points, or in terms of continuous harmonic motion driven by Brownian forces.[16] For small particles in water, frictional forces are large compared to inertial forces and the situation corresponds to strong overdamping. It has been shown that the theoretical spectrum based on these models does indeed consist of a narrow and a broad component, with parameters determined by a diffusion constant D and the ratio of harmonic to frictional forces. The intensity of the narrow line and the width of the broad line can be used to calculate the mean squared displacement and the diffusion constant, respectively, of the iron atoms participating in the diffusive motions.

## MÖSSBAUER SPECTROSCOPY OF MAGNETOTACTIC BACTERIA

Magnetotactic bacteria are various species of aquatic microorganisms that orient and swim along magnetic field lines.[4,17-19] All magnetotactic cells examined to date by electron microscopy contain iron-rich, electron-opaque particles.[4,18,20,21] In several

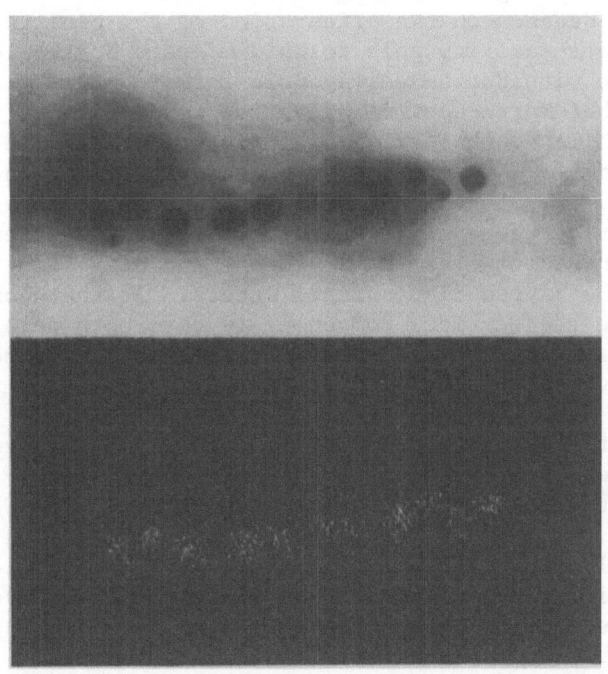

Fig. 1.  Transmission electron micrograph of a portion of $\underline{A}$. $\underline{magnetotacticum}$ showing 500 Å $Fe_3O_4$ particles (top).  An Fe x-ray pulse map of the same portion of the cell, show-ing that cellular iron is concentrated in the particles (bottom).  (After Ref. 28)

species of magnetotactic bacteria, and possibly all, the particles consist of magnetite, $Fe_3O_4$.[21,22] Cuboidal, rectangular parall-elipiped, and arrow-head shaped particles occur in different species with typical dimensions of 400 to 1200 Å. These are within the single-magnetic-domain size range of $Fe_3O_4$. In most species the particles are arranged in chains, which impart a magnetic moment of the cell, parallel to the axis of motility. The moment is suf-ficiently large that the bacterium is oriented in the geomagnetic field at ambient temperature as it swims, i.e., the chain of $Fe_3O_4$ particles functions as a biomagnetic compass.[23] The organism thus propels itself along the geomagnetic field lines. The direction of migration depends on the orientation of the biomagnetic compass. Those with North-seeking pole forward migrate North along the field lines. Those with South-seeking pole forward migrate South. It has been found that North-seeking bacteria predominate in the Northern Hemisphere while South-seeking bacteria predominate in the Southern Hemisphere.[24,25] The vertical component of the inclined geomagnetic field selects the predominant polarity in each hemisphere by pre-sumably favoring those cells whose polarity causes them to be directed downward towards the sediments and away from the toxic effects of the oxygen rich surface waters. At the geomagnetic equator where the vertical component is zero both polarities co-exist;[26] presumably, horizontally directed motion is equally bene-ficial to both polarities in reducing harmful upward migration.

In the freshwater magnetotactic spirillum, <u>A. magnetotacticum</u>, iron comprises 2% or more of the cellular dry weight.[27] Electron microscopy studies of this organism show that the $Fe_3O_4$ particles are cuboidal, 40 - 50 nm in width, and are arranged in a chain that longitudinally traverses the cell (Fig. 1). The particles are en-veloped by electron-transparent and electron-dense layers; a par-ticle and its enveloping membrane has been termed a magnetosome.[20]

Since <u>A. magnetotacticum</u> is cultured in a chemically defined medium in which iron is available as soluble ferric quinate[27] the presence of intracellular $Fe_3O_4$ implies a process of bacterial precipitation of this mineral, with control of particle size, number and location in the cell.

In order to elucidate the $Fe_3O_4$ biomineralization process, cells and cell fractions, some isotopically enriched in Fe-57, have been studied by Mössbauer spectroscopy.[28] Cells of a non-magnetotactic variant that accumulated iron but did not make $Fe_3O_4$ and of a cloned, nonmagnetotactic strain that accumulated less iron, were also studied. The results suggest that $Fe_3O_4$ is precipitated by reduction of a hydrous ferric-oxide precursor.

Mössbauer spectra of wet packed cells enriched in Fe-57 at 200 and 80 K are shown in Figs. 2 and 3a, respectively. The 200 K spectrum can be analyzed as a superposition of spectra corresponding

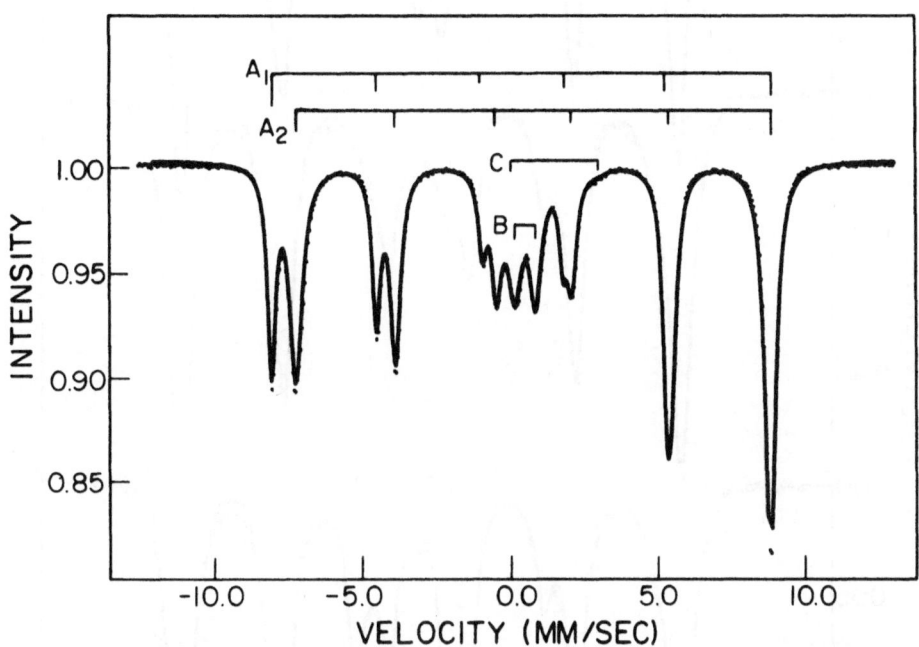

Fig. 2  Mössbauer spectrum of wet, packed cells of A. magnetotacticum at 200 K.  Subspectra $A_1$ and $A_2$ are due to $Fe_3O_4$;
spectrum B is a ferric doublet; spectrum C is a ferrous
doublet.  The solid line is a theoretical least-squares
fit to the data.  (Ref. 28).

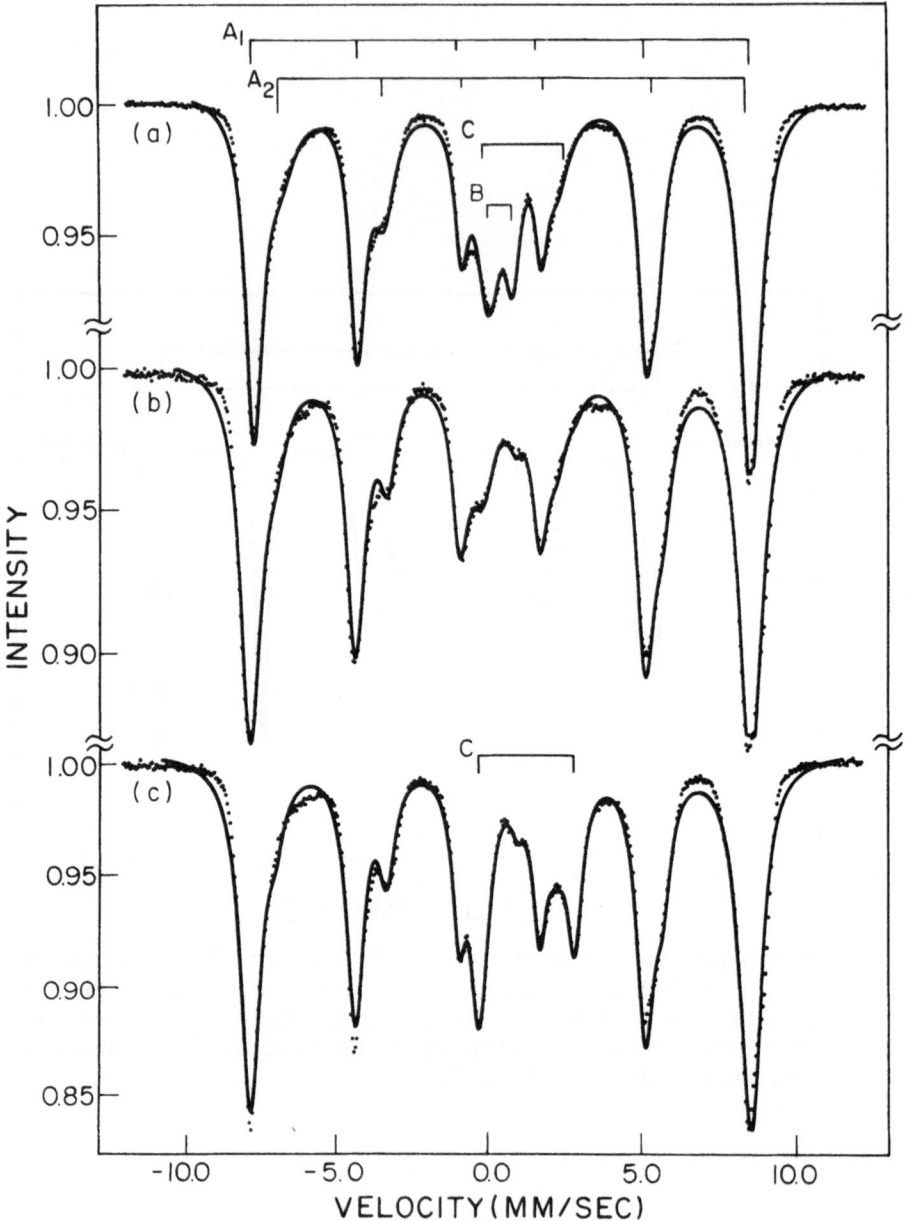

Fig. 3. Mössbauer spectra of <u>A. magnetotacticum</u> at (a) 80 and (b) 4.2 K. Note the reduction in the intensity of spectrum B at 4.2 K. (c) Cells at 4.2 K after anaerobic incubation above freezing temperature for 24 hours. Note enhancement of spectrum C at the expense of B. (Ref. 28).

to $Fe_3O_4$ (spectra $A_1$ and $A_2$), a broadened quadrupole doublet with parameters characteristic of ferric iron (spectrum B), and a weak quadrupole doublet with parameters corresponding to ferrous iron (spectrum C). Spectra $A_1$ and $A_2$ correspond to $Fe^{3+}$ in tetrahedral sites and $Fe^{2+}$ and $Fe^{3+}$ in octahedral sites in $Fe_3O_4$, respectively.[29]

Spectrum B was also observed in lyophilized cells and has isomer shift and quadrupole splitting parameters similar to iron in ferritin and in the mineral ferrihydrite, indicative of ferric iron with oxygen coordination. The relative intensity of B to $A_1 + A_2$ was somewhat variable from sample to sample, depending on growth conditions. At 80 K, spectra $A_1$ and $A_2$ correspond to $Fe_3O_4$ below the Verwey transition and the parameters of spectrum B and the relative intensity of B to $A_1 + A_2$ are relatively unchanged compared to the spectrum at 250 K. Between 80 and 4.2 K, however, the intensity of B decreased with decreasing temperature so that at 4.2 K only a residual doublet remained. A similar temperature dependence for spectrum B was also obtained in lyophilized cells.

The isomer shift and quadrupole splitting parameters of spectrum C correspond to high spin ferrous iron in coordination with oxygen or nitrogen. This spectrum was not observed with lyophilized cells, possibly as a result of oxidation during sample preparation. Wet, packed cells kept unfrozen under anaerobic conditions contained increased amounts of material responsible for spectrum C and correspondingly less material with spectral characteristics B (Fig. 3b). Thawing and aeration of these frozen cells resulted in increases in B spectral lines and concomitant decreases in C spectral lines. This indicates that the iron atoms responsible for spectrum C came from reduction of the iron atoms giving spectrum B. Unlike that of spectrum B, the intensity of spectrum C did not decrease between 80 and 4.2 K (Fig. 3c).

The decrease in the intensity of spectrum B between 80 and 4.2 K can be explained as the onset of magnetic hyperfine interactions at low temperature resulting in a concommitant decrease in the intensity of the central absorption doublet, similar to ferritin. However, in the present case, the magnetic hyperfine lines were obscured by the magnetite spectral lines ($A_1$ and $A_2$). To further resolve the nature of the materials responsible for spectrum B, the temperature dependent Mössbauer spectra of nonmagnetotactic cells which lacked the interfering magnetite were studied.

For $T \geq 80$ K, the spectrum of lyophilized nonmagnetotactic cells (Fig. 4) consisted primarily of the quadrupole doublet characteristic of ferric iron as denoted by spectrum B in Figs. 2 and 3. In addition, a very low intensity spectrum due to $Fe_3O_4$ (spectral lines $A_1 + A_2$ in Figs. 2 and 3) was observed. These latter spectral lines might have been due to a small fraction of magnetotactic cells in the sample or trace amounts of magnetite

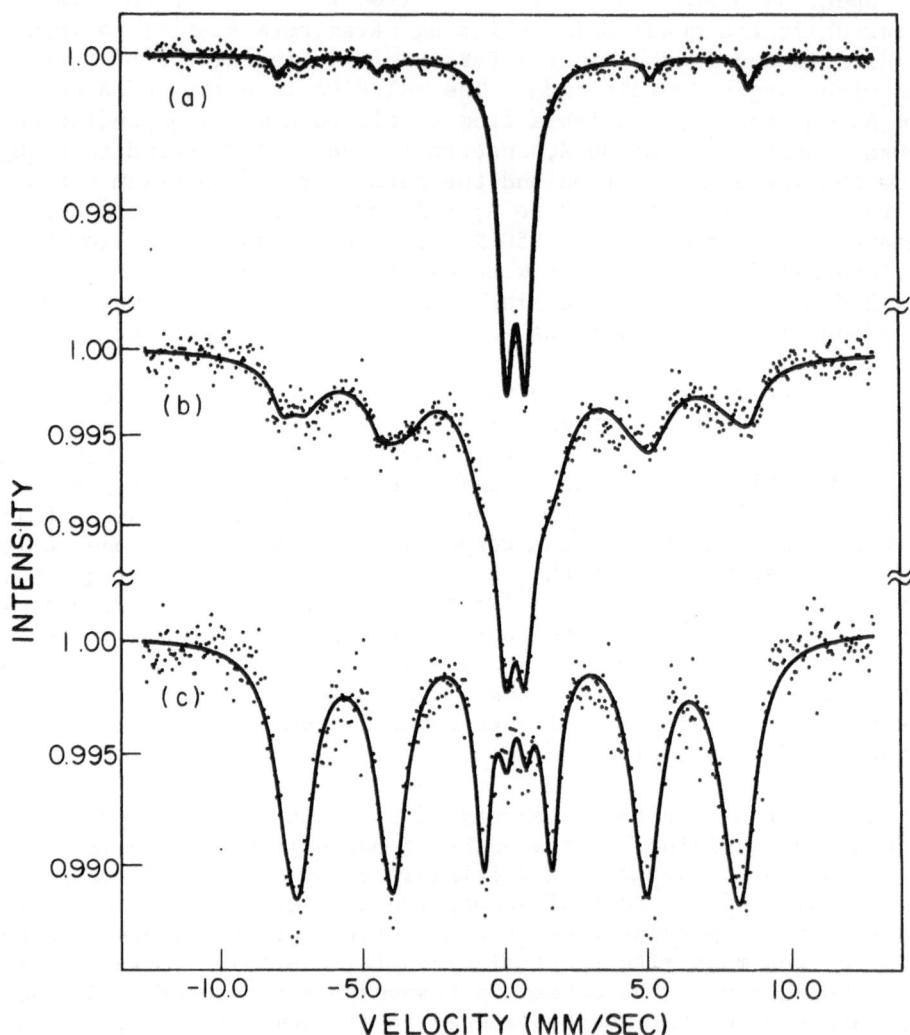

Fig. 4.  Mössbauer spectra of nonmagnetic cells at (a) 100 K; (b) 40 K; (c) 4.2 K.  (Ref. 28)

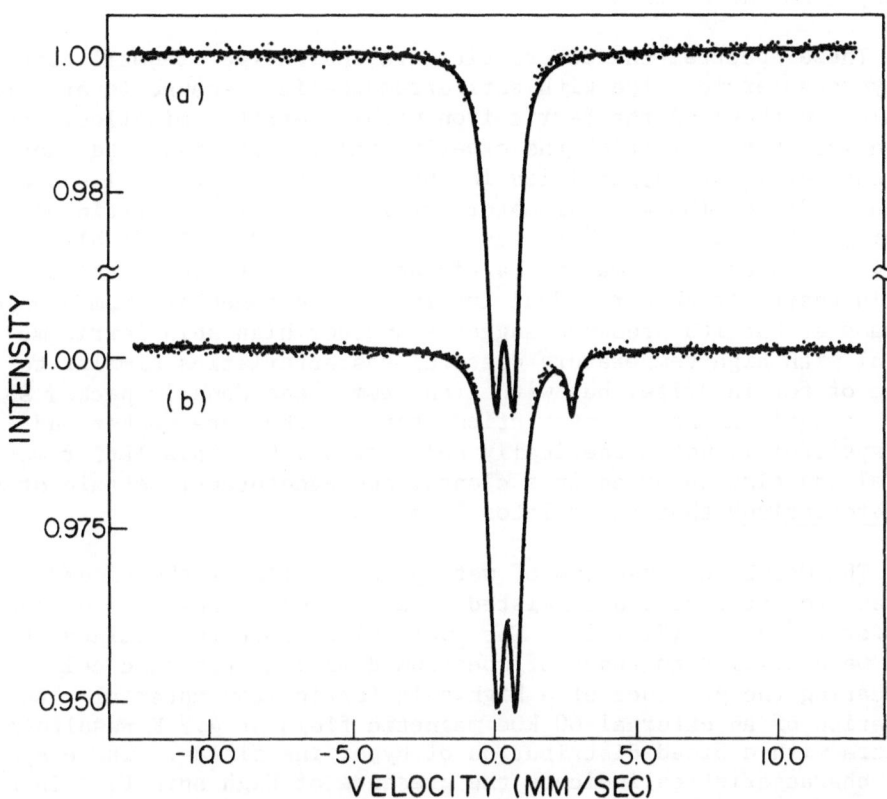

Fig. 5. Mössbauer spectra of a cloned, nonmagnetotactic strain at
(a) 200 K; (b) 200 K following incubation above freezing
temperature for 24 hours. (Ref. 28)

possibly present in the nonmagnetotactic cells. Below 80 K, the intensity of the quadrupole doublet decreased with decreasing temperature while the intensity of a six-line spectrum flanking the doublet increased. At 4.2 K the spectrum consisted primarily of the six broadened magnetic hyperfine lines, with a small residual doublet in the center. Spectral lines $A_1 + A_2$ were obscured by the six-line spectrum. Application of a longitudinal magnetic field of 60 kOe produced broadening of the six-line spectrum but with no appreciable shifts in the line position and no decreases in any line intensities.

These spectral characteristics are indicative of small particles of hydrous-ferric-oxide with antiferromagnetic exchange interactions similar to those of the ferric iron within ferritin micelles. By comparison with ferritin, the experimental results indicate that hydrous-ferric-oxide particles in the nonmagnetotactic cells are of the order of 100 Å in diameter, or less. Unlike ferritin or ferrihydrite, however, there was a residual quadrupole doublet in the 4.2 K spectrum of magnetotactic and nonmagnetotactic cells. The intensity of this residual doublet varied somewhat from sample to sample, but its presence suggests another high spin ferric material with high temperature spectral characteristics similar to those of ferrihydrite, but with iron atoms less densely packed so that magnetic exchange interactions between them are weaker and the spectrum is not magnetically split at 4.2 K. This latter material was also observed in a cloned, nonmagnetotactic strain of A. magnetotacticum that accumulates less iron.

The Mössbauer spectrum of wet, packed cells of the cloned, nonmagnetotactic strain consisted of a quadrupole absorption doublet for T ≥ 4.2 K (Fig. 5). The spectral parameters obtained at 80 K were similar to those of spectrum B in magnetotactic cells, indicating the presence of a high spin ferric iron material. Application of an external 60 kOe magnetic field at 4.2 K results in spectra with a broad distribution of hyperfine fields. These spectral characteristics indicate the presence of high spin $Fe^{3+}$ in a hydrous oxide with magnetic exchange interactions < 4 K, that is, where the iron atoms are less densely packed than in ferrihydrite. This material has similar spectral characteristics to the iron storage materials in E. coli.[14]

When these wet, packed cells were held above 275 K in an anaerobic environment, a ferrous spectrum similar to spectrum C appeared, in addition to the ferric-iron doublet. This indicates that the hydrous-ferric-oxide in cells of this strain can be reduced to ferrous iron as with cells of the other strain.

In summary, cells of A. magnetotacticum contain ferrous ions, a low-density hydrous-ferric-oxide, a high-density hydrous-ferric-oxide (ferrihydrite) and $Fe_3O_4$. Additional experiments with cell fractions show that ferrihydrite in the magnetotactic cells is associated with the magnetosomes.[28]

On the basis of the foregoing results it has been proposed[28] that A. magnetotacticum precipitates $Fe_3O_4$ in the sequence: $Fe^{3+}$ quinate $\rightarrow Fe^{2+} \rightarrow$ low density hydrous-ferric-oxide $\rightarrow$ ferrihydrite $\rightarrow Fe_3O_4$. In nonmagnetotactic cells the process stops with ferrihydrite. In cells of the cloned, nonmagnetotactic strain the process stops with low-density hydrous ferric oxide.

In the proposed process, iron enters the cell as $Fe^{3+}$ chelated by quinic acid. Reduction to $Fe^{2+}$ releases iron from the chelator. $Fe^{2+}$ is reoxidized and accumulated as the low density hydrous-iron-oxide. By analogy with the deposition of iron in the micellar cores of the protein ferritin,[7] this oxidation step might involve molecular oxygen, which is required for $Fe_3O_4$ precipitation in A. magnetotacticum.[30] Dehydration of the low-density hydrous-ferric-oxide results in ferrihydrite. Finally, partial reduction of ferrihydrite and further dehydration yields $Fe_3O_4$.

$Fe_3O_4$ is thermodynamically stable with respect to hematite and ferrihydrite at low $E_H$ and high pH.[31] However, rapid transformation of ferrihydrite to magnetite appears to involve more than simple reduction and dehydration. While the degree of crystallinity of ferrihydrite can vary, in crystalline samples it has a structure related to hematite, with hexagonal close-packed oxygen atoms and $Fe^{3+}$ octahedrally coordinated sites. $Fe_3O_4$ has a cubic, inverse spinel structure with $Fe^{3+}$ in octahedral and tetrahedral sites, and $Fe^{2+}$ in octahedral sites. This, plus the fact that the precipitation process requires spatial segregation of regions of differing $E_H$ and possibly pH, suggests that the process is organic matrix mediated. Thus the magnetosome envelope is probably an integral element in the precipitation process, functioning as a locus for enzymatic activities including control of $E_H$ and pH, as well as a structural element.

The $Fe_3O_4$ particles in A. magnetotacticum have also been studied by high resolution transmission electron microscopy.[5] The results show that many of the particles are well ordered single domain crystals with a distinct morphology. This is based on an octahedral prism truncated by {100} faces. The crystals are preferentially aligned with [111] direction parallel to the chain axis. This morphology is different from that of other magnetotactic bacterial particles.[32] Thus the morphology of $Fe_3O_4$ particles produced by magnetotactic bacteria appears to be species specific.

No other crystalline phases were detected. However, in some crystals, noncrystalline material was found contiguous with the $Fe_3O_4$. This suggests that the hydrous-ferric-oxide phase is amorphous ferrihydrite,[2] and that deposition of $Fe_3O_4$ occurs as a solution-reprecipitation process, possibly triggered by $Fe^{2+}$ ions.[32,33]

Finally, diffusive motions of the magnetosomes in A. magneto-tacticum have been observed in the Mössbauer spectrum of the whole cells above 275 K.[34] The Mössbauer spectrum of the whole cells at T > 275 K was dramatically different from that of the frozen cells (T < 265 K)(Fig. 6). At 275 K it consisted primarily of a broad line of width $\Gamma = 72 \pm 1$ mm/s. The width of the broad line increased with increasing temperature to $\Gamma = 139$ mm/s at T = 295 K (Fig. 7). However, the total spectral intensity was temperature independent and equal to the total spectral intensity of the sharp-line spectrum of the frozen cells. Some hysteresis in the solid-liquid transition was noted in spectra obtained at 270 K. If the sample temperature had been increased from 265 K, the sharp-line spectrum was observed. However, if the sample temperature had been decreased from 275 K the broad-line spectrum was obtained. For T > 275 K, computer analysis showed that the intensity of the sharp-line $Fe_3O_4$ spectrum superposed on the broad-line spectrum was less than 0.2%.

The temperature dependence of the additional quadrupole doublet depended on whether the iron was primarily $Fe^{3+}$ or $Fe^{2+}$. When the additional iron was $Fe^{3+}$, as evidenced by the parameters of the doublet in the T = 265 K spectrum, there was no residual doublet superposed on the broad-line spectrum at T > 275 K. However, when the additional iron was primarily $Fe^{2+}$, the low intensity, sharp line $Fe^{2+}$ doublet remained superposed on the broad-line spectrum.

The striking spectral change at 270 K can be explained by the onset of diffusive motions of the $Fe_3O_4$ particles in the bacteria as they are warmed through the solid-liquid phase transition of the cytoplasmic fluid at 270 K.[34] Evidence for this comes from the fact that for freeze-dried cells the sharp-line spectrum persists at 300 K and the broad-line spectrum is never observed.[22] The broad-line spectra were analyzed with an extention of the "bounded diffusion" model previously developed for iron-containing proteins in the whole cells.[16] The analysis yielded the diffusion constant D of the magnetosomes, the effective viscosity $\eta$ of the magnetosome environment, and the mean-squared translational displacements $<x^2> < 8.4$ Å and rotational displacement $<\theta^2>^{1/2} < 1.5$ Å. This implies that the particles are relatively fixed in the whole cells. The effective viscosity and diffusion constant are inversely proportional; at 295 K, $\eta = 10$ cP and $D = 96 \times 10^{-10}$ cm$^2$/s. $\eta$ has a temperature dependence similar to that of water.

The fact that the additional $Fe^{3+}$ quadrupole doublet in the spectrum broadened together with the $Fe_3O_4$ lines is consistent with the previous cell fractionation studies[28] that show the hydrous

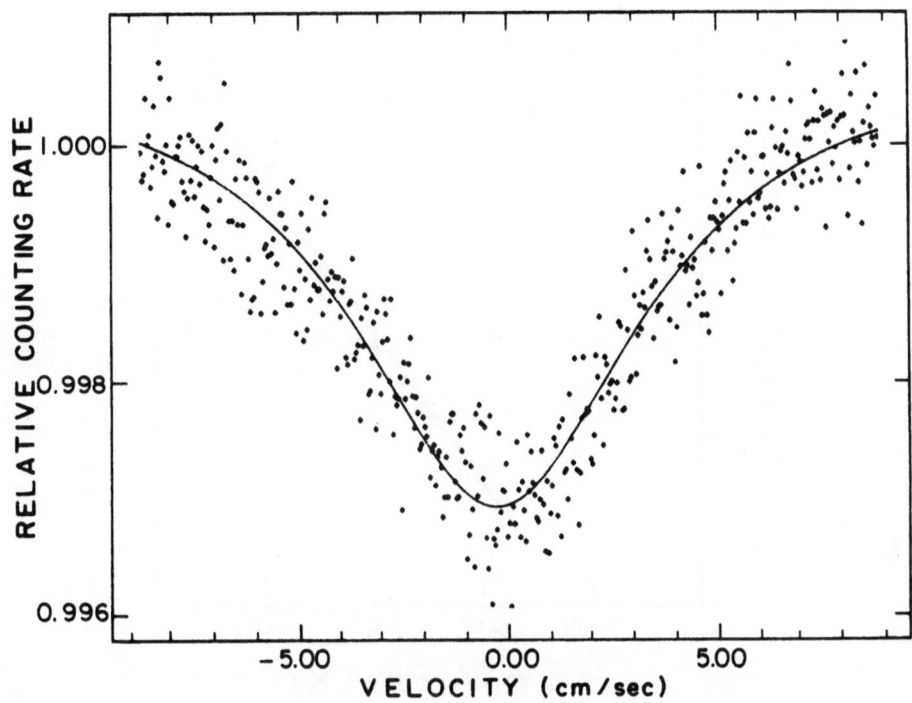

Fig. 6. Mössbauer spectrum of <u>A. magnetotacticum</u> at 275 K. (Ref. 34)

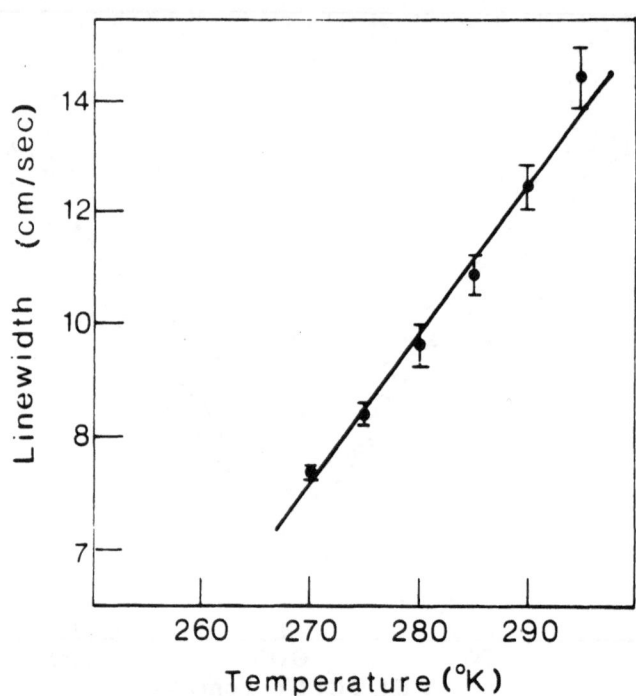

Fig. 7. Line width of the Mössbauer spectrum shown in Fig. 6 plotted as a function of temperature. (Ref. 34)

ferric-oxide to be primarily associated with the magnetosomes. Thus it undergoes the same diffusive motion as the magnetosomes. The fact that the sharp-line $Fe^{2+}$ spectrum remains even when the $Fe_3O_4$ lines have broadened shows that the $Fe^{2+}$ material is not associated with the magnetosomes. If it was, or if the $Fe^{2+}$ was dossolved in the cytoplasm, diffusive motion would broaden the sharp-line spectrum at T > 275 K, contrary to experiment. This suggests that the $Fe^{2+}$ is not associated either with the magnetosomes or with the cytoplasm in the cells. The $Fe^{2+}$ is very probably associated with the peptidoglycan layer of the cell wall. This association could occur during the conversion from the iron quimate complex outside the cell to ferric iron and ultimately $Fe_3O_4$ within the cell.

## CONCLUSION

Lowenstam[2] and Webb[35] have observed that biogenic iron oxides and oxyhydroxides are present in each of the five kingdoms of the biological world, with ferrihydrite the third more extensively formed mineral of biological origin and magnetite the fourth. Elucidation of the essentially bioinorganic processes of iron mineralization in organisms affords new opportunities for Mössbauer spectroscopy. We would like to understand how organisms accumulate and deposit iron minerals, with special emphasis on the mechanisms by which the deposition process is controlled.

## ACKNOWLEDGEMENT

We dedicate this paper to Solly G. Cohen and Shimon Ofer, our late friends and colleagues, and pioneers in Mössbauer spectroscopy. We thank R. Blakemore, W. O'Brien, S. Mann, I. Nowik, and E. R. Bauminger, for their contributions to the work reported here. This work was partially supported by the Office of Naval Research. The Francis Bitter National Magnet Laboratory is supported by the National Science Foundation.

## REFERENCES

1. A. Blaise, J. Chappert, and J. L. Giradet, 1965, Observation par mesures magnetiques et effet Mössbauer d'un antiferromagnetisme de grains fins dans la ferritine, <u>C.R. Acad. Sci. Paris</u> 261:2310-2313.
2. H. A. Lowenstam, 1981, Minerals formed by organisms, <u>Science</u> 211:1126-1130.
3. H. A. Lowenstam and S. Weiner, 1983, Mineralization by organisms and the evolution of biomineralzation, in <u>Biomineralisation and Biological Metal Accumulation,</u> P. Westbroek and E. W. deJong, Eds., Reidel, Boston, 191-203.

4. R. P. Blakemore, 1975, Magnetotactic bacteria, <u>Ann. Rev. Microbiol</u>. 36:217-238.

5. S. Mann, R. B. Frankel, and R. P. Blakemore, 1984, Structure, morphology and crystal growth of bacterial magnetite, <u>Nature</u> (submitted).

6. H. L. Ehrlich, <u>Geomicrobiology</u>, (M. Dekkar, New York, 1981). 165-200.

7. G. A. Clegg, J. E. Fitton, P. M. Harrison, and A. Treffry, 1980, Ferritin: Molecular structure and iron-storage mechanisms, <u>Prog. Biophys. Molec. Biol</u>. 36:56-80.

8. K. M. Towe and W. F. Bradley, 1967, Mineralogical constitution of colloidal hydrous ferric oxides, <u>J. Colloid. Interface Sci</u>. 24:384-392.

9. F. V. Chukrov, B. B. Zvyagin, A. I. Gorshkov, L. P. Yermilova, and V. V. Balshova, 1973, Ferrihydrite, <u>Int. Geol. Rev</u>. 16: 1131-1143.

10. J. F. Boas and B. Window, 1966, Mössbauer spectroscopy of ferritin, <u>Aust. J. Phys</u>. 19:573-576.

11. W. T. Oosterhuis, K. Spartalian, 1976, Biological iron transport and storage compounds, in <u>Applications of Mössbauer Spectroscopy</u>, Vol. I (R. L. Cohen, Ed.), Academic Press, New York, 141-170.

12. J. M. Williams, D. P. Danson, and Chr. Janot, 1978, A Mössbauer determination of the iron core particle size distribution in ferritin, <u>Phys. Med. Biol</u>. 23:835-851.

13. E. I. Stiefel and G. D. Watt, Azotobacter cytochrome $b_{557.5}$ is a bacterioferritin, <u>Nature,</u> 279:81-83.

14. E. R. Bauminger, S. G. Cohen, D. P. E. Dickson, A. Levy, S. Ofer, and J. Yariv, 1980, Mössbauer spectroscopy of <u>E. coli</u> and its iron storage protein, <u>Biochim. Biophys. Acta</u> 623: 237-242.

15. S. G. Cohen, E. R. Bauminger, I. Nowik, and S. Ofer, 1981, Dynamics of the iron-containing core in crystals of the iron-storage protein, ferritin, through Mossbauer spectroscopy. <u>Phys. Rev. Lett</u>. 46:1244.

16. I. Nowik, S. G. Cohen, E. R. Bauminger, and S. Ofer, 1983, Mössbauer absorption in overdamped harmonically bound particles in Brownian motion, <u>Phys. Rev. Lett</u>. 50:1528-1530.

17. R. P. Blakemore, 1975, Magnetotactic bacteria, <u>Science</u> 190: 377-379.

18. R. P. Blakemore and R. B. Frankel, 1981, Magnetic navigation in bacteria, <u>Sci. Am</u>. 245:58-65.

19. T. T. Moench and W. A. Konetzka, 1978, A novel method for the isolation and study of a magnetotactic bacterium, <u>Arch. Microbiol</u>. 119:203-212.

20. D. L. Balkwill, D. Maratea, and R. P. Blakemore, 1980, Ultrastructure of a magnetotactic spirillum, <u>J. Bacteriol</u>. 141: 1399-1408.

21. K. M. Towe, and T. T. Moench, 1981, Electron-optical characterization of bacterial magnetite, <u>Earth Planet. Sci. Lett</u>. 52:213-220.

22. R. B. Frankel, R. P. Blakemore, R. S. Wolfe, 1979, Magnetite in freshwater magnetotactic bacteria, Science 203:1355-1356.
23. R. B. Frankel and R. P. Blakemore, 1980, Navigational compass in magnetic bacteria, J. Magn. and Magn. Maters. 15-18:1562-1564.
24. R. P. Blakemore, R. B. Frankel, A. J. Kalmijn, 1981, South-seeking magnetotactic bacteria in the Southern Hemisphere, Nature (London) 286:384-385.
25. J. L. Kirschvink, 1980, South-seeking magnetic bacteria, J. Exp. Biol. 86:345-347.
26. R. B. Frankel, R. P. Blakemore, F. F. Torres de Araujo, D. M. S. Esquivel, and J. Danon, 1981, Magnetotactic bacteria at the geomagnetic equator, Science 212:1269-1270.
27. R. P. Blakemore, D. Maratea, R. S. Wolfe, 1979, Isolation and pure culture of a freshwater magnetic spirillum in chemically defined medium, J. Bacteriol. 140:720-729.
28. R. B. Frankel, G. C. Papaefthymiou, R. P. Blakemore, and W. O'Brien, 1983, $Fe_3O_4$ precipitation in magnetotactic bacteria Biochim. Biophys. Acta 763:147-159.
29. R. S. Hargrove, W. Kundig, 1970, Mössbauer measurements of magnetite below the Verwey transition, Solid State Commun. 8:303-308.
30. D. Bazylinski, R. P. Blakemore, R. B. Frankel, C. Rosenblatt, and K. Short, unpublished data.
31. R. M. Garrels and C. L. Christ, 1965, Solution, Minerals and Equilibria, Harper and Row, New York.
32. Y. Tamaura, K. Ito, and T. Katsura, 1983, Transformation of $\gamma$-FeO(OH) to $Fe_3O_4$ by adsorption of iron (II) ion on $\gamma$-FeO(OH) J. Chem. Soc. Dalton Trans., 189-194.
33. S. Mann, T. T. Moench, R. J. P. Williams, 1984, A high resolution electron microscope investigation of bacterial magnetite; Implications for crystal growth, Proc. Roy. Soc. (in press).
34. S. Ofer, I. Nowik, E. R. Bauminger, G. C. Papaefthymiou, R. B. Frankel, and R. P. Blakemore, 1984, Magnetosome dynamics in magnetotactic bacteria, Biophys. J. (in press).
35. J. Webb, 1983, A bioinorganic view of the biological mineralization of iron in Biomineralization and Biological Metal Accumulation, P. Westbroek and E. W. deJong, Eds., Reidel, New York, 413-422.

# MÖSSBAUER SPECTROSCOPY OF INTERCALATION COMPOUNDS

R. H. Herber and H. Eckert

Department of Chemistry
Rutgers University
New Brunswick, N.J. 08903

## I. INTRODUCTION

Intercalation compounds are formed by reversible topotactic
electron transfer reactions leading to the accommodation of guest
species (atoms or molecules) into the van der Waals gaps of
inorganic host matrices (1).

This solid state reaction, which frequently occurs readily at
ambient temperature, imposes several requirements on both reactants. The chemical bonding in the host is usually highly anisotropic, giving rise to a system of interconnected vacancies which
accept the guest material and facilitate its diffusion from the
surface to the interior. Also, a conduction band of sufficiently
low energy has to be present, which is involved in the reaction by
facilitating electron transfer according to the general reaction
schemes:

$$[M] + xA \rightarrow A_x^+ [M]^{x-} \qquad [1]$$

$$[M] + xB \rightarrow B_x^- [M]^{x+} \qquad [2]$$

Here, M denotes the matrix while A and B represent electron
donating and accepting guest species, respectively. In most well-
studied of the intercalation systems, reaction [1] prevails (1),
hence requiring guest species of low ionization potentials such as
alkali metals, metallocenes, and Lewis bases. On the other hand,
electron rich layered materials such as graphite (2), $Tl_2S$, and
$Ag_2F$ (3) also intercalate a variety of electron-deficient guest
materials, thus functioning as electron-donor matrices as
represented by eq. [2].

133

While, as implied by the term "topotactic", the structural features of the host lattices are basically maintained on intercalation, many physicochemical properties such as electron transport, optical, and magnetic characteristics may undergo dramatic changes. The opportunity to tailor the desired behavior by a judicious choice of the systems, as well as their compositional parameters, has stimulated a great deal of applied study (1,4). Thus, intercalation compounds have received technological attention as heterogeneous catalysts, secondary battery cathode materials, semiconductors and superconductors. Along with these studies, virtually all well-developed physicochemical techniques have been utilized in order to obtain further insight into the nature of the intercalation process (5). To this end, a large number of Mössbauer studies have been published which exploit the versatility of this technique to provide answers to questions belonging to quite different research areas. To provide a detailed overview of the application of nuclear gamma resonance techniques to the study of intercalation compounds, the present discussion is organized as follows: In Sections II and III is given an overview of the basic structural and chemical features of intercalation compounds, followed by a review of some general Mössbauer aspects in this class of materials. Section IV summarizes the basic structure and bonding inferences obtained mainly from the isomer shift and quadrupole coupling parameters. Section V deals with the lattice dynamical information (vibrational characteristics) obtained from studies of the Lamb-Mössbauer factors and/or the second order Doppler shift. Section VI is devoted to electron spin properties and special cooperative magnetic phenomena encountered in these systems. This paper is intended to be a comparative and critical survey of the available data. Whenever possible, the contribution which Mössbauer spectroscopy has made in the elucidation of a particular subject will be discussed within the context of the information obtained from complementary techniques such as NMR, magnetic susceptibility, x-ray, and neutron diffraction.

## II. STRUCTURE AND CHEMISTRY OF INTERCALATED MATERIALS

Since it is not appropriate in the context of this symposium to give a complete overview of all the matrices in which intercalation has been reported, the present discussion will be restricted to some of the most prominent systems in which significant Mossbauer work has been undertaken. Intercalation compounds are usually classified with respect to the structural dimensionality of their host lattices into:

  -three-dimensional systems
  -two-dimensional systems
  -one-dimensional systems
  -systems based on molecular host lattices

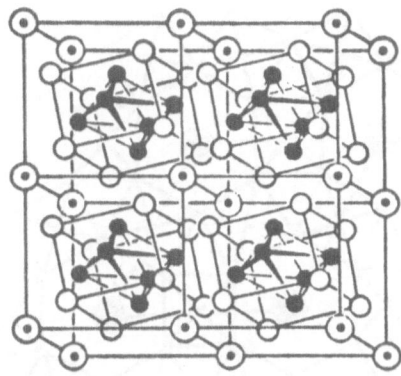

Fig. 1. Structure of Mo₆X₈ (X=S, Se)(from ref. (1)). Reproduced by
permission of Verlag Chemie.
● Mo
○ Chalcogen
◉ vacant site

3D systems display three-dimensionally connected host structures, in which one-dimensional arrays of vacancies give rise to
tunnel structures. An important member of this class, shown in
Fig. 1,is based on molybdenum cluster chalcogenides of the formula
$Mo_6X_8$ (X = S,Se) (6). These clusters are connected by covalent
bonds in such a way that interconnected channels, extending along
the three crystallographic axes, are formed. The intercalated
metal atom is situated in the middle of an asymmetric site surrounded by 8 chalcogen atoms. These materials, frequently referred
to as "Chevrel phases", have attracted a great deal of interest
because of their high temperature superconductive behavior (7).
Another important member of the 3D group is represented by tungsten
bronzes (8) and compounds of related structures. In fact, a
variety of different structures exist, depending on the nature and
the stoichiometric amount of the metal guest. The simplest structure, encountered in $Na_xWO_3$, is of the cubic perovskite type. More
complicated structures involve higher anisotropies, such as the
hexagonal structure (9), shown in Fig. 2, in which the vacancies
extend along the crystallographic c-direction, thus forming a
structure of isolated one-dimensional channels.

Intercalation compounds based on 2D systems have been most
widely studied. The structures of these materials are character-

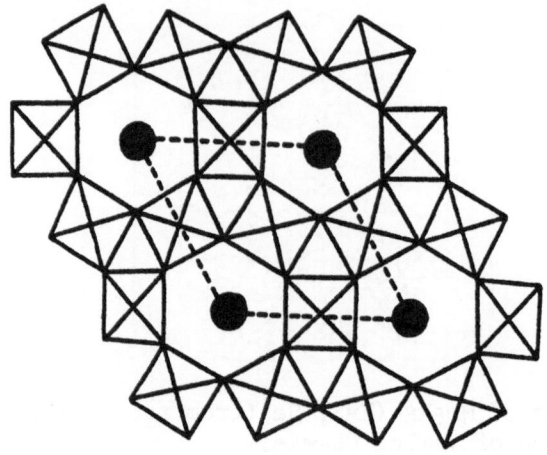

Fig. 2. Structure of hexagonal tungsten bronzes (projection onto the xy plane) (from ref. (8)). Reproduced by permission of the Royal Society of Chemistry.

ized by stacked layered matrix units. The best known member of this category is graphite which, under well defined intercalation conditions, gives rise to several ordered phases, differing in the arrangement of intercalated and unintercalated layers. If, for instance, every single layer is intercalated, the compound is called stage-1; while in a stage-2 compound every second layer is filled. Lower stoichiometry compounds of higher stages (stage 4 and 6) can be produced selectively (10).

Another important class of 2D materials is based on IVB and VB transition metal dichalcogenides (11). The structures of these compounds are related to hexagonal $CdI_2$, with the metal coordination inside the layers either octahedral (1T-type) or trigonal prismatic (2H-type). These units extend into the x, y plane while the bonding along the z-axis is of the van der Waals type. The van der Waals gaps between these layered matrix units represent an infinite two-dimensional periodic system of octahedral vacancies. the structure of the layered oxyhalides (FeOCl, VOCl) (12) (Fig. 3), which represent another important group of 2D materials, is slightly different, being made up of double layers extending into the orthorhombic x,z plane, rather than consisting of single layers. Within these units, the central metal is surrounded <u>cis</u>-octahedrally by 4 oxygen and 2 chlorine atoms. The van der Waals gaps are originally defined by distorted octahedra of 5 Cl atoms and one oxygen; on intercalation, however, changes in the interlayer stacking produce larger vacancies, thereby increasing the coordination number to 8.

1D systems are stabilized by chain units forming three-dimensional arrangements of vacancies. As with systems based on molecular lattices, these compounds have been studied to a much lesser extent, and within the context of the present review, only the $KFeS_2$ structure (13) will be briefly mentioned here. This compound (Fig. 4) consists of infinite zig-zag chains of edge-shared

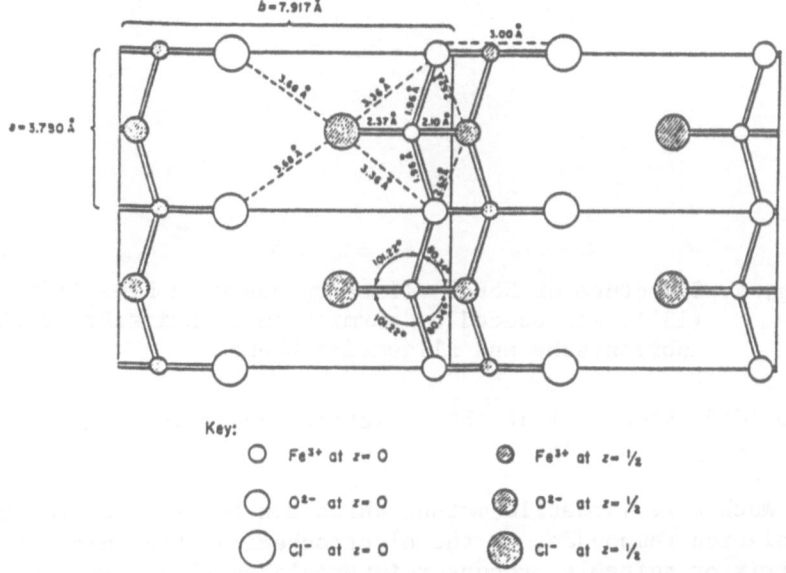

Key:

○ $Fe^{3+}$ at $z = 0$        ◉ $Fe^{3+}$ at $z = \frac{1}{2}$

◯ $O^{2-}$ at $z = 0$         ◉ $O^{2-}$ at $z = \frac{1}{2}$

◯ $Cl^-$ at $z = 0$          ◉ $Cl^-$ at $z = \frac{1}{2}$

Fig. 3. Structure of FeOCl (projection onto the xy plane) (from ref. (12)). Reproduced by permission of Acta Crystallographic.

Fe(III)S$_4$ tetrahedra. The interchain sites are occupied by K$^+$
ions. The structure contains, however, an equivalent number of un-
occupied sites, which can be electrochemically intercalated (14).

3D lattices are rigid and, hence, the size constraints of the
host vacancies impose severe restrictions upon the size of the
guest species which can be accommodated. In contrast, intercalation into systems of lower dimensionality is much more versatile,
because these lattices are flexible and intercalation usually
results in lattice expansions reflected by an increase of the unit
cell parameters. Since, however, the basic structural features of
the host lattices remain unaffected, the intercalation reaction is
referred to as being "topotactic". Another feature of this
reaction, the concomitant electron transfer, can occur by direct
reaction between guest and host, as in the case of FeOCl-Lewis base
intercalation (15). As will be shown later, Mössbauer spectroscopy

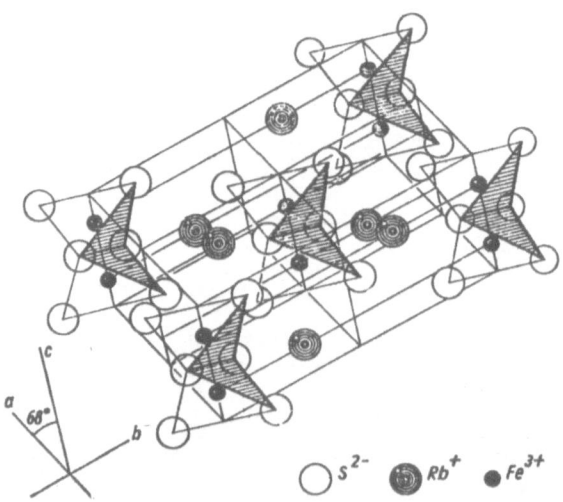

Fig. 4. Structure of RbFeS$_2$ (isomorphous with KFeS$_2$)(from ref.
(13)). Reproduced by permission of Zeitschrift für
Anorganische and Allgemeine Chemie.

has provided evidence that this reaction occurs via a molecular
redox disproportionation process (16).

A much more versatile method which can be used to prepare
intercalation compounds, is the electrochemical treatment of either
the matrix or suitable precursor intercalates (17). For most
systems, in which process [1] obtains, the host matrix is reduced
in the presence of the ionized guest species. The electrons are
either transferred by suitable reducing agents (I$^-$, S$_2$O$_3^=$, car-

banions, etc.) or provided at the cathode of an electrolytic cell. If, as in the case of $VS_2$ the host matrix itself is unstable, chemical $(I_2, O_2)$ or anodic oxidation of precursor intercalates (prepared by standard high temperature synthetic methods) is a useful pathway to obtain both the empty matrix as well as intercalates of lower stoichiometries.

## III. GENERAL ASPECTS OF MÖSSBAUER SPECTROSCOPY IN INTERCALATION COMPOUNDS

Fig. 5 gives a general overview of the potential usefulness of Mössbauer spectroscopy as a technique in the study of intercalation chemistry. Although this compilation is by no means complete, and, for the sake of clarity, many systems of minor importance have been omitted, this representation shows clearly that there is only a moderate overlap between those elements which are constituent members of intercalation matrices, and elements with which the Mössbauer effect has been developed to a well-established spectroscopy, and for which the hyperfine parameters can provide useful chemical information. In order to overcome this limitation, two approaches are possible:

--extended studies of less-common Mössbauer nuclei
  (such as $^{181}Ta$ and $^{183}W$)
--chemical modification of systems in order to
  introduce Mössbauer active nuclei which can serve
  as "reporter atoms" in the structure.

With respect to intercalation compounds, Mössbauer studies using nuclei different from the "easy-to-use" isotopes $^{57}Fe$, $^{119}Sn$, $^{151}Eu$ are still at a very early stage, whereas the second approach has been taken quite frequently. Introduction of substitutional Fe, especially into transition metal dichalcogenides (18) has considerably increased the number of materials available for study. It should be borne in mind, however, that the interpretation of the data in these cases may be quite involved, since the substitution effects of iron may be different in the host lattices and their intercalates, respectively. Also, quite commonly, in Mössbauer studies on compounds in which the probe atom is part of the host matrix, frequently it is noted that the structural and electronic changes induced by the guest species are not large enough to influence the interaction parameters significantly. This aspect limits the number of useful nuclei considerably, i.e., to isotopes on which excellent spectroscopic resolution can be obtained. If, on the other hand, the guest species itself contains a Mössbauer active isotope, the electronic state, as well as the bonding situation created by the matrix environment, are directly reflected in the spectra. Sometimes highly valuable complementary information can be obtained if both guest and host are studied.

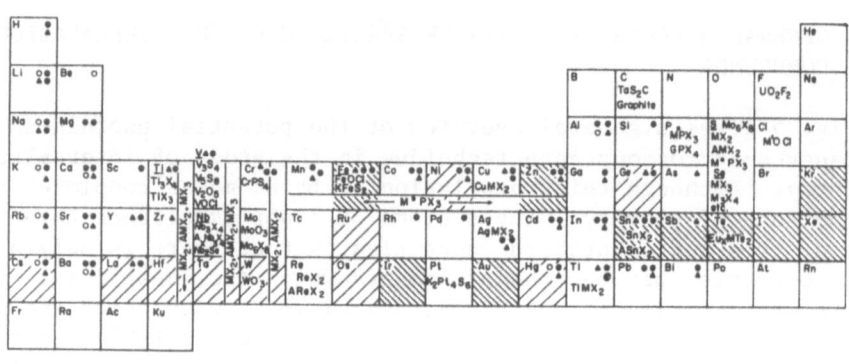

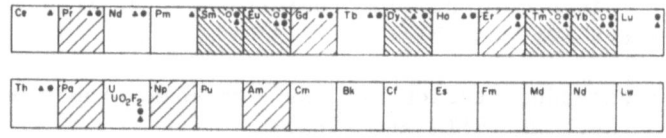

Fig. 5. Potential usefulness of Mossbauer methods in the investigation of intercalation compounds. Elements are grouped
into those

□ with no  Mossbauer isotope
▨ with an uncommon Mossbauer isotope
▧ with a common Mossbauer isotope

The most important host lattices are indicated: X=S,Se;
M=V,Nb,Ta; M*=Mn,Fe,Co,Ni,Cu,Zn; M'=Ti,V,Cr,Fe; A=Alkali
metal.
The respective element forms intercalation compounds with
O graphite;
⊗ $M_6S_8$; ● $TaS_2$; ▲ $WO_3$

Due to the severe constraints on the range of the materials available for such studies, these systems have to be specially designed. As will be shown later, successful results of such efforts have been obtained in the authors' laboratories.

Despite the limitations mentioned above the first publication of a $^{57}$Fe spectrum in graphite (19) has been followed by almost 200 papers. Although many of the results appear to be rather phenomenological, a great deal of chemically oriented systematic studies involving

> --the influence of different guest species upon a given host lattice
> --the behavior of a given guest species in different host lattices
> --the influence of structural and compositional parameters

have made substantial contributions towards a deeper understanding of the nature of intercalation compounds.


IV.  STUDIES OF STRUCTURE AND BONDING

1. Charge Density Waves

From x-ray, neutron, and electron diffraction work, it is known that many anisotropic materials which represent hosts for intercalation compounds exhibit intrapolytypic phase transitions associated with superlattice formation (20). Below a certain temperature these structural changes lead at first to a superlattice which is incommensurate with the basic lattice (i.e., its periodicity is different from that of the latter) and finally, at lower temperatures, a first order transition to a commensurate superlattice occurs. In both the commensurate and the incommensurate state, long period lattice distortions, frequently referred to as "charge density waves", exist. This manifests itself in site inequivalencies (21), and Mössbauer spectroscopy has proven to be an extremely sensitive tool to probe this phenomenon. At present, however, the results obtained from different studies do not yet present a consistent picture. In an $^{57}$Fe impurity study of commensurate 1T-TaSe$_2$ (22) three different sites have been identified, the IS and QS parameters of which are markedly different and scale roughly with the expected local charge density at the site in question. In sharp contradiction to this, complementary studies of the $^{181}$Ta Mössbauer effect (23) indicate two sites with identical quadrupole couplings and markedly different isomer shifts. Furthermore, it is worth mentioning in this context that temperature dependent TDPAC (time differential perturbed angular correlation) studies, utilizing the quadrupole moment of $^{181}$Ta in TaS$_2$, of the corresponding sulfide (24) have been carried out. These studies indicate a uniform distribution of quadrupole couplings, as long as the compound is in the incommensurate state, while in the commensurate state 2 distinct sites with ~11% difference in the quadrupole couplings can be distinguished. It would be very interesting to test this result by complementary [181]Ta Mössbauer studies. The spectra of the 2H structures of $TaS_2$ and $TaSe_2$ have so far been studied only above their CDW phase transitions (25) and, consistent with expectation, one unique site is observed. No results on their corresponding intercalation compounds have as yet been reported. These investigations seem to be very intriguing, however, since (a) intercalation itself is expected to introduce periodic lattice distortions at low temperatures (26) and (b) TDPAC results (27) in tantalum dichalcogenide intercalates as well as [93]Nb NMR data on the related niobium disulfide systems (28) indicate that intercalation results in a significant decrease of the electric field gradient at the site of the transition metal atom.

## 2.   Charge Transfer Between Guest and Host

Mössbauer spectroscopy has contributed a great deal of experimental evidence that the intercalation reaction is essentially accompanied by topotactic charge transfer. Although equations [1] and [2] imply the latter to be quantitative, recent studies, primarily solid state NMR results, indicate the existence of many borderline cases in which there is either only fractional charge transfer (29) or essentially none at all (30). With respect to elucidating the details of the charge transfer process, Mössbauer spectroscopy is an excellent spectroscopic tool since isomer shifts and quadrupole splittings respond sensitively to changes in electron densities and geometries, and hence permit not only the assignment of distinct valency states but also give quantitative information concerning the amount of charge transfer. Within the frame of this subsection three types of intercalation complexes are reviewed:

> --metal cation insertion compounds of electron-accepting
>   host matrices
> --Lewis base intercalation compounds formed by electron-
>   accepting host matrices, mainly FeOCl
> --Lewis acid intercalation complexes formed by electron
>   donating host matrices, mainly graphite

Metal cation insertion compounds. Although most of the basic information on these systems has been obtained by intercalating Mössbauer-active nuclei, a good deal of host-lattice studies have also been reported for matrices which either contain Mössbauer active atoms or which can be doped with such species to serve as "reporter nuclei". With respect to the latter, experiments have not always been fruitful, because the transferred electrons may not influence the s-electron densities or local symmetries seen by the matrix atoms, at least not to an extent which is significant enough to be detected within the resolving power of the particular method employed. For instance, no inferences can be drawn by comparing $^{182}W$ literature data of tungsten bronzes (31) with other studies on the unintercalated matrix (32). Similarly the informational content of $^{119}Sn$ spectroscopy in layered $A_xSnS_2$ intercalates (A = alkali metal) (33) suffers from the lack of resolution as well as the complexity of the chemical system (34). In a study of the 3D system $Li_xFeV_3O_8$ (35) the vanishing influence of lithium content on the isomer shift of the $^{57}Fe$ resonance has been taken as evidence that the electrons are transferred to the vanadium atoms rather than to iron. In this case the stoichiometric dependencies of the quadrupole splittings and linewidths were qualitatively explained in terms of a more nearly ideal local symmetry of the iron site, due to the successive convergence of the metal ionic radii in the structure, resulting from the reduction of vanadium from the pentavalent state to +IV, and subsequently to the +III state.

An analogous reasoning is suggested by $^{57}Fe$ results of $FePS_3$ intercalated (36) by Li and cobaltocene (37). The reaction clearly creates site inequivalencies, but this does not affect the Fe(II) valency and high spin state. This conclusion is consistent with $^{31}P$ NMR data (38,39) which indicate that the electron donation might occur to antibonding phosphorus or sulfur orbitals rather than to the iron atoms. Several host lattice studies have been undertaken with solid state cathode materials based on IVB and VB transition metal dichalcogenides. The clearest picture has been obtained on the system $Li_xFe_yTi_{1-y}S_2$ (40). In the unlithiated compounds the IS is virtually independent of the substitutional content (y) and falls well into the range of IS data reported for low spin Fe(III) in octahedral coordination (41). On lithiation, a dramatic increase of both IS and QS is observed, the numerical values of which are compatible with high spin Fe(II). These results have been consistently interpreted as follows: Since the conduction band of $TiS_2$ is empty, substitution of Ti by Fe produces holes in the valence band established by the sulfur 3p lone pairs. Therefore, if the amount of iron is y, the sulfur is oxidized to an average valence state of $-(2-y/2)$. Lithiation, at first, results in a filling of these electron holes and, subsequently, converts Fe(III) (low spin) into Fe(II) (high spin). If this simple model obtains, one would expect the transition Fe(III) → Fe(II) to be shifted to higher Li contents when y is increased (since this produces more holes to be filled in the first place). This, however, is not observed since the situation is slightly complicated by the fact that the bottom of the Ti d band is below the Fe(II) energy level at low y contents, and a corresponding portion of this band is filled before the Fe becomes reduced. At higher iron contents this contribution decreases. Analogous studies have been carried out with systems based on VB dichalogenides. As shown by Mössbauer spectroscopy (22) substitution of the Ta atoms in 1T-TaS$_2$ by iron rapidly destroys the CDW present. In VS$_2$ the substitution of V by Fe probably has the same effect (18). A very important practical consequence is the suppression of intermediate phases in the Li$_x$VS$_2$ system which improves the reversibility of cathodes based on this material (42). Although the compositional trends of the Mössbauer parameters in the Li$_x$Fe$_y$V$_{1-y}$S$_2$ system (43,44) are strikingly similar to those of the above mentioned titanium disulfide system here the presence of low spin Fe(III) seems to be ruled out by magnetic susceptibility data. Therefore, in order to account for the observed effects it was suggested that in the VB transition metal disulfides the iron is divalent throughout the whole compositional region, and undergoes a low spin → high spin transition as the Li content is increased.

Electrochemical lithiation of KFeS$_2$ has been shown (45) to convert the tetrahedrally coordinated Fe(III) centers successively into Fe(II) with increasing charge transfer. The Fe(II) site produced is clearly distinguished in the Mössbauer spectrum (IS = 0.54 mm/s, QS = 1.02 mm/s) at room temperature and dominates the spectrum at the nominal composition LiKFeS$_2$. Upon deintercalation the original spectrum of KFeS$_2$ reappears. At intermediate lithium concentrations complex spectra are observed, indicating at least three additional phases or sites. At a given nominal lithium content, the relative amounts of these sites are different for the discharge (lithiation) and the recharge (delithiation) process, respectively. These results show that, although the intercalation process, in toto, is reversible, the charging and recharging processes are mechanistically different.

A key experiment in order to characterize the guest → host charge transfer is to study the Mössbauer hyperfine parameters of intercalant nuclei. Table 1 summarizes a comprehensive survey of the IS data obtained. The (in most cases) unambiguous valence assignments derived from these data have proven the validity of the charge transfer concept represented by eq. [1]. Beyond this, Table I clearly shows that, with the exception of Fe in V$_2$O$_5$ (46), both iron and tin are divalent in all host matrices. Eu is divalent in graphite (47) and Chevrel phases (48-49) while the trivalent state prevails in tungsten bronzes (50,51). If, however, both Sn and Eu are present in mixed tungsten bronzes, the Eu becomes successively divalent with increasing Sn content (52). This behavior was suggested to reflect charge contraints related to an upper limit of ~ 0.65 electrons per W in tetragonal tungsten bronzes. Similar mixed valence states for Eu have been reported in several niobates and tantalates having the tetragonal tungsten bronze structure (53,54). With respect to the bronzes, Fe in Fe$_{0.33}$V$_2$O$_5$ as well as

Eu in general, violate the rule for polyvalent elements being present in the lowest stable oxidation state (55).

As another mixed valence compound, $YbMo_6S_8$ has been identified by [170]Yb absorption spectroscopy (56). The spectra were fitted to a near 1:1 ratio of divalent and tervalent Yb. Mössbauer emission spectra of [170]$YbMo_6S_8$ created by neutron irradiation of [169]$TmMo_6S_8$ have been interpreted similarly. The latter result seems, however surprising in view of results of a [169]Tm absorption experiment carried out in the same study. This experiment indicates that only a single Tm(III) site exists in $TmMo_6S_8$. Since $t_{1/2}$ of [170]Yb is only 1.57ns (57) one would expect, therefore, an emission spectrum without chemical after effects, i.e., it should reflect the unique tervalence of the Tm atoms in $TmMo_6S_8$ rather than the mixed valence of Yb encountered in $YbMo_6S_8$.

Beyond simple valence assignments the numerical IS values carry information about the degree of ionicity, i.e., they make it possible to decide how complete is the guest → host charge transfer. A compilation of solid state NMR data utilizing mainly [7]Li, [23]Na, [133]Cs, and [205]Tl chemical shifts of the respective guest atoms (58) clearly indicates that: (a) bronzes represent the most ideal systems with respect to eq. [1], while in sulfide-and selenide based systems there is substantial covalency, and (b), increasing guest metal content decreases the ionic bonding character of the guest species.

In general, these statements are reinforced by the available Mössbauer data. This can be seen from the eventual occurrence of higher oxidation states Fe(III), Eu(III) only in oxide bronzes and the fact that the isomer shift of divalent iron is substantially larger in $WO_3$ (59,60) than in the sulfide and selenide materials (61-68). Moreover, for Cs-graphite intercalation compounds, the [133]Cs Mössbauer isomer shift (69) obeys the same compositional trend as the [133]Cs NMR Knight shift (70) indicating ~50% ionization in the stage-1, and complete ionization in the stage-4 compound. In transition metal dichalcogenides, the isomer shifts of intercalated [57]Fe fall well into the range reported for high spin Fe(II) in octahedral sulfur coordination (71), thus indicating a comparable amount of covalency. For iron, however, an interesting anomaly is observed in all layered host systems: in contrast to the trends usually observed, the ionicity, as inferred from the IS magnitude, appears to _increase_ with increasing iron content.

The bonding of tin in intercalated host matrices has been thoroughly reviewed elsewhere (72) and only a short synopsis is given here. The spectra of tin intercalates of transition metal dichalcogenides $Sn_xMX_2$ (x<1; M = Nb,Ta; X = S,Se) are characterized by the absence of an observable quadrupole interaction and isomer

Table I: Mossbauer isomer shifts for metal insertion compounds, x represents the total guest content per host matrix formula unit. References are Fe metal for 57Fe, BaSnO3 for 119Sn, and EuF3 for 151Eu.

**$WO_3$**

| Element | x | T/K | IS | Ref. |
|---|---|---|---|---|
| Fe | 0.0018 | 300 | 1.16 | (59) |
| | 0.005 | 300 | 1.15 | (59) |
| | 0.0102 | 300 | 1.16 | (59) |
| | 0.0193 | 300 | 1.16 | (59) |
| | 0.025 | 300 | 1.16 | (60) |
| Sn | 0.01 | 78 | -3.4 | (55) |
| | 0.10 | 78 | -3.4 | (55) |
| | 0.31 | 78 | -3.4 | (55) |
| | 0.40 | 78 | -3.4 | (55) |
| Eu | 0.01 | 80 | +0.24 | (50) |
| | 0.02 | 80 | +0.14 | (50) |
| | 0.04 | 80 | -0.17 | (50) |
| | 0.05 | 80 | -0.05 | (50) |
| | 0.06 | 80 | -0.04 | (50) |
| | 0.07 | 80 | -0.07 | (50) |
| | 0.08 | 80 | 0 | (50) |
| | 0.1 | 80 | 0 | (50) |
| | 0.2 | 80 | +0.03 | (50) |
| | 0.1 | 300 | 0 | (51) |

**$Mo_6S_8$**

| Element | x | T/K | IS | Ref. |
|---|---|---|---|---|
| Fe | 1.0 | * | * | (66) |
| | 1.3 | * | * | (65) |
| | 2.0 | 90 | 0.81 | (67) |
| | | 300 | 0.77 | (67) |
| Sn | 1.0 | 4.2 | 3.77 | (86) |
| | 1.0 | 80 | 3.72 | (86) |
| | 1.0 | 300 | 3.62 | (86) |
| | 1.0 | 4.2 | 3.76 | (87) |
| | 1.0 | 80 | 3.71 | (87) |
| | 1.0 | 300 | 3.61 | (87) |
| Eu | 1.0 | 4.2 | -12.8 | (49) |
| | 0.94 | 0.55 | -11 | (48) |

**$NbS_2$**

| Element | x | T/K | IS | Ref. |
|---|---|---|---|---|
| Fe | 0.25 | 4.2 | 0.73 | (61) |
| | 0.25 | 78 | 0.73 | (61) |
| | 0.25 | 300 | 0.61 | (61) |
| | 0.33 | 4.2 | 0.84 | (61) |
| | 0.33 | 78 | 0.80 | (61) |
| | 0.33 | 300 | 0.68 | (61) |
| | 0.33 | 78 | 0.92 | (63) |
| | 0.33 | 300 | 0.79 | (63) |
| | 0.25 | 78 | 0.75 | (69) |
| | 0.33 | 78 | 0.79 | (64) |
| | 0.5 | 78 | 0.86 | (64) |
| Sn | 0.56 | 300 | 3.62 | (28) |
| | 0.33 | 300 | 3.78 | (76) |
| | 1.0 | 300 | 3.12 | (77) |
| | 0.33 | 78 | 3.68 | (77) |
| | 1.0 | 78 | 3.13 | (77) |

**$TaS_2$**

| Element | x | T/K | IS | Ref. |
|---|---|---|---|---|
| Fe | 0.2 | 300 | 0.8 | (62) |
| | 0.26 | 300 | 0.82 | (62) |
| | 0.28 | 300 | 0.84 | (62) |
| | 0.39 | 300 | 0.86 | (62) |
| Sn | 0.33 | 78 | 3.85 | (73) |
| | 0.33 | 300 | 3.94 | (73) |
| | 1.0 | 78 | 3.19 | (73) |
| | 1.0 | 300 | 3.15 | (73) |
| | 0.17 | 300 | 4.08 | (76) |
| | 0.33 | 300 | 3.94 | (76) |
| | 1.0 | 300 | 3.14 | (76) |

**Others**

| Element | Host | x | T/K | IS | Ref. |
|---|---|---|---|---|---|
| Fe | $V_2O_5$ | 0.33 | 300 | 0.225 | (45) |
| | | 0.33 | 78 | 0.329 | (45) |
| | $TiS_2$ | 0.25 | 78 | 0.86 | (63) |
| | | 0.25 | 300 | 0.73 | (63) |
| | | 0.33 | 78 | 0.90 | (63) |
| | | 0.33 | 300 | 0.76 | (63) |
| | | 0.5 | 78 | 0.91 | (63) |
| | | 0.5 | 300 | 0.79 | (63) |
| Sn | $VS_2$ | 0.25 | 4.2 | 0.83 | (68) |
| | | 0.25 | 300 | 0.57 | (68) |
| | | 0.5 | 4.2 | 0.71 | (68) |
| | | 0.5 | 300 | 0.89 | (68) |
| | $NbSe_2$ | 0.5 | 300 | 3.76 | (76) |
| | | 1.0 | 300 | 2.86 | (76) |
| Eu | Graphite | 0.17 | 300 | -11.7 | (46) |

*not determined

shifts falling into the range 3.6-4mm/s (73-78). Since, in general, $^{119}$Sn isomer shifts are not a straightforward measure of bond ionicities in divalent tin compounds (79), the chemical information derived has to be considered with great care. It can be noted, however, that for tin in transition metal disulfide hosts a rough correlation with the $^{119}$Sn NMR Knight shift (28,80) obtains, indicating increasing covalency with increasing tin content. The origin of the two-line spectrum observed in SnTaS$_2$ (73-75) as well as SnNbS$_2$ (77), which had been a point of extensive discussion, was settled by a recent single crystal study (75). In these compounds a single site with an unusual linear S-Sn-S array (81,82) is responsible for a large quadrupole coupling and an isomer shift which is substantially lower as compared to the phases containing smaller amounts of tin. It should be noted here that the d$z^2$ band of the host can only accept one electron per VB metal (83). Thus, the donation cannot be complete for Sn contents above 0.5 and the excess electron density has to be either localized at the tin atom (which would make the latter paramagnetic) or stored in an s/p band formed by the intercalated metal. Whereas the first possibility has been ruled out in an applied field study (78), the second explanation is favored in view of (a) metallic behavior, (b) short Sn-Sn bonding distances, and (c) a strong $^{119}$Sn NMR Knight shift (80) which is comparable to the respective value in Sn metal. Hence the low IS value probably reflects the peculiar bonding situation encountered here, being ionic/covalent in one dimension, while metallic in two dimensions. The fact that, in spite of the Mössbauer results, two sites are seen in an XPS experiment (84) may indicate that valence fluctuations are present, which are fast on the timescale of the former ($10^{-10}$s) but slow on the timescale of the latter ($10^{-17}$s) experiment.

Another point of controversy, which again casts some doubt on the usefulness of Sn(II) isomer shifts to yield refined charge transfer information is the bonding state of Sn in Chevrel phases. Although in these compounds the tin isomer shifts (85-87) are comparable to the data for the previously discussed transition metal dichalcogenide intercalates, the $^{119}$Sn NMR Knight shifts in SnMo$_6$S$_8$ (88) and SnMo$_6$Se$_8$ (89) are unusually strong, suggesting metallic rather than ionic bonding in these compounds. No attempts have been made so far to settle this striking discrepancy.

Lewis base intercalation complexes formed by electron accepting host matrices. A number of layered host matrices are known to form intercalation compounds by direct reaction with the guest species. Extensive studies of the system TaS$_2$-pyridine have shown that this reaction involves a redox disproportionation of the guest molecules according to the sequence (90):

$$\text{py} \rightarrow \text{bipy} + 2\text{H}^+ + 2\text{e}^- \qquad\qquad [3]$$

$$py + H^+ \rightarrow pyH^+ \qquad\qquad [4]$$

$$TaS_2 + npyH^+ + ne- \rightarrow TaS_2{}^{n-}(pyH^+)n \qquad\qquad [5]$$

where "py" = $C_5H_5N$ and "bipy" is 4,4' bipyridine, $C_{10}H_8N_2$. Accordingly, the matrix undergoes a reduction, with the additional electron being delocalized in the band structure of the host. The excess charge is neutralized by a complementary amount of protonated species, which, rather than neutral molecules, are localized within the van der Waals layer. It should be noted, however, (vide infra) that neutral molecules can be inserted into the layer structure of the host to serve - for example - as molecules of solvation or as hydrogen bonding partners of the cationic species involved. The above model has been supported by some [1]H and [2]D NMR studies (91) and, more recently, by neutron diffraction experiments (92), which have unambiguously revealed that the guest molecules adopt an orientation in which the lone pairs are directed parallel to the layers. For $TaS_2(py)_{0.5}$ a charge transfer of ~0.2 electrons per Ta was estimated from magnetic susceptibility data (93). These results are corroborated by TDPAC results which also speak in favor of an ionic bonding bonding concept for $TaS_2$-Lewis base intercalates (27).

Although being chemically rather different from the transition metal dichalcogenides, the layered compound iron oxide chloride (FeOCl) shows a quite analogous intercalation chemistry. Thus, a large number of different Lewis bases including $NH_3$ and amines (94), pyridines (95,100), anilines (96), organophosphorus (97) and -sulfur compounds (98) have been successfully intercalated under comparatively mild reaction conditions. In addition, alkali metal insertion (99,100) as well as intercalation of several organometallic guest molecules (101-104), have been demonstrated. Mössbauer studies have been undertaken on these systems utilizing the uniquely favorable properties of the [57]Fe nucleus as a probe for the guest → host interaction. Early results obtained by a Japanese group (105) were based on samples which may have been inadequately characterized, and have been re-examined in later studies (95). In the following, the recent results obtained in the authors' laboratories (16,106) will be discussed in some detail. The different spectroscopic behavior of the intercalation compounds compared to unintercalated FeOCl is depicted in Fig.'s 6 and 7. Whereas in the unintercalated matrix only Fe(III) (high spin) is present and no anomalous temperature dependent effects are observed, all Lewis base intercalation compounds studied so far show the additional presence of Fe(II) (high spin) at low temperatures (see Fig. 8). Above 130 K this component can no longer be readily resolved from the spectra; in the temperature range 78 < T < 300 K the quadrupole splitting of the majority doublet shows a characteristic sigmoid decrease with increasing temperature. Lines are particularly broad in the temperature region 150 < T < 200 K

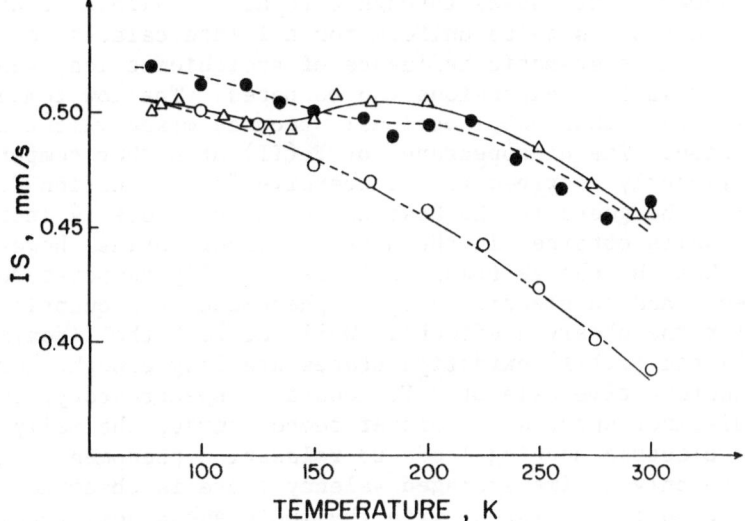

Fig. 6. Temperature dependence of the isomer shift (vs. Fe metal)
of the majority doublet in
○ FeOCl
● FeOCl (phenylguanidine)0.1
△ FeOCl (α-picoline)0.2

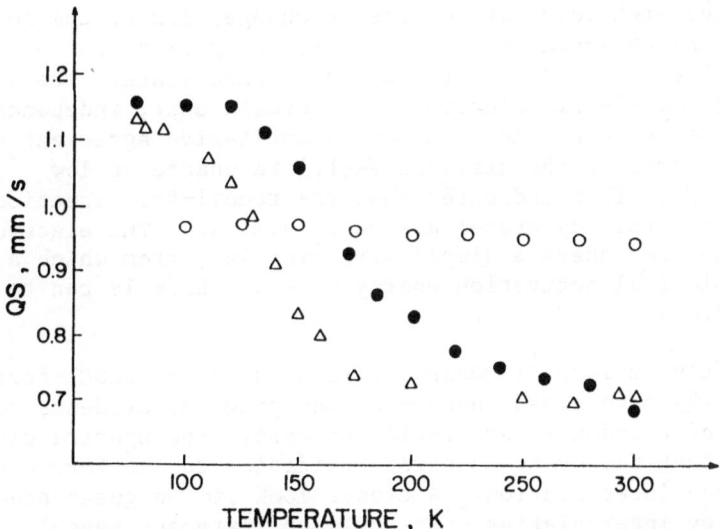

Fig. 7. Temperature dependence of the quadrupole splitting of the
majority doublet in
○ FeOCl
● FeOCl (phenylguanidine)0.1
△ FeOCl (α-picoline)0.2

and the isomer shift passes through a relative maximum at about 200 K. This behavior is quite uniform for all intercalates so far studied, and no systematic influence of stoichiometries, guest basicity, or lattice expansions can be noted. The low temperature spectra indicate that FeOCl obviously becomes mixed valent upon intercalation. The disappearance of Fe(II) at higher temperatures has been recently ascribed to a successive "back-donation" of charge from the guest to the host as the temperature is increased (107). Results obtained in the authors' laboratories, however, indicate that the charge transfer is essentially temperature independent, and an electron hopping phenomenon can quantitatively account for the observed effects. While at 78 K the lifetimes of the Fe(II) and Fe(III) oxidation states are long enough, compared to the inherent timescale of $^{57}$Fe Mössbauer spectroscopy, to give rise to distinct spectra, at higher temperatures, thermally activated electron hopping leads to relaxation phenomena. At room temperature only a time-averaged valency state is observed. This interpretation is consistent with the ionic guest-host bonding model, as well as the chemical intercalation process depicted by eqs. [3-5]. In addition, subsequent FTIR studies have contributed further evidence for this model, since in the FeOCl-pyridine system py, pyH$^+$ and bipy can all be shown to be present (108).

As indicated by Fig. 8, computer-simulated spectra generated by the application of a simple relaxation model (109) closely reproduce the experimental lineshapes.

In the high temperature (fast exchange) limit, the numerical value of the observed quadrupole splitting permits an evaluation of the Fe(II) contribution to the mixed valence state. The latter amounts to ~0.10-0.12 electrons per formula unit, independently of the guest species and in almost quantitative agreement with the fractional area of the distinct Fe(II) resonance at low temperatures. This indicates that the recoil-free fractions for iron in both valency states are quite similar. The electron relaxation time obeys a simple Arrhenius law, from which a phenomenological activation energy $E_a \approx 5.0$ kJ/mole can be extracted.

Recently, bilateral Mössbauer results in the FeOCl-ferrocene system (110) have added another strong piece of evidence for the validity of a redox intercalation process. The spectra clearly indicate that ferrocene undergoes oxidation to the ferricenium cation upon intercalation. A closer look at the guest species obtained by intercalating $^{57}$Fe enriched ferrocene reveals (111) that a small amount of the unoxidized guest molecule is also present, the fraction of which appears to increase at lower temperatures. These results may indicate that the Fe atom of the

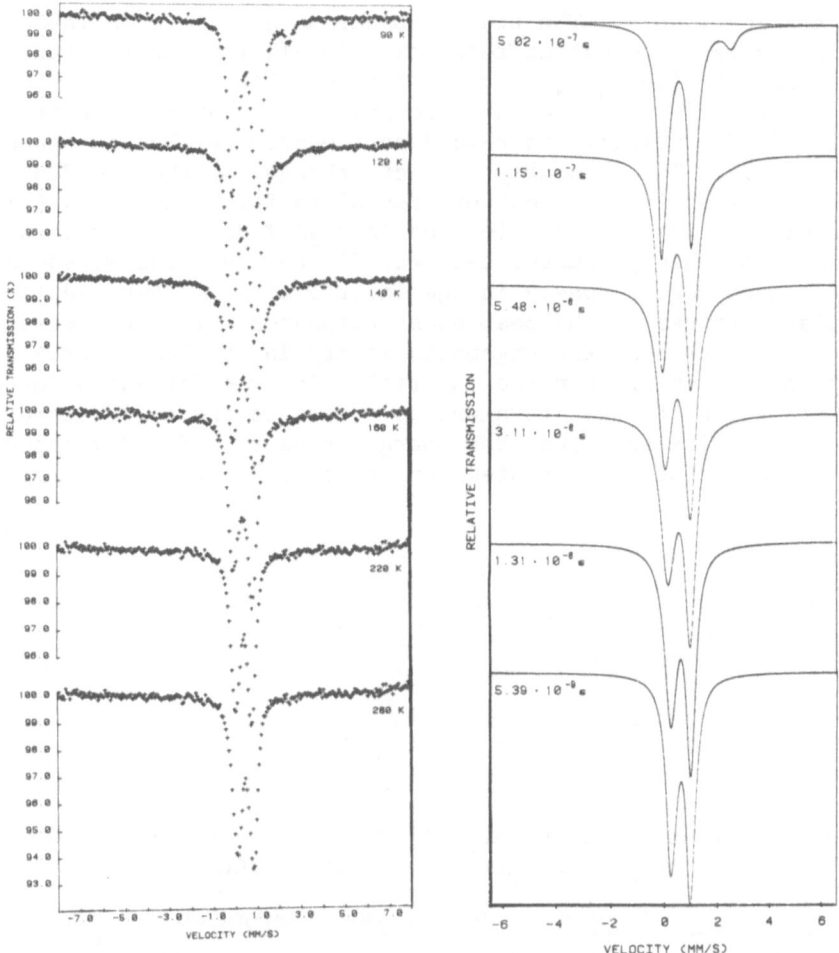

Fig. 8. Temperature dependent experimental spectra for FeOCl(Krypto-fix-22)$_{1/18}$ and computer simulated relaxation spectra, assuming 10% Fe(II) and 90% Fe(III).

intercalated ferrocene molecule takes part in the hopping process, and some fraction of the itinerant electrons become trapped on the guest species at low temperatures.

Although these studies have proven very useful, the observed lineshapes are quite complicated due to the superposition of both guest and host spectra. In order to avoid these problems without abandoning the high informational content of bilateral guest/host Mössbauer studies, recent efforts have been undertaken to inter-calate organotin compounds into FeOCl (112). A series of four FeOCl organotin intercalates (cf. Table II) has been studied. The iron Mössbauer data are consistent with those observed in simple

Lewis base intercalates and indicate transfer of about 0.1 electrons per FeOCl unit.  The tin Mössbauer data, listed in Table II,
exhibit several interesting features.  In all the intercalates,
except that of {(CH$_3$)$_2$N}$_4$Sn, the presence of two different tin
atoms can be inferred from the spectra.  The quadrupole splitting
of these is dramatically increased as compared to the neat guest
species (113).  This probably reflects changes of the coordination
number from four (in the neat organotin) to five and/or six (within
the FeOCl layers).  In addition the ln f $\underline{vs}$ T data (for a detailed
discussion of this parameter see Sec. V) show that the molecules
are held much more strongly in the intercalates as compared to the
molecular lattices of the neat guest compounds.  In other words,
the forces which bind the organotin moiety in the host matrix are
appreciably stronger than the intermolecular van der Waals forces
which stabilize the neat covalent molecules in the solid state.
All these results show that the charge transfer model developed
above is valid for these systems and that at least part of the

Table II. Mössbauer parameters of organotin guest species in the
neat state and intercalated into FeOCl.

| | (CH$_3$)$_3$SnPy | | (CH$_3$)$_3$SnN(CH$_3$)$_2$ | | (CH$_3$)$_2$Sn[N(CH$_3$)$_2$]$_2$ | | Sn[N(CH$_3$)$_2$]$_4$ |
|---|---|---|---|---|---|---|---|
| IS(78K), neat mms$^{-1}$ | 1.30 | | 1.24 | | 1.16 | | 0.85 |
| IS(78K), inter. mms$^{-1}$ | 1.29 | 1.42 | 1.12 | 1.41 | 0.77 | 1.27 | 0.16 |
| QS(78K), neat mms$^{-1}$ | 0.61 | | 0.92 | | 1.22 | | <0.5 |
| QS(78K), inter. mms$^{-1}$ | 0.71 | 3.38 | 2.61 | 3.18 | 1.84 | 2.84 | 0.67 |
| $-10^3$dlnA/dT, neat k$^{-1}$ | 34.0 | | 28.5 | | 21.5 | | 19.5 |
| $-10^3$dlnA/dT, inter. k$^{-1}$ | 20.8 | 18.5 | 12.7 (average) | | 12.3 (average) | | 7.6 |

organotin molecules are ionized. The exceptional behavior of the
{(CH$_3$)$_2$N}$_4$Sn intercalate, in which a single site with a
surprisingly small isomer shift is found, may be accounted for as
follows: Although the intercalated stoichiometry is only 1/18, the
[57]Fe Mössbauer data indicate that the charge transfer in this
compound is approximately twice as large. This has been found
previously for FeOCl intercalated by crown ethers (16) and implies
that the intercalated species must be doubly ionized. Qualitatively the drastic decrease of the IS upon intercalation may
reflect this unusual situation, although, in order to reach a more
quantitative understanding, additional experiments reflecting
variations of the molecular structure of organotin guest species
will be necessary.

Lewis acid intercalation complexes formed by electron donating
host matrices. The general reaction scheme [2], involving oxidation
of the host matrix, while the guest species is reduced, has been
shown by a variety of physicochemical techniques (2) to obtain
for intercalation compounds of graphite with Lewis acids. Mössbauer studies, quite generally, have reinforced this view and have
provided quantitative information concerning the amount of charge
transfer. Thus, [121]Sb studies at 4.2 K clearly indicate the presence of (~40%) reduced Sb(III) species in graphite intercalated
with antimony pentahalides (114,115) while analytical data confirm
that the Sb/halogen ratio remains 1:5 during this process. These
studies also show that, although SbCl$_3$ alone cannot be intercalated, it can be intercalated from a SbCl$_5$/SbCl$_3$ mixture (115).

In contrast, no reduced species have been identified in [119]Sn
studies (at 80 K) of intercalated SnCl$_4$ and trimethyltinchloride,
which have been recently successfully introduced into graphite by
means of a photochemical technique (116). Also, no Eu(II) was
detected within the temperature region 4.2K < T < 300 K in an [151]Eu
investigation of the graphite-EuCl$_3$/AlCl$_3$ system (117). With respect to these apparently controversial results, a look at the
extensive literature published on the intercalation system formed
with FeCl$_3$ is quite instructive. From resistivity data, Dzurus and
Henning (118) postulated a charge transfer model in which ca. 25%
of the Fe atoms are in the Fe(II) state. The room temperature [57]Fe
Mössbauer spectrum, however, consists of a single peak with an
isomer shift characteristic of Fe(III), which has been repeatedly
(119-121) taken as evidence against the above model. Such reasoning is, however, based on a static picture, involving the assumption that the timescale of the Mössbauer effect is fast compared to

the correlation time of the valence fluctuations between Fe(II) and Fe(III), an assumption that has been proven to be incorrect by detailed subsequent work. A closer look at the temperature dependence shows, that while pristine and intercalated $FeCl_3$ have identical isomer shifts at 80 K, there is a significant difference between them at room temperature (122), the IS of intercalated $FeCl_3$ being about 0.09 mm/s larger. In addition, the low temperature spectra reveal the presence of Fe(II) (122-125) while the latter is absent at room temperature. These results seem to be qualitatively consistent with a thermally activated dynamic electron hopping process which occurs rapidly relative to the Mössbauer timescale at room temperature and, hence, renders the observation of distinct valency states impossible (125). Attempts to treat this problem quantitatively by relaxation theory (126), however, show poor agreement between theory and experiment, although it is noted that electron hopping cannot be ruled out on the basis of the Mössbauer results (123). A "quasi-static" model, involving a "bound to free transition" (the nature of which seems not clear at present) of the transferred electrons was noted to yield better agreement.

In addition, the situation is complicated by the presence of another site typical of Fe(III) characterized by a rather large, temperature independent quadrupole splitting of 1.07 mm/s (see Fig. 9). This site, which had been repeatedly ignored in prior publications, is attributed to $FeCl_3$ molecules next to a vacancy site. The number of these vacancy sites is strongly stage dependent, in accord with a monotonic decrease of the in-plane density as the stage is increased (127). A careful investigation of sample dependence reveals that at low temperature, the overall sum of the number of Fe(II) sites, together with the number of sites which are nearest neighbors to vacancies, amounts to ca. 25% of the total area. This universal behavior strongly supports the concept of a constant amount of charge transfer. Apparently, the chlorine atoms surrounding the iron vacancies act as the primary acceptor site for the electrons donated by the host matrix, while only those electrons which remain after the number of available vacancy sites is exhausted, are donated to Fe(III) and are trapped to yield localized Fe(II) states a low temperature. This explanation may also account for the above mentioned lack of observation of lower metal oxidation states in the Mössbauer spectra of $SnCl_4$ and $EuCl_3$ graphite intercalates. The charge transfer model developed above is only in agreement with the proposal of Dzurus and Henning (118) if the in-plane density is high, i.e. there are no vacancies present. For those samples the narrow linewidth of the Fe(II) absorption suggests that these sites are not randomly situated (128).

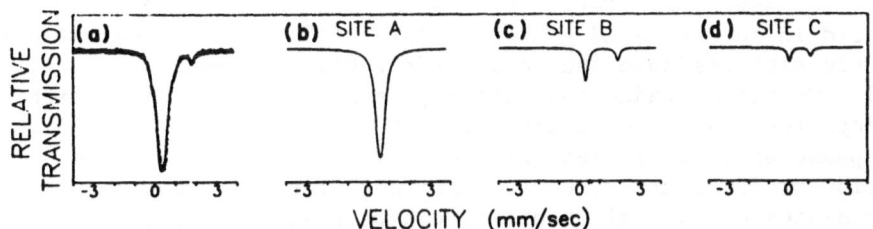

Fig. 9. Decomposition of the 10.0K $^{57}$Fe Mossbauer spectrum of
stage 1 graphite-FeCl$_3$ into three sites (from ref. (123)).
Reproduced by permission of the American Physical Society.

The concept of a constant charge transfer is strongly
reinforced by experiments conducted on graphite intercalated with
FeCl$_3$ and AlCl$_3$ simultaneously (129). Below an FeCl$_3$/AlCl$_3$ ratio
of 1 the room temperature Mössbauer spectrum gives clear evidence
of Fe(II), the fraction of which increases with decreasing iron
content. In these compounds the presence of AlCl$_3$, which is
unstable in lower valency states, is not only responsible for the
selective reduction of Fe(III) but apparently also quenches the
valence fluctuations at room temperature. Two Fe(II) sites with
identical isomer shifts but different quadrupole splittings
(~1.2mm/s and ~2.0mm/s at 80 K) can be distinguished. These
results are similar to the values obtained for graphite-FeCl$_2$
compounds (120,122,124,130) which can be prepared from the
respective FeCl$_3$ intercalates by heating above 350°C, or treatment
with reducing agents. Despite extensive discussion, the exact
nature of these Fe(II) sites has not yet been established. The
inner quadrupole doublet closely reproduces the spectral parameters
of neat FeCl$_2$ and shows a distinct angular dependence of the
intensity ratio, consistent with a fixed 90° orientation of the $V_{zz}$
axis with respect to the hexagonal c-axis of the material. In
contrast, the outer doublet is angularly independent indicating
that these sites are randomly oriented (122). The fraction of the
latter increases with increasing flake size which led to the
suggestion that the larger QS is due to Fe(II) centers in the
neighborhood of trapped Cl$_2$ molecules (131).

Reduction of $FeCl_3C_{7.1}$ (stage-1) with $Fe(CO)_5$ increases the total iron content of the intercalate, leading to a stoichiometry $FeCl_{1.96}C_{4.7}$. The Mössbauer spectrum of this compound (132) reveals an additional Fe(II) site which was subsequently also detected in intercalated grafoil samples (133). The spectral data for this site resemble the values for $FeCl_2$ adsorbed on grafoil (134). Continued reduction with $H_2$, metallic potassium or lithium biphenyl results in small particles of $\alpha$-Fe which sinter into the bulk phase after prolonged treatment (135-137). While these reaction products are rather sensitive to thermal treatment, metal intercalates of unusual stability can be prepared (138) by reduction of $FeCl_3$ cointercalates with $MoCl_5$ or $WCl_6$ (139). $^{57}Fe$ Mössbauer spectra confirm that the majority iron atoms are in a paramagnetic, zerovalent state (139), a unique situation in intercalation chemistry, which is probably only possible as a consequence of the redoxamphoteric behavior of the graphite matrix.

Source experiments on a $CoCl_2$ graphite compound, prepared with $^{57}Co$, (140) reveal a behavior quite analogous to that of the iron halide intercalates. The emission spectrum shows two distinct cobalt(II) sites with quadrupole splittings similar to those of $FeCl_2$ in graphite. Treatment with lithium biphenyl leads to the appearance of a six-line hyperfine pattern typical of metallic cobalt.

In contrast to the graphite-iron chloride intercalates, the corresponding system based on the bromides is decidedly nonstoichiometric, and the Fe/halogen ratio can be varied continuously over a wide range. As far as intercalates with stoichiometric Fe/halogen ratios are concerned, the Mössbauer results (141) are completely analogous to those observed for the chloride compounds. In sharp contrast, the spectrum of the intermediate compound $FeBr_{2.7}$-graphite is fundamentally different, and cannot be accounted for in terms of a superposition of the Fe(III) and Fe(II) sites observed at the stoichiometric compositions. More systematic studies involving the influence of the stoichiometry, as well as detailed temperature dependent data, seem to be desirable in order to confirm or rule out the possibility of charge hopping in this mixed valence compound.

Host $\rightarrow$ guest charge donation in accord with reaction mechanism [2] is also suggested by Mössbauer results on $FeCl_3$ intercalates of 1D systems, such as polyacetylene and polyphenylene (142), as well as molecular lattices based on polynuclear aromatic hydrocarbons (143). Results on the latter, as well as the polyphenylene intercalates, show a close resemblance to graphite based systems, displaying two Fe(II) sites and/or some additional Fe(III) in the Mössbauer spectra. Quite in contrast, only a single

Fe(II) site with a QS = 2.38 mm/s (at room temperature) is present in $FeCl_3$-polyacetylene intercalates. On the basis of analytical data, Mössbauer spectra, as well as physicochemical investigation by a variety of techniques, the following reaction mechanism was formulated:

$$(CH)_x + 0.2xFeCl_3 \rightarrow [(CH)^{0.2+}(FeCl_4)^{2-}_{0.1}]_x + 0.1xFeCl_2 \qquad [6]$$

The $FeCl_2$ is removed by the rinsing solvent and not observed in the Mössbauer spectra.


V. LATTICE DYNAMICAL PROPERTIES

1. Debye temperatures and effective vibrating masses

The area under the resonance curve is related not only to the number of resonant atoms of a given kind, but also to the probability of a γ-ray being absorbed without recoil. The latter probability is called the recoil-free fraction f, which, in turn, depends on the vibrational dynamics of the nucleus under consideration. Using the Debye-Waller formalism it can be shown that f decreases with increasing mean square displacement $\langle x^2 \rangle$ according to the relationship:

$$f = \exp\left\{-\frac{E_\gamma^2 \langle x^2 \rangle}{hc^2}\right\} \qquad [7]$$

Here $\langle x^2 \rangle$ is dependent on the strength of the chemical bonding of the nucleus to its environment and is strongly temperature dependent. Hence, a study of the temperature dependence of the recoil-free fraction affords a valuable tool for studying the lattice dynamics of crystals, provided the vibrational spectrum of the solid under consideration is known. Most solids can be described in terms of the Debye formalism, the application of which leads to the expression

$$f = \exp-\left[\frac{6E_r}{k\theta_D}\left\{1/4 + \left(\frac{T}{\theta_D}\right)^2 \int_0^{\theta_D/T} \frac{xdx}{e^x-1}\right\}\right] \qquad [8]$$

where $\theta_D$, the so-called Debye temperature, is a measure of the average vibrational frequency present in the vibrational spectrum of the solid. For $T > \theta_D/2$, eq. [8] can be simplified, by neglecting higher order terms, into

$$f = \exp\left[-\frac{6E_r T}{k\theta_D^2}\right] \qquad [9]$$

Here $E_r$ represents the recoil energy, given by

$$E_r = \frac{E_\gamma^2}{2Mc^2} \qquad [10]$$

where M is the effective vibrating mass of the nucleus, expressed in amu.

Thus, $\theta_D$ can be evaluated from the slope of the linear ln f vs. T plot in the high temperature region. In practice, it is not generally necessary to determine the absolute value of f, but sufficient to evaluate the temperature dependence of the area under the resonance curve, assuming that the sample is sufficiently thin, so that saturation effects can be neglected. The analysis assumes, moreover, that the effective vibrating mass is identical to the bare mass of the vibrating nucleus, an assumption which is questionable when the nucleus is held by strongly covalent bonding forces. Fortunately, the actual vibrating mass can be estimated from the relativistic dependence of the isomer shift

$$\frac{dIS}{dT} = -\frac{3}{2}\frac{E_\gamma k}{Mc^2} \qquad [11]$$

so that, by combining eq's [9] and [11] a more realistic Debye temperature can be obtained. The significance of these results, however, depends upon the applicability of the Debye model to the solid under consideration. It has been pointed out frequently, that polyatomic lattices are usually poorly approximated by such formalism. For this reason it is usually more appropriate to speak of a "Mössbauer lattice temperature", represented by the symbol $\theta_M$.

Moreover, the highly anisotropic bonding forces present in intercalation compounds can be expected to give rise to further gross deviations from a Debye model like behavior. Nevertheless, the temperature dependence of ln f is frequently found linear in intercalation compounds, in agreement with a Debye model. The slopes of these plots provide a qualitative measure for the tightness of binding which is especially useful within the context of comparative studies. Depending on whether the Mössbauer-active atom belongs to the host matrix or the guest species, respectively, the influence of various guests on the properties of a given host can be studied and/or the binding of a given guest species in different hosts can be compared.

Influence of the guest species upon the lattice dynamics of a given host matrix. The only systematic study carried out so far deals with Lewis base intercalates of FeOCl. Table 3 summarizes the Mössbauer lattice temperatures obtained for various guest species. In these studies the $\theta_M$ values were calculated by using the point mass assumption, since the isomer shift shows an intrinsic (non-relativistic) temperature dependence due to electron hopping in the temperature region 150K < T < 250 K and, hence, cannot be used in order to assign effective masses. To this end, measurements above room temperature are necessary and are currently being undertaken. Depite this limitation the data indicate unambiguously a significant softening of the FeOCl lattice as a result of intercalation. There is a quite uniform behavior among all intercalates, and no systematic influence of the structure and the chemical properties of the guest species can be noted nor does there appear to be a systematic trend with the expansion of the crystallographic b axis of the matrix. The lattice softening, however, is qualitatively in agreement with expectation since the density of the lattice decreases upon intercalation, due to the already noted b axis expansion of the unit cell. This behavior is also reflected in FTIR studies, which show that the interlayer vibration of the intercalation compounds is significantly red-shifted as compared to the host compound (144). It would be interesting to determine whether the spread of $\theta_M$ values among the various intercalates correlates with this red-shift; for the present, it can only be noted that the influence of the exact values of $\Delta b$, as well as the intercalated stoichiometry, appear to be minor.

Binding of the guest species. Linear temperature dependencies of ln f have been found for $FeCl_3$ in graphite (122), for Eu in europium tungsten bronze (46), as well as Fe and Sn in transition metal dichalcogenide hosts (63, 73, 75). The results obtained for the latter indicate that, except for $Sn_{1/3}NbS_2$ the effective vibrating mass as obtained from the temperature dependence of the IS is very close to the bare mass of the intercalated metal atom.

While these results can be consistently interpreted from the Debye model, tin-containing Chevrel phases show significantly different behavior. Attempts to fit the data to a phonon spectrum consisting of a Debye continuum with $\theta_D = 74$ K and an Einstein peak with $\theta_E = 155$ (intensity ratio 1:2) have been only moderately successful (86). In all investigations a pronounced change in both dIS/dT and dlnf/dT at about 80 K was noted indicating an apparent change in the Debye temperature. While the possibility that this might reflect a structural phase transition has been ruled out by temperature dependent x-ray diffraction data, neutron scattering results (146) suggest that this peculiarity reflects a phonon softening phenomenon linked to the librational modes of the $Mo_6S_8$ cubes. Based on comparative studies of $SnMo_6S_8$ and $SnMo_6Se_8$ (145),

Table III. Mössbauer lattice temperature $\theta_M$ for FeOCl and its intercalates, X represents the guest stoichiometry while 1/2b is the layer separation in pm.

| Guest | x | $\frac{1}{2}b$/pm | $\theta_M$/K(±10) |
|---|---|---|---|
| - | - | 791 | 341 |
| isopropylamine | 0.33 | 1154 | 235 |
| butylamine | 0.25 | 1171 | 210 |
| pyridine | 0.33 | 1345 | 200 |
| 2-picoline | 0.19 | 1390 | 201 |
| 2-picoline, N-oxide | 0.36 | 1820 | 220 |
| 2,6 Lutidine | 0.17 | 1480 | 211 |
| pyrazine | 0.28 | 1322 | 219 |
| imidazole | 0.25 | 1316 | 222 |
| N,N-dimethylamino-pyridine | 0.16 | 1341 | 215 |
| benzamidine | 0.11 | 1330 | 227 |
| phenylguanidine | 0.11 | 1340 | 185 |
| quinuclidine | 0.15 | 1336 | 216 |
| triethylenediamine | 0.15 | 1306 | 220 |
| hexamethylenetetramine | 0.09 | 1230 | 210 |
| piperazine | 0.14 | 1200 | 232 |
| | | 1300 | |
| kryptofix-22 | 0.05 | 1220 | 196 |
| thiourea | 0.15 | 1240 | 205 |
| triethylphosphine | 0.16 | 1187 | 203 |
| trimethylphosphite | 0.16 | 1438 | 219 |

it was suggested that this phonon softening is related to the superconducting transition temperatures of these materials.

## 2. Vibrational anisotropies

Due to their anisotropic bonding forces, intercalation compounds are expected to represent well-suited model systems for the study of vibrational anisotropies by means of the Mössbauer effect. For an axially symmetric quadrupole interaction in nuclei such as $^{119}Sn$, $^{57}Fe$ and $^{197}Au$, the recoil-free fractions for the respective doublet components depend on the mean square displacements $\langle z^2 \rangle$ and $\langle x^2 \rangle$ parallel and perpendicular to the EFG principal axis.

$$f_{3/_2 \rightarrow 1/_2} \sim \exp{-(k^2 \langle x^2 \rangle)} \int_0^1 (1 + \cos^2\theta) \exp\{-(\langle z^2 \rangle - \langle x^2 \rangle) \cos^2\theta\} d\cos\theta \quad [12]$$

$$f_{1/_2 \rightarrow 1/_2} \sim \exp{-(k^2 \langle x^2 \rangle)} \int_0^1 (\frac{5}{3} - \cos^2\theta) \exp\{-(\langle z^2 \rangle - \langle x^2 \rangle) \cos^2\theta\} d\cos\theta \quad [13]$$

If vibrational anisotropy is present, $\langle z^2 \rangle - \langle x^2 \rangle \neq 0$ and hence an intensity asymmetry of the quadrupole doublet is noted in the spectrum. This asymmetry increases with increasing temperature. Despite this significant behavior, frequently interference by other sources of asymmetry renders the observation of vibrational aniso-tropies difficult. For example, as a consequence of the platelike habit of many 2D materials, a temperature independent intensity asymmetry results from non-random orientation of the crystalline material with respect to the γ-beam. (This "preferential orientation effect" can be identified by measuring the sample in different geometries). Intensity asymmetries can also arise from an accidential superposition of two doublets with slightly different IS and QS (in this case, the two components seen in the spectra, have equal areas, in contrast with the effect produced by a vibrational anisotropy). Although the latter can thus be clearly distinguished from other sources of asymmetry, it is clear that, in order to analyze the Mössbauer spectrum in terms of vibrational anisotropies, the interferences discussed above have to be absent or at least estimable from other measurements, a requirement which limits considerably the number of potential materials which can be studied. Two independent investigations (86,87) have confirmed that the temperature dependence of the quadrupole doublet inten-sity ratio gives clear evidence of vibrational anisotropy in $SnMo_6S_8$. Applied magnetic field experiments (87) show unambigu-ously that the sign of $V_{zz}$ is positive in this compound, hence permitting the conclusion that $\langle x^2 \rangle > \langle z^2 \rangle$.

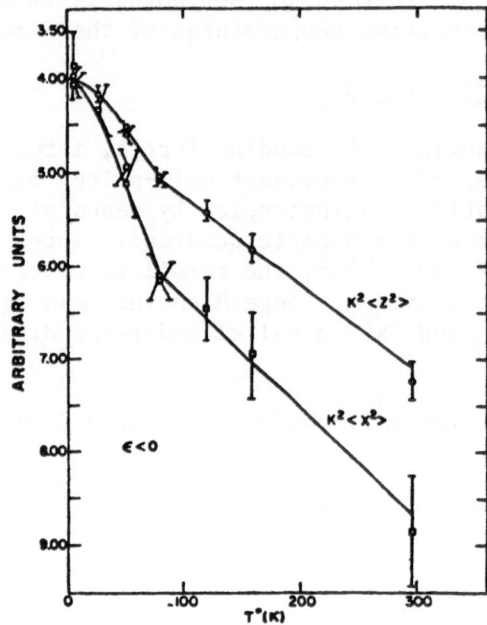

Figure 10. Temperature dependent vibrational anisotropy in SnMo$_6$S$_8$ (from ref. (87)). Reproduced by permission of the American Physical Society.

This result can be understood from the nearest neighbor surrounding of Sn, consisting of 6 sulfur atoms at a distance of 320 pm while there are two atoms at a closer distance of 270 pm on the threefold rhombohedral axis, which represents the $V_{zz}$ direction. Beyond this general conclusion the detailed analyses of both references are very much at variance with each other. Bolz et al. (86) note, however, that the large anisotropic amplitude displacements of the Sn atoms are incompatible with their experimentally measured Lamb Mössbauer factors, concluding that simple relationships, such as [7], may not be valid in this compound.

Utilizing the 81 KeV transition of the $^{133}$Cs nucleus, Perlow et. al. demonstrated a pronounced vibrational anisotropy for the Cs ion in the single stage crystalline graphite intercalation compound

162

$C_8Cs$ which has been studied (69) at 4.2 K. The recoil free fraction in this material is strongly orientation dependent, increasing by a factor of 20 when the angle between the optical axis and the crystallographic $c$-direction (which coincides with $V_{zz}$) is changed from 90 to 0 degrees. The best fit parameters are $\langle x^2 \rangle = 35.8 Å^2$ and $\langle z^2 \rangle = 18.9 Å^2$, in agreement with the expectation that the guest ion displacement within the layers is larger than along the $c$-axis direction.

## VI. MAGNETIC PHENOMENA

Mössbauer spectroscopy has proven to be a highly valuable tool in the elucidation of cooperative magnetic phenomena. The spectra can provide information about the number of magnetic sublattices, the magnetic ordering temperature, the value of the magnetic hyperfine field and its temperature dependence, the so-called Brillouin function. Moreover, oriented single crystal studies allow one to determine the preferential orientation of the magnetic spins with respect to the crystallographic axes. On the other hand Mössbauer spectra alone do not provide information about the detailed nature of the interaction and, in addition, the observation of a magnetically split hyperfine spectrum is only a necessary, but not a sufficient condition for the existence of cooperative ordering. Especially in systems where the magnetic spins are rather dilute (a situation frequently encountered in intercalation compounds), similar patterns, which may be mistaken as evidence for magnetic ordering, can be produced by slow electron spin relaxation phenomena. Therefore complementary information obtained from independent physicochemical investigations is desirable and, to this end, Mössbauer spectra are usually discussed in the context of magnetic susceptibility and/or neutron diffraction data.

In the following, the available data relating to the influence of intercalation upon the magnetic properties of a given host lattice will be briefly discussed while the second part of this section will deal with cooperative magnetic effects associated with the intercalated species.

### 1. Magnetic phenomena associated with the host lattice

Magnetic properties of Fe(II) substituted in transition metal dichalcogenides and their intercalates. Magnetic susceptibility data indicate that substitutional iron present in $Fe_yTa_{1-y}S_2$ compounds undergoes a thermally activated low spin → high spin transition in the temperature range 200 K < T < 500 K. This behavior is reflected in the Mössbauer data which have been quantitatively analyzed for y = 0.1 (147). Single site spectra are observed, which show that this transformation is dynamic, i.e. the spin state in the transition region fluctuates rapidly as compared to the timescale of the Mössbauer effect; a rather unique situation, as far as spin-conversions of Fe(II) compounds are concerned. The above phenomenon results in a strong non-linearity of the IS <u>vs</u>. T curve, showing almost no temperature dependence within the transition region. In addition an unusual temperature dependence of the quadrupole splitting is observed, which can be interpreted in terms of a magnetic modulation of the lattice contribution to the electric field gradient (148). While thorough studies have been undertaken on the host matrix 1T-TaS$_2$ with several doping levels, the effect of intercalation on the above phenomena has not been the subject of detailed temperature dependent work so far. In general, however, for the lithiated compounds based on TaS$_2$ as well as VS$_2$, magnetic susceptibility data and Mössbauer spectra at 4.2 K are consistent with an increasing evolution of a magnetic moment as the lithium content in the sample is increased (43).

Effect of intercalation on magnetic ordering parameters in FeOCl. Magnetic susceptibility data show that both FeOCl (149-151) and its Lewis base intercalates (151) order antiferromagnetically with a pronounced 2 D behavior. In addition, detailed temperature dependent Mössbauer investigations (16,107,152) indicate that the evolution of the magnetic hyperfine spectra from the paramagnetic doublets occurs over a wide temperature range, which renders the assignment of a Neel temperature somewhat ambiguous. While, however, in unintercalated FeOCl the magnetic splitting becomes evident below 90 K, in most Lewis base intercalates the onset of spin ordering is depressed below liquid nitrogen temperature and varies from 55 to 80 K for different guest species. Fig. 11 contrasts the hyperfine patterns observed near 4.2 K in FeOCl and its piperazine intercalate. The spectra arise from the simultaneous presence of both quadrupole and magnetic hyperfine interactions. Also, for the intercalated matrices, the Mössbauer line patterns give evidence for at least two magnetic sublattices, the spectral components of which, however, cannot be readily resolved. The extracted hyperfine fields, resulting from a simple six-line least squares fitting procedure, in which only the positions of the outermost resonance maxima are considered in the calculation, are approximately the same for all guest species as well as the parent FeOCl, amounting to ~43 ± 3 T. Interesting variations are found, however, for the line positions. If the lines of the majority hyperfine pattern are denoted from low to high velocities by indices 1 to 6, one finds that the parameter

$$\varepsilon = (IS_6 - IS_5) - (IS_2 - IS_1) \qquad [14]$$

164

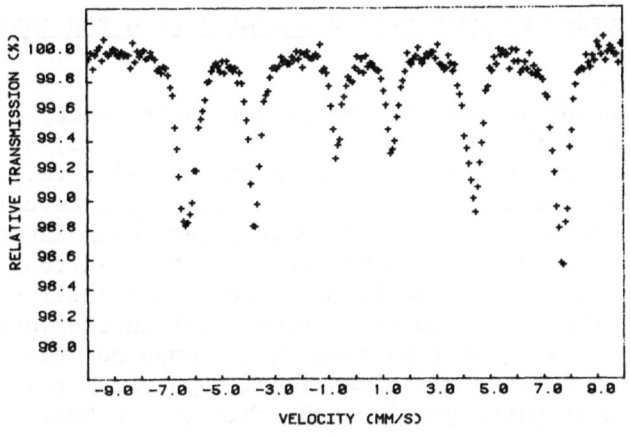

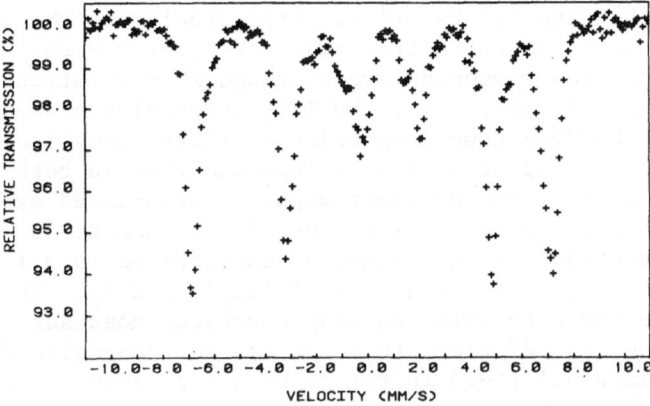

Figure 11. $^{57}$Fe Mössbauer spectra of FeOCl and FeOCl(piperazine)$_{1/6}$ at 8.0K.

which relates to the quadrupole interaction by the relationship

$$\varepsilon = \frac{e^2 q \ Q^*}{4} \ \frac{3\cos^2\theta - 1}{2} \qquad\qquad [15]$$

varies among different intercalates. Since the quadrupole coupling constant at low temperatures is approximately constant for all intercalates ($QS = \frac{e^2 q \ Q^*}{2} = 1.15 \pm 0.05$ mm/s), the variations of $\varepsilon$ indicate differences in the angle $\theta$ which specifies the orientation of the $V_{zz}$ axis with respect to the direction of the magnetic hyperfine field. An analogous phenomenon has been reported for FeOCl ferrocene intercalates, where, in addition, a strong influence of sample preparation was noted and irreversible sample aging effects were studied (111,153).

## 2. Magnetic ordering phenomena associated with the intercalated guest species

As pointed out in Sec. IV, Mössbauer spectroscopy has added substantial proof for the redox intercalation model, which implies that the guest species are present in an ionized form. Accordingly, if transition metal or rare earth atoms are intercalated, the resulting compounds frequently display interesting magnetic properties as the result of unpaired d- or f- electrons localized on the guest ions. Large scale magnetic susceptibility studies, especially of the first row transition metal intercalates of $NbS_2$ and $TaS_2$ have shown (154,155) that these compounds can be described as structurally ordered two-dimensional magnetic arrays situated between the Pauli paramagnetic sandwiches of the host materials. Orientation dependent data reveal, that the ordering is either ferromagnetic (for M = V,Cr,Mn) with the easy axis of magnetization pointing within the layers, (i.e. perpendicular to the crystallographic c-direction) or antiferromagnetic (for M = Co, Ni). The corresponding iron compounds are distinguished by strongly anomalous behavior. First of all, the $TaS_2$ intercalates are ferromagnetic while the $NbS_2$ compounds order antiferromagnetically (155). Moreover, the susceptibility data indicate that in both the ferromagnetic as well as the antiferromagnetic iron-based systems the direction of the preferred alignment of the moments is along the c-axis, in contrast to the situation encountered in all other first row transition metal intercalates of $TaS_2$ and $NbS_2$. This result is strongly confirmed by oriented single crystal Mössbauer studies. The data show, in addition, that for the ferromagnetic $Fe_xTaS_2$ system the internal field increases with increasing x (62) while the opposite behavior is observed for the antiferromagnetic niobium disulfide intercalates (61). For the latter, the spectra also indicate that increasing the iron content from 0.25 to 0.33 depresses the Neel temperature by nearly 100 K and inverts the sign of the $V_{zz}$ component with respect to the magnetic field axis. No quantitative accounting for these surprisingly drastic effects of stoichiometry has so far been reported. Likewise, it is not clear at the present time why changing the host compound from $NbS_2$ to its homolog $TaS_2$ results in such fundamental differences of the magnetic behavior.

Cooperative magnetic ordering of the ionized guest species was also observed in the stage one compound $EuC_6$ (47) as well as the stage one and stage two graphite intercalation compounds of $FeCl_3$, (156-162) (see Table IV). In both cases the ordering is antiferromagnetic, with the orientation of the internal magnetic field perpendicular to the c-axis, i.e.: within the basal plane. The iron atoms in graphite-$FeCl_3$ (stage 1) can, to a first approximation, be described as a two dimensional array with isotropic exchange. Mössbauer studies indicate that intercalating

Table IV. Magnetic ordering phenomena associated with the intercalated guest species. $T_O$ represents the magnetic
ordering temperature. $H_{hf}$ is the magnitude of the
internal field, as determined by Mössbauer spectroscopy.

| Compound | $T_O$/K | $H_{hf}$(T) | Type of Ordering | Ref. |
|---|---|---|---|---|
| $Fe_{0.2}TaS_2$ | * | 8.0 | ferromagnetic | (62) |
| $Fe_{0.26}TaS_2$ | * | 11.2 | H ∥ C | (62) |
| $Fe_{0.28}TaS_2$ | * | 14.8 | | (62) |
| $Fe_{0.34}TaS_2$ | * | 17.7 | | (62) |
| $Fe_{0.25}NbS_2$ | 138 | 17.7 | antiferromagnetic | (61) |
| $Fe_{0.33}NbS_2$ | 47 | 5.0 | H ∥ C | (61) |
| $Fe_{0.5}TiS_2$ | 140 | 3.0 | ferromagnetic | (63) |
| $Fe_{0.25}VS_2$ | * | 0 | | (68) |
| $Fe_{0.5}VS_2$ | * | 12.0 | antiferromagnetic | (68) |
| $Fe_{0.5}CrSe_2$ | 218 | 8.9 | antiferromagnetic H ⊥ C | (163) |
| $EuC_6$ | 34 | 10.7 | antiferromagnetic H ⊥ C | (47) |
| $FeCl_3$ | 8.6 | 49.5 | antiferromagnetic | (162) |
| $FeCl_3$-graphite (stage-1) | 4.2 | 49.3 | antiferromagnetic H ⊥ C | (160) |
| $FeCl_3$-graphite (stage-2) | 2.0 | 49.3 | | (160) |

*Not determined.

$FeCl_3$ into the graphite matrix does not change the hyperfine field,
while it depresses the Neel temperature from 8.6 K (158) to 4.2 K
(160). While this result is in good agreement with magnetic
susceptibility (157) as well as neutron diffraction data (159), the
results on the stage 2 compounds reported in the literature
(156,157) have been inconsistent, possibly due to inadequate sample
characterization. A reinvestigation indicates a magnetic ordering
temperature of 2.0 K (160,161), which, however, has not as yet been
confirmed by a subsequent neutron diffraction study (159).
Intercalates of higher stages do not show any magnetic ordering
down to 65 mK (161) and these results consistently point out the
importance of three-dimensional (interlayer) interactions in
graphite intercalation compounds, which become successively
weakened as the stage is increased. Moreover a closer look at the
spectra reveals that, at all stages, an additional site exists,
which contributes a six-line hyperfine pattern arising from slow
Fe(III) spin relaxation rather than magnetic ordering (160,161).

This site is ascribed to the periphery of the magnetic domains. In addition, especially in intercalates with low in-plane densities, the iron sites next to vacancies (as discussed in Sec. IV) contribute another relaxation pattern at low temperatures. For stage 2 compounds, a critical concentration of these sites (~11 %) has been noted to "turn on" a pronounced spin glass behavior in magnetic susceptibility measurements (162).

Recently the possibility of a coexistence of superconductivity and magnetic order within the same volume element was demonstrated for a mixed Chevrel phase containing both Sn and Eu. Bolz et al. (48) investigated the Mössbauer spectra of $Eu_{0.63}Sn_{0.38}Mo_6S_{6.61}$ in the superconducting state. The spectra reveal a magnetic hyperfine pattern below 0.5 K which is believed not to be due to a relaxation effect. On the other hand there is no detectable transferred hyperfine field at the tin nuclei to within 0.1 T. The results are in close agreement to those obtained with the non-superconducting compound $^{119}Sn_{0.015}Eu_{0.93}Mo_6S_{6.61}$. Accordingly, the authors conclude that the order is not a long range one, but rather of a spin glass type. The authors note, that the coexistence of super-conductivity and magnetic order in their sample indicates very weak interaction between the "superconducting" Mo electrons and the "magnetic" Eu electrons.


SUMMARY AND ACKNOWLEDGMENTS

In this review, we have tried to present an overview of the present status of research on intercalation compounds - both of electron acceptor and electron donor host matrices - which has been influenced by the application of Mössbauer Effect spectroscopic techniques. As is usual in studies focussing on the chemical properties of materials, the most "popular" nuclear gamma ray resonance effect nuclides, $^{57}Fe$ and $^{119}Sn$, have been most widely employed in such studies, but other Mössbauer effect reporter atoms will find increasing use in such studies. One of the major advantages of the application of the Mössbauer technique to the study of intercalation systems is the ready feasibility of carrying out such studies over a wide temperature range. This makes it possible to obtain data which is complementary to studies on superconductivity phenomena and low temperature heat capacity determinations on the one hand, and to room temperature X-ray crystallographic studies and high temperature conductance data, inter alia, on the other. Moreover, the lattice and electron dynamical and magnetic hyperfine interaction results which can be extracted from Mössbauer effect studies provide a unique opportunity to study intercalation phenomena on the nuclear excited state lifetime time scale. The results summarized in the present chapter can only be considered as a prologue to the scientific studies on intercalation compounds yet to come.

It is appropriate to record here the many fruitful discussions and informative exchanges which we have had with our co-workers and colleagues, including Profs. M. Katada and Y. Maeda, Drs. A. Salmon and A. Lechtenböhmer, as well as J.E. Phillips, R.S. Bannwart, R.A. Cassell, and R.F. Davis. One of us (H.E.) likes to thank his teachers Prof. W. Müller-Warmuth and Prof. R. Schöllhorn for the invaluable academic preparation and for drawing his attention to solid state spectroscopy and intercalation compounds. Much of the research described in this review and cited as originating in the authors' laboratory has been supported by the Division of Materials Research of the National Science Foundation under grant DMR-8102940, and well as by a grant from the Center for Computer and Information Services of Rutgers University, and by a grant from the "Deutsche Forschungsgemeinschaft" West Germany.

## VII. REFERENCES

1a. R. Schöllhorn, Angew. Chem. 92:1015 (1980) and references therein.

1b. F. Levy, "Intercalated Layered Materials", D. Reidel Publishing Comp. Dordrecht/Holland (1979) and references therein.

1c. M.S. Whittingham and A.J. Jacobson, "Intercalation Chemistry", Academic Press, New York (1982) and references therein.

2. J.E. Fischer, ref. (1b), p. 481 and references therein; N. Bartlett and B.W. McQuillan, ref. (1c), p. 19.

3. A.P. Mil'ner, Yu. R. Zabrodskii, V.M. Koshkin, V.V. Kukol', E.B. Yagubskii, and G.D. Guseinov, Zh. Strukt. Khim. 21:93 (1980).

4. M.S. Whittingham and L.B. Ebert, ref. (1b), p. 533.

5. C.F. van Brüggen, C. Haas, and H.W. Myron, "Layered Materials and Intercalates", Proc. Int. Conf. 1979, North Holland, Amsterdam (1980).

6. R. Chevrel, M. Sergent, and J. Prigent, J. Solid State Chem. 3:515 (1971).

7. Ø. Fischer, Appl. Phys. 16:1 (1978).

8. P.G. Dickens and M.S. Whittingham, Quart. Rev. 22:30 (1968).

9. P.E. Biersted, T.A. Bither, and F.J. Barnell, Solid State Commun. 4:25 (1966).

10. F. Hulliger, in "Structural Chemistry of Layer-type Phases", F. Levy, ed., D. Reidel Publishing Comp., Dordrecht/Holland (1976).

11. J.A. Wilson and A.D. Yoffe, Adv. Phys. 18:193 (1969).

12. M.D. Lind, Acta Crystallogr. B26:1058 (1970).

13. W. Bronger, Z. Anorg. Allg. Chem. 359:225 (1968).

14. A.J. Jacobson, M.S. Whittingham, and S.M. Rich, J. Electrochem. Soc. 126:887 (1979).

15. S. Yamanaka, T. Nagashima, and M. Tanaka, Thermochim. Acta 19:236 (1977).

16. H. Eckert and R.H. Herber, J. Chem. Phys., in press; R.H. Herber and H. Eckert, to be published.

17. G.V. Subba Rao and J.C. Tsang, Mat. Res. Bull. 9:921 (1974); M.S. Whittingham, J.C.S. Chem. Commun. (1974), 328.

18. D.W. Murphy, F.J. DiSalvo, and J.N. Carides, J. Solid State Chem. 29:339 (1979).

19. B.V. Liengme, M.W. Bartlett, J.G. Hooley, and J.R. Sams, Phys. Lett. A25:127 (1967).

20. F.J. DiSalvo, Physics Today (1979), 32.

21. J.A. Wilson, F.J. DiSalvo, and S. Mahajan, Adv. Phys. 24:117 (1975).

22. M. Eibschütz and F.J. DiSalvo, Phys. Rev. B15:5181 (1977).

23. L. Pfeiffer, M. Eibschütz, and D. Salomon, Hyperfine Int. 4:803 (1978).

24. T. Butz, A. Vasquez, and A. Lerf, J. Phys. C:12:4509 (1979).

25. M. Eibschütz, D. Salomon, and F.J. DiSalvo, Phys. Lett. A93 (1983), 259.

26. C.B. Carter and P.M. Williams, Phil. Mag. 26:393 (1972).

27. T. Butz, A. Vasques, H. Saitovich, R. Mühlberger, and A. Lerf, Physica B99:69 (1980).

28. N. Karnezos, L.B. Welsh, and M.W. Shafer, Phys. Rev. B11:1808 (1975).

29. R. Schöllhorn, Comments in Inorg. Chem., in press.

30. W. Schramm, R. Schöllhorn, H. Eckert, and W. Müller-Warmuth, Mat. Res. Bull. 18:1283 (1983).

31. A.G. Maddock, R.H. Platt, A.F. Williams, and R. Gancedo, J.C.S. Dalton (1974), 1314; L.R. Conroy and G.J. Perlow, Phys. Lett. A31:400 (1970).

32. M. Drosg and E. Ujlaki, Acta Phys. Austr. 23:47 (1966).

33. R.H. Herber and R.F. Davis, J. Inorg. Nucl. Chem. 42:1577 (1980).

34. A. Leblanc and J. Rouxel, C.R. Acad. Sci. (Paris) 274:786 (1972).

35. M. Eibschütz, D.W. Murphy, S.M. Zahurak, and P.A. Christian, Appl. Phys. Lett. 39:664 (1981); Solid State Ionics 5:339 (1981).

36. A.H. Thompson and M.S. Whittingham, Mat, Res. Bull. 12:741 (1977).

37. R.H. Herber and A. Lechtenböhmer, to be published.

38. C. Berthier, Y. Chabre, and M. Minier, Solid State Commun. 28:327 (1978).

39. Ref. (1c), p. 267.

40. J.M. Tarascon, F.J. DiSalvo, M. Eibschutz, D.W. Murphy, and J.V. Waszczak, Phys. Rev. B28:6397 (1983).

41. G.A. Fatseas and J.B. Goodenough, J. Solid State Chem. 33:219 (1980).

42. F.J. DiSalvo, M. Eibschütz, C. Cros, D.W. Murphy, and J.V. Waszczak, Phys. Rev. B19:3441 (1979).

43. M. Eibschütz, D.W. Murphy, and F.J. DiSalvo, Physica B99:145 (1980).

170

44. M. Eibschütz, Adv. Chem. Ser. 194:523 (1981).

45. A.J. Jacobson and L.E. McCandlish, J. Solid State Chem. 29:355 (1979).

46. J. Korecki, A. Polaczek, and M. Pekala, Nucl. Instr. Meth. 199:209 (1982).

47. G. Kaindl, J. Feldhaus, U. Ladewig, and K.H. Frank, Phys. Rev. Lett. 50:123 (1983).

48. J. Bolz, G. Crecelius, H. Maletta, and F. Pobell, J. Low Temp. Phys. 28:61 (1977).

49. M.M. Abd-Elmeguid and H. Micklitz, J. Phys. C15:L479 (1982).

50. C.S. Dimbylow, I.J. McColm, C.M.P. Barton, N.N. Greenwood, and G.E. Turner, J. Solid State Chem. 10:128 (1974).

51. A. Polaczek, K. Krop, J. Korecki, and J. Arabski, Acta Phys. Pol. A60:55 (1981).

52. I.J. McColm, R. Steadman, C.S. Dimbylow, N.N. Greenwood, and G.E. Turner, J. Solid State Chem. 19:161 (1976).

53. N.N. Greenwood, F. Viegas, and F. Studer, J. Solid State Chem. 31:347 (1980).

54. F. Studer, G. Allais, and B. Raveau, J. Phys. Chem. Solids 41:1187 (1980).

55. I.J. McColm, R. Steadman, and A. Howe, J. Solid State Chem. 2:555 (1970); L.E. Conroy and M.J. Sienko, J. Am. Chem. Soc. 79:4048 (1957).

56. P. Bonville, R. Chevrel, J.A. Hodges, P. Imbert, G. Jehanno, and M. Sergent, Int. Conf. Appl. Mössb. Effect (Indian Nat. Sci. Acad. N.D. 1981), 171; Rev. Phys. Appl. 15:1139 (1980).

57. N.N. Greenwood and T.C. Gibb, "Mössbauer Spectroscopy", Chapman and Hall, London 91971).

58. H. Eckert, Dissertation Münster/W. Germany (1982) and references therein.

59. I.J. McColm, R.I.D. Tilley, C.M.P. Barton, and N.N. Greenwood, J. Solid State Chem. 16:265 (1976).

60. J.P. Doumerc, G. Schiffmacher, P. Caro, and M. Pouchard, C.R. Acad. Sci. (Paris) 282:295 (1976).

61. O. Gorochov, A. LeBlanc-Soreau, J. Rouxel, P. Imbert, and G. Jehanno, Phil. Mag. B43:621 (1981).

62. M. Eibschütz, F.J. DiSalvo, G.W. Hull Jr., and S. Mahajan, Appl. Phys. Lett. 27:464 (1975); M. Eibschütz, S. Mahajan, F.J. DiSalvo, G.W. Hull Jr., and J.V. Waszczak, J. Appl. Phys. 52:2098 (1981).

63. M. Katada and R.H. Herber, J. Solid State Chem. 33:361 (1980).

64. M. Katada, K. Sato, Y. Hirasawa, and H. Sano, Radiochem. Radioanal. Lett. 54:293 (1982).

65. R. Rangel, J. Bolz, and F. Pobell, Vortr. Dt. Phys. Ges. (1979), 372.

66. J.M. Friedt, B.D. Dunlap, G.K. Shenoy, A.T. Aldred, F.Y. Fradin, and C.W. Kimball, Physica B107:61 (1981).

67. H. Eckert, R.H. Herber, and R. Schöllhorn, unpublished work.

68. Y. Oka, K. Kosuge, and S. Kachi, Mat. Res. Bull. 12:1117 (1977).
69. L.E. Campbell, G.L. Montet, G.J. Perlow, Phys. Rev. B15:3318 (1977).
70. H. Estrade-Szwarckopf, J. Conard, P. Lauginie, J. van der Klink, D. Guerard, and P. Lagrange, in "Physics of Intercalation Compounds, Proc. Int. Conf.", L. Pietronero, E. Tosatti, eds. (1981), p. 274.
71. J.B. Goodenough and G.A. Fatseas, J. Solid State Chem. 41:1 (1982); J.T. Hoggins and H. Steinfink, Inorg. Chem. 15:1682 (1976).
72. R.H. Herber, Accts. Chem. Res. 15:216 (1982).
73. R.H. Herber and R.F. Davis, J. Chem. Phys. 63:3668 (1975).
74. F.J. DiSalvo, G.W. Hull Jr., L.H. Schwartz, J.M. Voorhoeve, and J.V. Waszczak, J. Chem. Phys. 59:1922 (1973).
75. R.H. Herber and R.F. Davis, J. Chem. Phys. 65:3773 (1976).
76. P.S. Gentile, D.A. Driscoll, and A.J. Hockman, Inorg. Chim. Acta 35:249 (1979).
77. R.H. Herber and M. Katada, J. Solid State Chem. 27:137 (1979).
78. R.H. Herber, F.J. DiSalvo, and R.B. Frankel, Inorg. Chem. 19:3135 (1980).
79. P.A. Flinn, in "Mössbauer Isomer Shifts", G.K. Shenoy and F.E. Wagner, eds., North Holland Publishing Comp. (1978), p. 593.
80. A.C. Gossard, F.J. DiSalvo, and H. Yasuoka, Phys. Rev. B9:3965 (1974).
81. R. Eppinga and G.A. Wiegers, Mat. Res. Bull. 12:1057 (1977).
82. R. Eppinga and G.A. Wiegers, Physica B99:121 (1980).
83. Ref. (1b), p. 229.
84. R. Eppinga, G.A. Sawatzky, C. Haas, C.F. van Brüggen, J. Phys. C9:3371 (1976).
85. H.A. Wagner and H.C. Freyhardt, Physica B107:657 (1981).
86. J. Bolz, J. Hauck, and F. Pobell, Z. Phys. B25:351 (1976).
87. C.W. Kimball, L. Weber, G. van Landuyt, F.Y. Fradin, B.D. Dunlap, and G.K. Shenoy, Phys. Rev. Lett. 36:412 (1976).
88. N.E. Alekseevskii and E.G. Nikolaev, Pis'ma Zh. Eksp. Teor. Fiz. 31:770 (1980); 34:350 (1981).
89. M. Matsumura, N. Sano, T. Taniguchi, and K. Asayama, J. Phys. Soc. Jpn. 50:3937 (1981).
90. R. Schöllhorn, H.D. Zagefka, T. Butz, and A. Lerf, Mat. Res. Bull. 14:369 (1979).
91. F.R. Gamble and B.G. Silbernagel, J. Chem. Phys. 63:2544 (1975); M. Molitor, W. Müller-Warmuth, H.W. Spiess, and R. Schöllhorn, Z. Naturforsch. A38:237 (1983).
92. C. Riekel, D. Hohlwein, and R. Schöllhorn, J.C.S. Chem. Commun. (1976), 863; C. Riekel and C.O. Fischer, J. Solid State Chem. 29:181 (1979).
93. D.C. Johnston, Solid State Commun. 43:533 (1982).

94. P. Hagenmüller, J. Portier, B. Barbe, and P. Bouclier, Z. Anorg. Allg. Chem. 355:209 (1967).

95. S. Kikkawa, J. Solid State Chem. 31:249 (1980); S. Kikkawa, F. Kanamaru, and M. Koizumi, Physica B105:249 (1981); Bull. Chem. Soc. Jpn. 52:963 (1979).

96. Y. Maeda, M. Yamashita, H. Ohshio, N. Tsutsumi, and Y. Takashima, Bull. Chem. Soc. Jpn. 55:3138 (1982).

97. R.H. Herber and Y. Maeda, Physica B105:243 (1981); Inorg. Chem. 19:3411 (1980).

98. M.R. Antonio and B.A. Averill, J.C.S. Chem. Commun. (1981), 382; J. de Phys. 44:C3:1373 (1983).

99. H. Meyer and A. Weiss, Mat. Res. Bull. 13:913 (1978); M. Armand, L. Coic, P. Palvadeau, and J. Rouxel, J. Power Sources 3:137 (1978).

100. P. Palvadeau, L. Coic, and J. Rouxel, Mat. Res. Bull. 13:221 (1978).

101. T.R. Halbert and J. Scanlon, Mat. Res. Bull. 14:415 (1979).

102. H. Schäfer-Stahl and R. Abele, Z. Anorg. Allg. Chem. 465:147 (1980).

103. H. Stahl, Inorg. Nucl. Chem. Lett. 16:271 (1980); H. Schäfer-Stahl and R. Abele, Angew. Chem. Int. Ed. Engl. 19:477 (1980).

104. H. Schäfer-Stahl, Mat. Res. Bull. 15:1157 (1980).

105. F. Kanamaru, M. Shimada, M. Koizumi, M. Takano, and T. Takada, J. Solid State Chem. 7:297 (1973); F. Kanamaru, S. Yamanaka, M. Koizumi, and S. Nagai, Chem. Lett. (1974), 373.

106. R.H. Herber and Y. Maeda, Inorg. Chem. 20:1409 (1981); R.H. Herber and R.A. Cassell, J. Chem. Phys. 75:4669 (1981); Inorg. Chem. 21:3713 (1982).

107. J. Rouxel and P. Palvadeau, Rev. Chim. Miner. 19:317 (1982); G.A. Fatseas, P. Palvadeau, and J.P. Venien, Stud. Inorg. Chem. 3:627 (1983); J. Solid State Chem. 51:17 (1984).

108. A. Salmon, H. Eckert, and R.H. Herber, to be published.

109. H.H. Wickman, M.P. Klein, and D.A. Shirley, Phys. Rev. 152:345 (1966).

110. P. Palvadeau, L. Coic, J. Rouxel, F. Menil, and L. Fournes, Mat. Res. Bull. 16:1055 (1981).

111. H. Schäfer-Stahl, Mat. Res. Bull. 17:1437 (1982).

112. J.E. Phillips and R.H. Herber, to be published.

113. J.E. Phillips and R.H. Herber, J. Organomet., in press.

114. J.G. Ballard and T. Birchall, J.C.S. Dalton (1976), 1859.

115. P. Boolchand, W.J. Bresser, D. McDaniel, K. Sisson, V. Yeh, and P.C. Eklund, Solid State Commun. 40:1049 (1981).

116. P. Bowen, W. Jones, J. M. Thomas, and R. Schlögl, J.C.S. Chem. Commun. (1981), 677.

117. P. Boolchand, G. Lemon, W. Bresser, D. McDaniel, R.E. Heinz, P.C. Eklund, E. Stumpp, and G. Niefeld, Mater. Res. Soc. Symp. Proc. 20:393 (1983).

118. M.L. Dzurus and G.R. Henning, J. Am. Chem. Soc. 79:1051 (1957).

119. A.G. Freeman, J.C.S. Chem. Commun. (1968), 193.
120. F.D. Grigutsch, D. Hohlwein, and A. Knappworst, Z. Phys. Chem. 65:322 (1969).
121. J.G. Hooley, M.W. Bartlett, B.V. Liengme, and J.R. Sams, Carbon 6:681 (1968).
122. R.H. Herber and M. Katada, J. Inorg. Nucl. Chem. 41:1097 (1979).
123. S.E. Millman and G. Kirczenow, Phys. Rev. B28:5019 (1983).
124. Yu. N. Novikov, M.E. Vol'pin, V.E. Prusakov, R.A. Stukan, V.I. Gol'danskii, V.A. Semion, and Yu. T. Struchkov, Zh. Strukt. Khim. 11:1039 (1970).
125. D. Hohlwein, P.W. Readman, A. Chamberod, and J.M.D. Coey, Phys. Status Solidi 64:305 (1974).
126. M. Blume, Phys. Rev. 174:351 (1968).
127. S.E. Millman, Phys. Lett. A92:441 (1982).
128. S.E. Millman, and G. Kirczenow, Solid State Commun. 44:1217 (1982).
129. T. Tominaga, T. Sakai, and T. Kimura, Chem. Lett. (1974), 853; Bull. Chem. Soc. Jpn. 49:2755 (1976).
130. K. Ohhashi and I. Tsujikawa, J. Phys. Soc. Jpn. 37:63 (1974).
131. J.G. Hooley, J.R. Sams, and B.V. Liengme, Carbon 8:467 (1970).
132. B. Pritzlaff and H. Stahl, Carbon 15:399 (1977).
133. K. Ohhashi, T. Shinjo, T. Takada, and I. Tsujikawa, J. Phys. Colloq. C2 40:269 (1979).
134. H. Shechter, J.G. Dash, M. Mor, R. Ingalls, and S. Bukshpan, Phys. Rev. B14:1876 (1976).
135. M.J. Tricker, E.L. Evans, P. Cadman, N.C. Davies, and B. Bach, Carbon 12:499 (1974).
136. P.P. Vaishnava and P.A. Montano, J. Phys. Chem. Solids 43:809 (1982).
137. R.A. Stukan, V.A. Prusakov, Yu. N. Novikov, M.E. Vol'pin, and V.I. Gol'danskii, Zh. Strukt. Khim. 12:622 (1971); A.V. Nefed'ev, R.A. Stukan, V.A. Postnikov, V.B. Shur, Yu. N. Novikov, and M.E. Vol'pin, Izv. Akad. Nauk SSSR, Ser. Khim. 10:2376 (1976).
138. A.V. Nefed'ev, N.D. Lapkina, R.A. Stukan, Yu. T. Struchkov, O.L. Lependina, Yu. N. Novikov, and M.C. Vol'pin, Zh. Strukt. Khim. 20:835 (1979).
139. A.V. Nefed'ev, R.A. Stukan, V.A. Makarov, V.A. Kondakov, A.T. Shuvaev, N.D. Lapkina, Yu. N. Novikov, and M.E. Vol'pin, Zh. Strukt. Khim. 21:68 (1980).
140. V.E. Prusa kov, Yu. N. Novikov, R.A. Stukan, M.E. Vol'pin, V.I. Gol'danskii, Dokl. Akad. Nauk SSSR 207:1394 (1972).
141. H. Stahl, Z. Anorg. Allg. Chem. 434:201 (1977).
142. A. Pron, D. Billaud, I. Kulszewicz, C. Budrowski, J. Przyluski, and J. Suwalski, Mat. Res. Bull. 16:1229 (1981).
143. V.G. Jadhao, R.M. Singru, G.M. Joshi, K.P.R. Pisharody, and C.N.R. Rao, Z. Phys. Chem. 92:139 (1974).
144. R.H. Herber and A. Rein, J. Chem. Phys. 73:6345 (1980).

145. C.W. Kimball, G.L. van Landuit, C.D. Barnet, G.K. Shenoy, B.D. Dunlap, and F.Y. Fradin, J. Phys. Colloq. C2 40:671 (1979).

146. S.D. Bader, G.S. Knapp, S.K. Sinha, P. Schweiss, and B. Renker, Phys. Rev. Lett. 37:344 (1976).

147. M. Eibschütz and F.J. DiSalvo, Phys. Rev. Lett. 36:104 (1976); M. Eibschutz, M.E. Lines, and F.J. DiSalvo, Phys. Rev. B15:103 (1977).

148. M. Eibschütz, M.E. Lines, Phys. Rev. Lett. 39:726 (1977).

149. T.R. Halbert, D.C. Johnston, L.E. McCandlish, A.H. Thompson, J.C. Scanlon, and J.A. Dumesic, Physica B99:128 (1980).

150. H. Bizette and A. Adam, C.R. Acad. Sci. (Paris), B275:911 (1972).

151. R. Bannwart and R.H. Herber, to be published.

152. R.H. Herber, H. Eckert, Y. Maeda, R.A. Cassell, J.E. Phillips, unpublished results.

153. H. Schäfer-Stahl, Mat. Res. Bull. 15:1091 (1980).

154. A.R. Beal, in ref. (1b), p. 251, and references therein.

155. S.S.P. Parkin and R.H. Friend, Phil. Mag. B41:65 (1980); R.H. Friend, A.R. Beal, and A.D. Yoffe, Phil. Mag. B35:1269 (1977).

156. K. Ohhashi and I. Tsujikawa, J. Phys. Soc. Jpn. 36:422 (1974).

157. Yu. S. Karimov, A.V. Zvarykina, and Yu. N. Novikov, Fiz. Tverd. Tela 13:2836 (1971).

158. C.W. Kocher, Phys. Lett. A24:93 (1967); J.P. Stampfel, W.T. Oosterhuis, B. Window, and F. de S. Barros, Phys. Rev. B8:4371 (1973).

159. Ch. Simon, F. Batallan, I. Rosenman, J. Schweitzer, H. Lauter, and R. Vangelisti, J. Phys. Lett. 44:641 (1983).

160. S.E. Millman, M.R. Corson, and G.R. Hoy, Phys. Rev. B25 (1982), 6595.

161. M.R. Corson, S.E. Millman, G.R. Hoy, and H. Mazurek, Solid State Commun. 42:667 (1982).

162. S.E. Millman and G.O. Zimmerman, J. Phys. C16:L89 (1983).

163. S.R. Hong and H.N. Ok, Phys. Rev. B11:4176 (1975).

# HOT-ATOM CHEMISTRY AND TRAPPED SPECIES

Hirotoshi Sano

Department of Chemistry, Faculty of Science
Tokyo Metropolitan University
Fukasawa, Setagaya, Tokyo

## INTRODUCTION

Hot-atom chemistry is a unique field of radiochemistry investigating either kinetically or electronically excited chemical species resulting from nuclear transformations. The importance of hot-atom chemistry lies not only in the elucidation of the fundamental mechanisms of chemical consequences associated with nuclear events, but also in its applications to various fields. The first of them may be energy-related researches, because the knowledge of hot-atom chemistry provides us with microscopic information about locally caused radiation damages in materials.

In spite of its importance, there had long been recognized difficulties that genuine chemical states caused by nuclear events could hardly be observed by conventional hot-atom chemical techniques, because of limited experimental procedures available for radioactive tracer amounts and of possible perturbations given to the nucleogenic species by the applied analyticochemical procedures involved in conventional methods. In solids, for instance, there are many examples in which the yields of a certain species are vulnerable to experimental conditions, such as the type of radiochemical technique used for chemical separation, the length of duration of and of storage before chemical separation. Those difficulties had made some hot-atom chemists pessimistic about the prospects of ever fully understanding the mechanisms of hot-atom chemical reactions in the solid state. Similar problems had also been experienced in radiation chemistry as far as local or in situ radiolytic processes were studied by using non-destructive techniques.

Fortunately, however, the situation has much improved since the discovery of the Mössbauer effect,[1] because the gamma-rays emitted from the transformed nuclei accompany abundant Mössbauer spectroscopic information concerning the chemical state of the produced nuclide.

Soon after the discovery of Mössbauer effect, Mössbauer emission spectroscopic studies on $^{57}$Co-labelled cobalt(II) oxide were reported suggesting the presence of anomalous charge state ($^{57}$Fe$^{3+}$ in the case of $^{57}$CoO) of the produced Mössbauer atoms in a lifetime shorter than the Mössbauer nuclear level (e. g., about $10^{-7}$ s in $^{57}$Fe).[2] Additional more sophisticated studies were needed,[3-7] until it was commonly accepted that chemical states detected by Mössbauer emission spectroscopy are mostly stable on a time-scale comparable to the lifetime of the Mössbauer nuclear level. Another significant conclusion that the presence of lattice defects plays an essential role in determining the fate of the Mössbauer atoms produced in nuclear transformations is very suggestive to the understanding of hot-atom chemistry.

Even in stoichiometric compounds labelled with radioactive nuclides providing a Mössbauer nuclear level, after-effects of nuclear transformation were observed in a number of compounds and most nuclear decays. For instance, $^{57}$Co-labelled trisacetylacetonatocobalt(III) was the first stoichiometric insulator source compound, in which both divalent and trivalent $^{57}$Fe species were observed in the Mössbauer emission spectrum.[8]

Attempts to observe a similar after-effect with $^{119}$Sn, which is as popular a Mössbauer nuclide as $^{57}$Fe, had not been successful, until a divalent $^{119}$Sn species was observed in the Mössbauer emission spectrum of a $^{119m}$Sn-labelled tetravalent tin oxalate compound, $K_6Sn_2(C_2O_4)_7 4H_2O$.[9] Since then, other aliovalent $^{119}$Sn species, for instance, divalent $^{119}$Sn species in $^{119m}$Sn-labelled tin(IV) compounds and tetravalent $^{119}$Sn species in $^{119m}$Sn-labelled tin(II) compounds have become widely observed.

In addition to those most frequently studied after-effect on the EC decay of $^{57}$Co-labelled compounds and on the IT(IC) decay of $^{119m}$Sn-labelled compounds, a number of Mössbauer emission spectroscopic studies have also been reported on various nuclear events, such as $^{83}$Br($\beta^-$)$^{83}$Kr,[10] $^{99}$Rh(EC)$^{99}$Ru,[11] $^{119}$Sb(EC)$^{119}$Sn,[12-14] $^{119m}$Te (EC,EC)$^{119}$Sn,[15] $^{125}$Sb($\beta^-$)$^{125}$Te,[16] $^{125}$I(EC)$^{125}$Te,[16] $^{151}$Ga(EC)$^{151}$Eu,[17] $^{161}$Tb($\beta^-$)$^{161}$Dy,[18] $^{193}$Os($\beta^-$)$^{193}$Ir,[19] $^{197}$Pt($\beta^-$)$^{197}$Au,[20] $^{197}$Hg(EC) $^{197}$Au,[21] $^{237}$U($\beta^-$)$^{237}$Np[22] and $^{241}$Am($\alpha$)$^{237}$Np.[23] Readers who would like to get further information about more detailed progress in this field and experimental techniques may be advised to refer to relevant reviews,[24-26] books[27-33] or original references cited there, because of the limited space in this chapter.

AFTER-EFFECTS ASSOCIATED WITH NUCLEAR DECAY

Although one of the goals of hot-atom chemistry is to describe the sequential processes initiated immediately after a nuclear event, the lack of suitable experimental techniques has not completely satisfied the investigators yet. However, various kinds of efforts have been made devoted to filling the lack of knowledge and to extend our understanding.

Based on the theoretical points of view, the formation of highly charged ionic species is widely accepted for the earliest stage, causing Coulombic explosion, electronic excitation, nucleogenic radiolysis of species, which have been, at least, partially proved by the results obtained in charge and mass spectrometry in the gas phase. After parts of nucleogenic energy have been consumed for vibrational excitation and local heating, fast relaxation of electronic and molecular dynamic state is considered to proceed in the solid state.

Subsequent slow relaxation processes, e. g., in the time-scale of $10^{-9}$ to $10^{-7}$ s, have occasionally been observed by means of time-differential Mössbauer emission spectroscopy (TDMES) in $^{57}$Co-labelled compounds. One of the typical examples is the studies on the change of charge state of $^{57}$Fe species produced by EC decay in $^{57}$Co-labelled $Co_3[Fe(CN)_6]_2$.[34] Based on the comparison of the data obtained by TDMES with those of conventional Mössbauer emission spectroscopy or time-integral Mössbauer emission spectroscopy (TIMES), the following process is proposed:

$$^{57}Co^{2+}Co^{2+}{}_2[Fe^{III}(CN)_6]_2 \xrightarrow{EC} {}^{57}Fe^{2+}Co^{2+}{}_2[Fe^{III}(CN)_6]_2$$
$$\text{(after ca. } 10^{-8} \text{ s)}$$

$$\longrightarrow {}^{57}Fe^{3+}Co^{2+}{}_2[Fe^{II}(CN)_6][Fe^{III}(CN)_6] \quad \text{(after ca. } 10^{-7} \text{ s)}$$

Besides such charge state relaxation processes, dynamic changes regarding electronic spin state relaxation have also been demonstrated especially in so-called intermediate-ligand field coordination compounds.[35-37]

Information about chemical states, such as charge state, electronic spin state, which are produced after the time-scale of a Mössbauer nuclear level, has widely been supplied by using TIMES, as described in the following paragraphs for typical examples.

In most cases an anomalous electronic state is observed in the

Mössbauer emission spectra of the daughter nuclides. Based on the systematic studies of the accumulated results, one has consolidated a conclusion that the charge state of nucleogenic atoms appears to be determined by the soft-hardness or oxidation-reduction process and excitation-deexcitation process in the interaction with species surrounding the decaying atoms, such as ligands, anions and radicals produced through the after-effect, which can be called nucleogenic local radiolytic processes.[24-33,38-42]

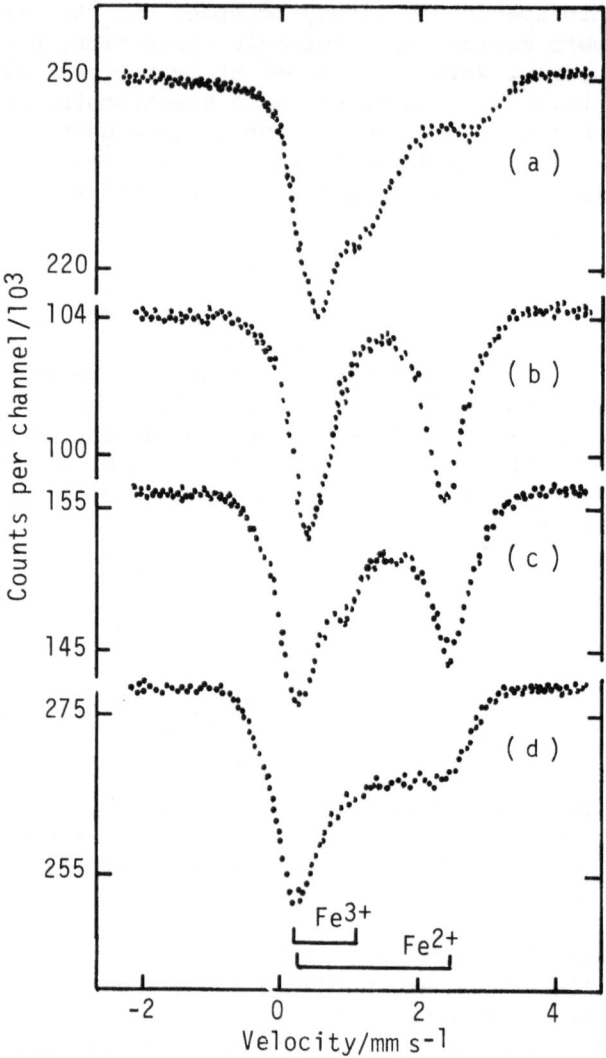

Fig. 1. Mössbauer emission spectra of $^{57}$Co-labelled (a) $K_3[Co(CO_3)_3]\cdot 3H_2O$, (b) $K_3[Co(C_2O_4)_3]3H_2O$, (c) $[Co(gly)_3]2H_2O$, and (d) $[Co(en)_3](NO_3)_3$, all at 78 K. Velocity scale is normalized with respect to metallic iron foil at room temperature.[43]

A comparison of some Mössbauer emission spectra is shown in Fig. 1. It is found that the yield of $^{57}Fe$ species depends on the kind of ligand in $^{57}Co$-labelled cobalt(III) coordination compounds. The yield of $^{57}Fe(II)$ species in $^{57}Co$-labelled trisglycinatocobalt(III) may simply be expected to fall in between the yields in $^{57}Co$-labelled potassium trisoxalatocobaltate(III) and tris ethylenediaminecobalt(III) nitrate, because the chemical structure of glycinate consists of an each half of oxalate and ethylenediamine ligand. Table 1 summarizes the yields of divalent $^{60}Co$ species produced in the $^{59}Co(n,\gamma)^{60}Co$ hot-atom reaction, the yields of divalent $^{57}Fe$ species obtained in the EC decay by Mössbauer emission spectroscopy and the $G(Co^{2+})$ values in gamma-ray irradiation of these compounds.

Table 1. Yields of $^{60}Co(II)$ species in $(n,\gamma)$ reaction and of $^{57}Fe(II)$ species in EC decay and $G(Co^{2+})$ values in gamma-ray irradiation of cobalt(III) coordination compounds[43]

| Compound / Reaction | $K_3[Co(ox)_3]3H_2O$ | $[Co(gly)_3]2H_2O$ | $[Co(en)_3](NO_3)_3$ |
|---|---|---|---|
| $^{60}Co(II)$ yield in $(n,\gamma)$ reaction | 83 % | 82 % | 69 % |
| $^{57}Fe(II)$ yield in EC decay | 85 % | 67 % | 53 % |
| $G(Co^{2+})$ values in $\gamma$-ray radiolysis | 11 | 1.5 | 0.46 |

There is a similar trend found in each reaction, although a simple averaged value is not obtained for $[Co(gly)_3]2H_2O$ except for EC decay, probably because of additional factors involved in the destructive analysis for the $(n,\gamma)$ hot-atom chemistry and the radiation chemistry. The results suggest that the carboxyl group plays an intrinsic role in the reduction of the nucleogenic atoms, such as $^{57}Fe$ and $^{60}Co$, and that the first step is the rupture of the C–C bond in oxalato or glycinato ligands caused by local radiolysis producing $\cdot CO_2^-$ radicals as follows:

$$O_2C-CO_2^- \xrightarrow{\phantom{xx}} 2 \cdot CO_2^- \longrightarrow 2\ CO_2 + 2\ e^-$$

$$H_2N-CH_2-CO_2^- \xrightarrow{\phantom{xx}} H_2N-\dot{C}H_2 + \cdot CO_2^- \longrightarrow \tfrac{1}{2}\ H_2NCH_2CH_2NH_2 + CO_2 + e^-$$

The $\cdot CO_2^-$ radicals may be used to reduce the trivalent species to divalent species giving $CO_2$. The yield of divalent $^{57}Fe$ species is found to be relatively low in $K_3[Co(CO_3)_3]3H_2O$ which has no C-C bond in the $CO_3^{2-}$ ligand,[43] as seen in Fig. 1.

## RANGE INVOLVED IN AFTER-EFFECT

It is frequently observed that the chemical species neighboring the decaying atoms play a role in determining the chemical state of the decayed atoms, as demonstrated in the preceding section. Although chemical species surrounding the decaying atoms are expected to be radiolyzed by conversion and Auger electrons and X-rays causing electron defect and then bond rupture, what extent of range is involved in the local radiolysis has remained unsolved experimentally. As seen in Fig. 2, it is found from the comparison of the re-

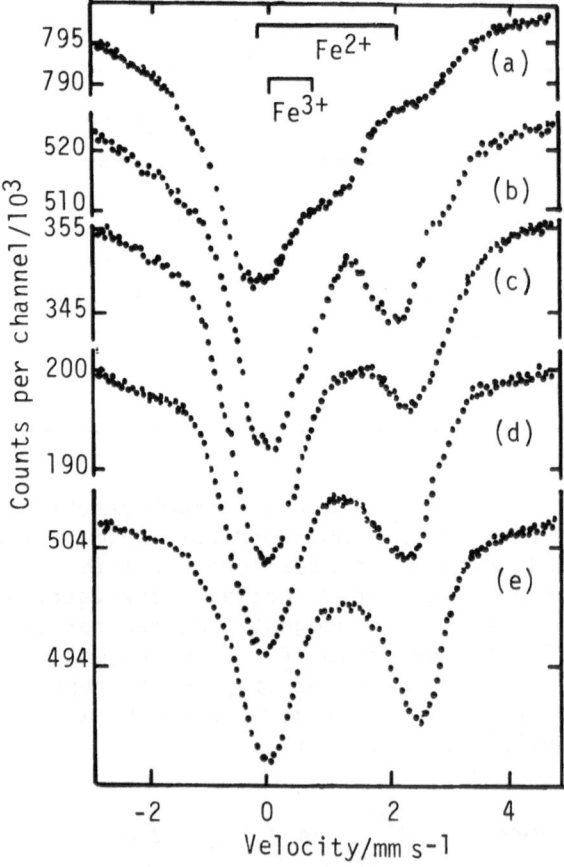

Fig. 2. Mössbauer emission spectra at 78 K of $^{57}$Co-labelled (a) [Co-$(NH_3)_6](NO_3)_3$, (b) $[Co(NH_3)_6]_2(C_2O_4)_34H_2O$, (c) $[Co(NH_3)_6][Cr-(C_2O_4)_3]3H_2O$, (d) $[Co(NH_3)_6][Fe(C_2O_4)_3]3H_2O$, and (e) $K_3[Co-(C_2O_4)_3]3H_2O$.[44]

sults in several $^{57}$Co-labelled hexamminecobalt(III) compounds that oxalate species included as counter anions or as counter complex anions still reduce remarkably the produced $^{57}$Fe atoms to divalent state.[44] This implies that the chemical consequences involve not only the species directly adjacent to the decaying atom but also the species which sits a little distant away from the decaying atoms, while the decayed atom, $^{57}$Fe, gains the maximum recoil energy of 3.4 eV due to neutrino emission following the EC decay which is too insufficient to displace the $^{57}$Fe atom from the original lattice site of the $^{57}$Co precursor atom.

The same conclusion is also derived from the comparison of the emission spectra of $^{57}$Co-labelled trisacetylacetonatocobalt(III)

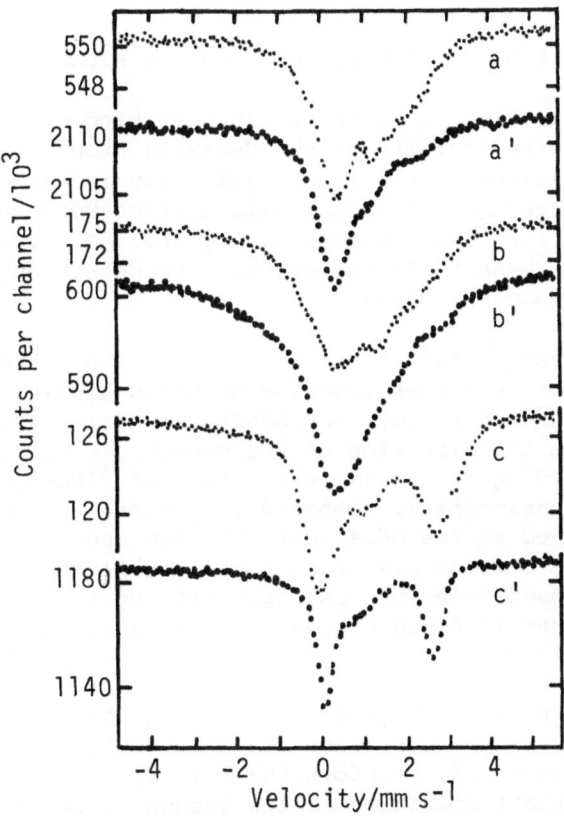

Fig. 3. Mössbauer and absorption spectra at 78 K of (a and a') ($^{57}$Co,Co)(acac)$_3$ and γ-ray irradiated ($^{57}$Fe,Co)(acac)$_3$, (b and b') ($^{57}$Co,Mn)(acac)$_3$ and γ-ray irradiated ($^{57}$Fe,Mn)(acac)$_3$, and (c and c') ($^{57}$Co,Fe)(acac)$_3$ and γ-ray irradiated ($^{57}$Fe,Fe)(acac)$_3$, respectively.[41]

doped in each trisacetylacetonatocobalt(III), manganese(III) and iron(III) host matrices with the absorption spectra of $^{57}$Fe-enriched trisacetylacetonatoiron(III) doped in the same acetylacetonate host matrices, respectively, which is subjected to a gamma-ray irradiation (ca. $10^8$ to $10^9$ rad) prior to the Mössbauer absorption spectroscopic measurement.[41]

Figure 3 shows that each set of the emission and absorption spectrum is different in shape from the other set depending upon the host matrix but that the emission spectrum shows quite similar line shape to the absorption spectrum within a set. The results also suggest that the chemical species surrounding the molecule, in which the decaying atom resides, have an effect on the chemical state of the decayed atoms. It is worth mentioning that there seems to be a similarity between the chemical consequences associated with the nuclear decay and the gamma-ray radiolysis and that there seems no magnetic hyperfine component in the spectra of $^{57}$Co-doped Co(acac)$_3$ and gamma-ray irradiated $^{57}$Fe-doped Co(acac)$_3$.

RADICAL SPECIES PRODUCED AROUND THE DECAYING ATOMS

The next question which arises is what kind of chemical species are produced in the vicinity of the decaying atoms. One of the hypothetical explanations is the local radiolysis initiated by the nuclear decay proposed based on the similarities between the retention value in conventional hot-atom chemistry and the G-value,[45,50] although there has been no direct evidence to detect any radiolytic products in the nuclear decay.

It is well-known that the magnetic hyperfine structure (hfs) of a Mössbauer spectrum arises from the magnetic dipole of a Mössbauer nucleus in a magnetic field. The Mössbauer nucleus starts its Larmor precession about the direction of the magnetic field and the nuclear levels split, giving six peaks in the case of $^{57}$Fe-Mössbauer spectrum. In most paramagnetic compounds, however, the magnetic hfs cannot be observed in the Mössbauer spectrum, because the neighboring paramagnetic species give rise to a fast fluctuation of the magnetic field through spin-spin and spin-lattice interaction in such a way that the magnetic field averages out to zero during the Larmor precession time.

On the other hand, it is also known that the magnetic hfs appears if the paramagnetic species are diluted by a large amount of diamagnetic species and therefore the spin-spin interaction decreases and if the temperature is low enough to suppress the spin-lattice interaction. As found in Fig. 4, the absorption spectra of paramagnetic ($S = 5/2$) $^{57}$Fe-enriched Fe(acac)$_3$ doped in diamagnetic Co(acac)$_3$, Fe(oxin)$_3$ doped in Al(oxin)$_3$ and Fe(dbm)$_3$ doped in Co(dbm)$_3$ show magnetic hyperfine relaxation wings at liquid nitrogen

temperature (even at dry-ice temperature). If the paramagnetic
Mössbauer species sits in a paramagnetic host material, the magnetic
hyperfine relaxation component tends to disappear as seen in Fig. 5,
where absorption spectra of $^{57}$Fe-enriched Fe(acac)$_3$ doped Cr(acac)$_3$
(S = 3/2) and Mn(acac)$_3$ (S = 2) both show only a broad single peak
similar to the spectrum of neat Fe(acac)$_3$,[32,46,47] although the line
broadening still reflects a slight contribution of spin-spin relaxation.

In contrast to these Mössbauer absorption spectroscopic data,
there is essentially no magnetic hyperfine relaxation component observed in most emission spectra of diamagnetic trivalent $^{57}$Co-label-
led cobalt(III) coordination compounds, in which the $^{57}$Fe species
produced by the EC decay appears predominantly in the high-spin (S =
5/2) electronic state.

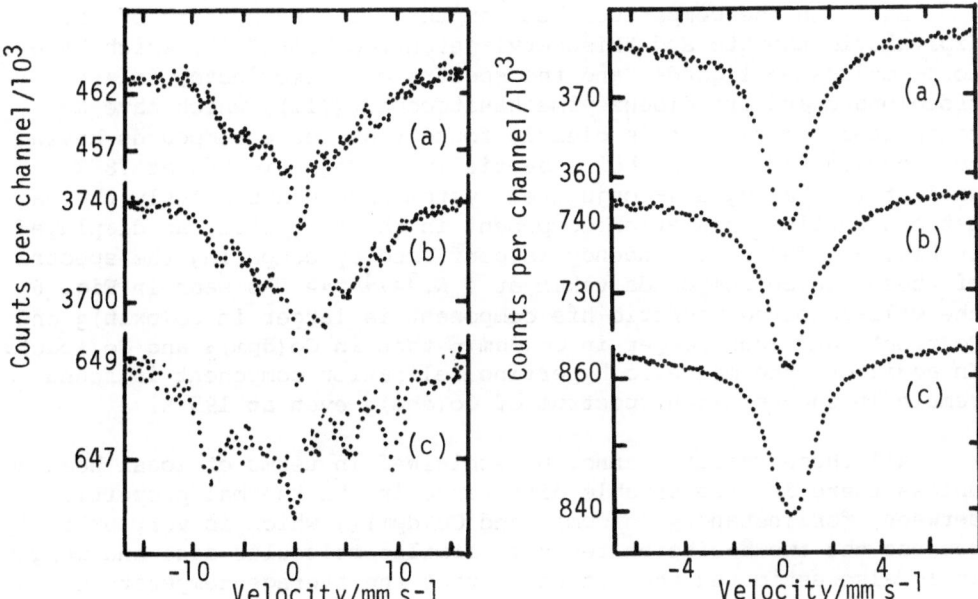

Fig. 4.  Mössbauer absorption
spectra at 78 K of
(a) ($^{57}$Fe,Al)(oxin)$_3$,
(b) ($^{57}$Fe,Co)(acac)$_3$,
and (c) ($^{57}$Fe,Co)(d-
bm)$_3$.[32,46]

Fig. 5.  Mössbauer absorption
spectra at 78 K of
(a) ($^{57}$Fe,Cr)(acac)$_3$,
(b) ($^{57}$Fe,Mn)(acac)$_3$,
and (c) ($^{57}$Fe,Fe)(acac)$_3$.[32,46]

One of the most plausible interpretations for the absence of the magnetic hfs even at a higher temperature as 78 K may be to assume that some paramagnetic species are produced in the vicinity of the decaying atom of such an amount that the spin-spin interaction between the paramagnetic species and high-spin decayed atoms causes the magnetic field to average out to zero.

It is reasonable to conclude that there is a sufficient amount of paramagnetic radicals produced from ligands or anions in the vicinity of the decaying atom through the local radiolytic process initiated by the EC decay, because the concentration of decayed atoms, $^{57}$Fe(II) or (III), is estimated to be too low to give rise to spin-spin interaction between nucleogenic iron atoms.

In order to prove the conclusion experimentally, the Mössbauer emission spectra of several kinds of $^{57}$Co-labelled tris-β-diketonato, trisoxinato and tristropolonatocobalt(III) complexes were studied. One can expect that $^{57}$Co-labelled cobalt(III) complexes which contain π-conjugated ligands should give rise to magnetic hfs in the emission spectrum, because it is well-known in radiation chemistry that a π-conjugated system has a remarkable tendency to reduce radiolysis, i. e., the formation of paramagnetic radicals.

Based on the comparison between the emission spectra of tris-dipivaloylmethanato and trisacetylacetonatocobalt(III), which have no π-conjugated ligands, and the spectra of trisoxinato, tris-tropolonato and trisdibenzoylmethanatocobalt(III), which have π-conjugated ligands, it is clearly found that those compounds having no π-conjugated system show essentially no magnetic hfs at 78 K, while those having a π-conjugated system show unambiguously the magnetic hyperfine relaxation component in their spectra, as displayed in Fig. 6. The same tendency is confirmed by comparing the spectra of these source compounds taken at 4 K.[32,46-48] As seen in Fig. 6, the well-resolved magnetic hfs component is larger in Co(oxin)$_3$ and Co(trop)$_3$ and much larger in Co(dbm)$_3$ than in Co(dpm)$_3$ and Co(acac)$_3$. In addition, the magnetic hyperfine relaxation component is found to remain in the emission spectrum of Co(dbm)$_3$ even at 195 K.

All these results cannot be explained in terms of local heating, unless there is a remarkable difference in the thermal properties between, for instance, Co(dbm)$_3$ and Co(dpm)$_3$, which is very unlikely because the two β-diketonates have similar molecular size and weight. It is also estimated theoretically that the average temperature increase is as low as a negligibly low value after the Mössbauer lifetime.[46]

Further experiments were carried out to answer the question how far the radical formation is involved outside the $^{57}$Co-labelled species and interacts with the decayed atoms, $^{57}$Fe. In the emission spectrum of polystyrene film homogeneously doped with $^{57}$Co-labelled

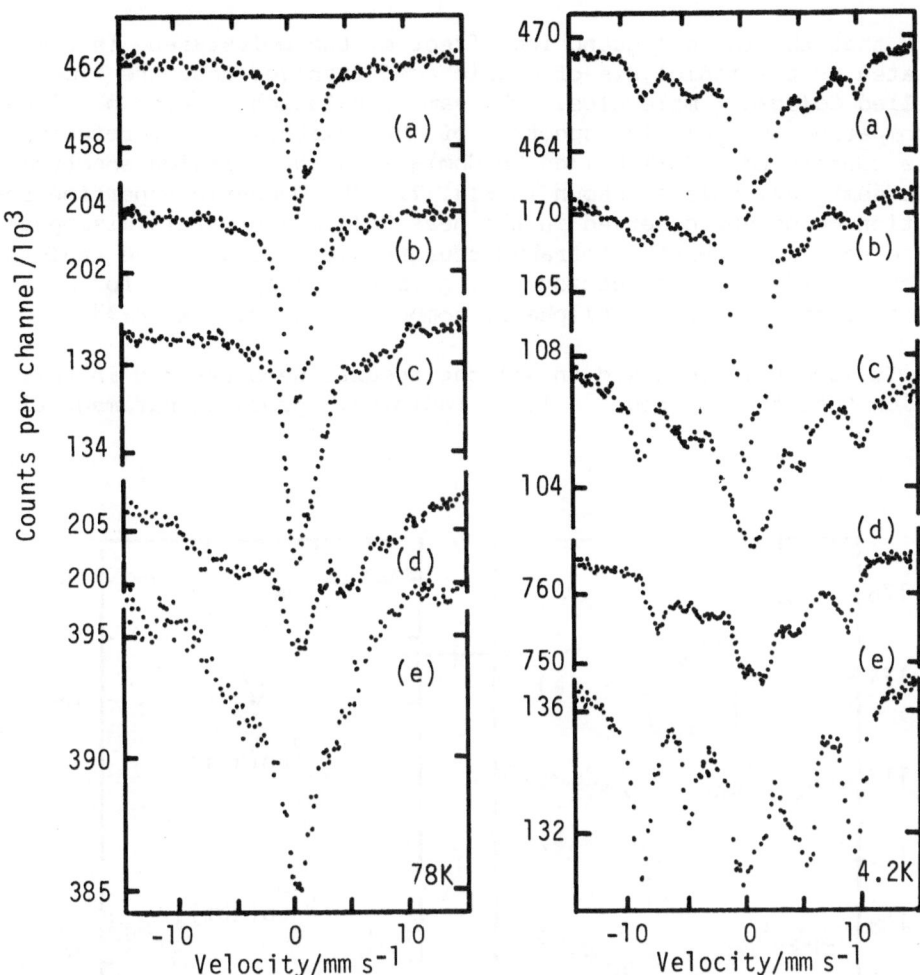

Fig. 6. Mössbauer emission spectra of [57]Co-labelled (a) Co(acac)$_3$, (b) Co(dpm)$_3$, (c) Co(oxin)$_3$, (d) Co(tropol)$_3$, and (e) Co(dbm)$_3$ at 78 K and 4.2 K.[32,46,47]

Co(acac)$_3$, the component of magnetic hyperfine relaxation revives even at 78 K, as found in Fig. 7, although it was hardly observed in the emission spectrum of Co(acac)$_3$ crystals at 78 K. This indicates that not a sufficient amount of radicals is produced in the polystyrene host, because of the presence of the π-conjugated system in the polystyrene network.[49]

The magnetic hyperfine relaxation component is found to be suppressed by adding an amount of oxalic acid into the polystyrene film containing [57]Co-labelled Co(acac)$_3$ as also seen in Fig. 7. This suggests that the radical quenching effect of the polystyrene is compensated by the radiolysis of oxalic acid involved near the [57]Co-labelled Co(acac)$_3$ molecules. The same conclusion is also obtained by comparing the emission spectrum of the 2-methyltetrahydrofurane glass containing [57]Co-labelled Co(dbm)$_3$ with the emission spectrum of Co(dbm)$_3$ crystals as shown in Fig. 7. The magnetic hyperfine relaxation component observed in the neat Co(dbm)$_3$ crystals disappears in the matrix of 2-methyltetrahydrofurane glass because the radical quenching effect of the benzoyl group is also compensated for by the radical formation out of aliphatic compound host matrix.[32,49]

It is concluded based on all the results demonstrated in this section that the nucleogenic local radiolysis produces paramagnetic

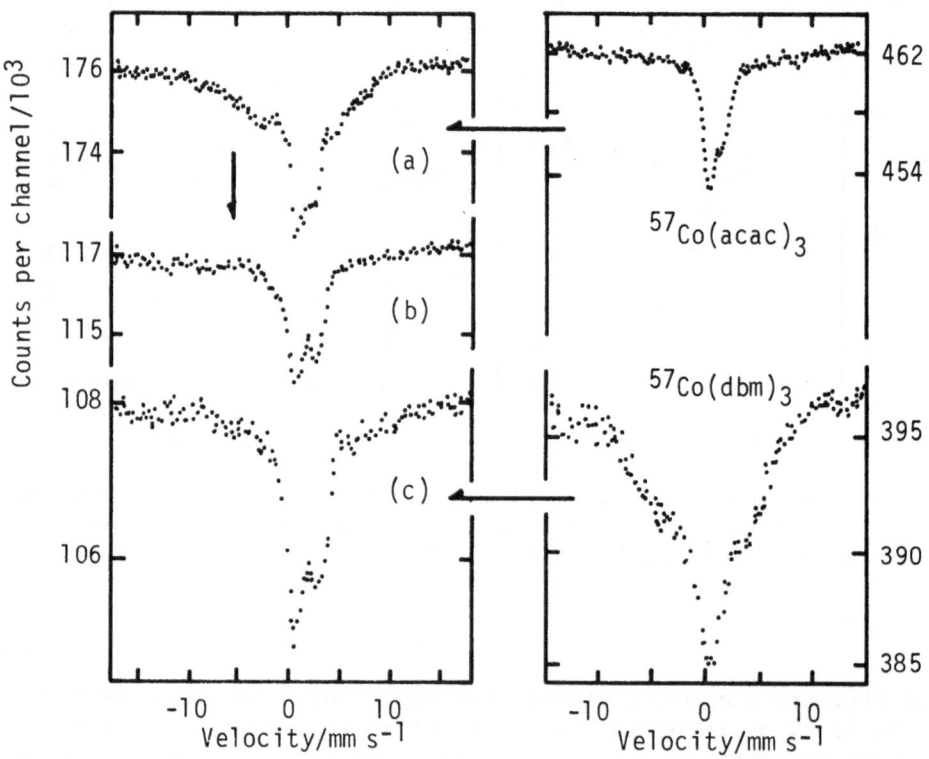

Fig. 7. Mössbauer emission spectra at 78 K of (a) [57]Co-labelled Co(acac)$_3$ doped in polystyrene film, (b) [57]Co-labelled Co(acac)$_3$ doped in polystyrene film including oxalic acid, and (c) [57]Co-labelled Co(dbm)$_3$ doped in 2-methyltetrahydrofurane glass.[32,49] For the sake of comparison the emission spectra of [57]Co-labelled Co(acac)$_3$ and Co(dbm)$_3$ shown in Fig. 6 are reproduced.

radicals involving not only the ligand species but also the species
in a more distant vicinity of decaying atoms.

## LOCAL RADIATION DAMAGE PRODUCED BY NUCLEAR EVENTS

Examples which show similarities among the after-effects in EC
decay observed by Mössbauer emission spectroscopy, conventional hot-
atom chemistry in $^{59}Co(n,\gamma)^{60}Co$ reaction and radiation chemistry of
cobalt(III) coordination compounds were already mentioned in Table
1. A roughly good correlation between the yield of $^{60}Co(II)$ species
estimated in the $^{59}Co(n,\gamma)^{60}Co$ reaction and $^{57}Fe(II)$ species in the
EC decay of $^{57}Co$ for the same cobalt(III) coordination compounds is
found in Fig. 8 as well, with an exception in the case of hexammine-

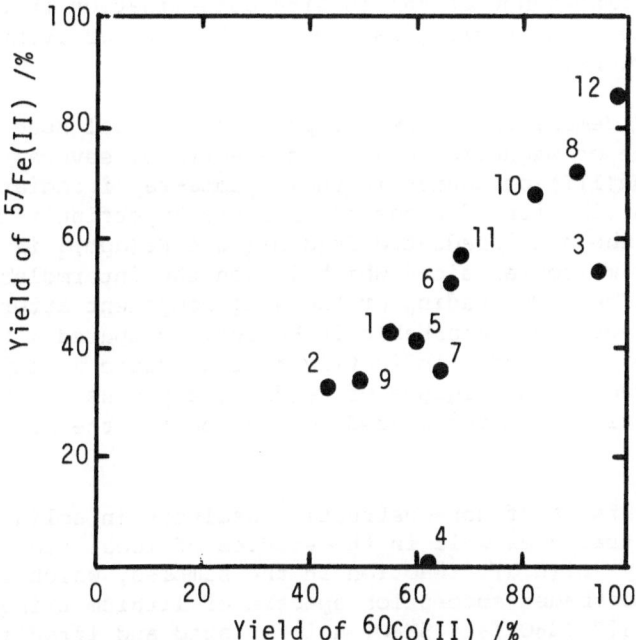

1. $[Co(NH_3)_6]Cl_3$, 2. $[Co(NH_3)_6]_2(CO_3)_3$, 3. $[Co-$
$(NH_3)_6]_2(SO_4)_3 5H_2O$, 4. $[Co(NH_3)_6]_2(CrO_4)_3$, 5. $[Co-$
$(NH_3)_6]CO_3Cl$, 6. $[Co(NH_3)_6]SO_3Cl$, 7. $[Co(NH_3)_6]-$
$(NO_3)_3$, 8. $[Co(NH_3)_6]_2(C_2O_4)_3 4H_2O$, 9. $[Co(NH_3)_5-$
$OH_2](NO_3)_3$, 10. $[Co(NH_3)_5OH_2]_2(C_2O_4)_3$, 11. $[Co-$
$(en)_3](NO_3)_3$, and 12. $K_3[Co(C_2O_4)_3]3H_2O$.

Fig. 8. Comparison of the radiochemical yield of $^{60}Co(II)$ species
in $^{59}Co(n,\gamma)^{60}Co$ hot-atom chemistry with the yield of
$^{57}Fe(II)$ species produced after EC decay for the same
Co(III) coordination compounds.[50]

cobalt(III) chromate.[50]  The large yield of $^{60}$Co(II) species as compared to that of $^{57}$Fe(II) in [Co(NH$_3$)$_6$]$_2$(CrO$_4$)$_3$ may be ascribed to the reactions, such as,

$$\mathrm{Cr^{VI}O_4}^{2-} + \mathrm{NH_3} \longrightarrow \tfrac{1}{2}\,\mathrm{Cr^{III}_2O_3} + \tfrac{1}{2}\,\mathrm{H_2O} + 2\,\mathrm{OH^-} \quad \text{and/or}$$

$$\mathrm{Cr^{VI}O_4}^{2-} + \tfrac{2}{3}\,\mathrm{NH_3} \longrightarrow \mathrm{Cr^{IV}O_2} + \tfrac{1}{3}\,\mathrm{N_2} + 2\,\mathrm{OH^-}$$

In the case of the conventional hot-atom chemistry, the produced ligand-deficient $^{60}$Co(III) species may easily be reduced to $^{60}$Co(II) species during the chemical analysis, increasing the apparent yield of $^{60}$Co(II) species, while the ligand-deficient $^{57}$Fe (III) species produced in a similar way can be determined non-destructively as $^{57}$Fe(III) species in Mössbauer spectroscopy.  This suggests the importance of the in situ determination in the investigations of the local radiolysis produced by nuclear events of external radiations.

Figure 9 demonstrates how the produced radical species can affect the shape of magnetic relaxation spectra of several tris-β-diketonatoiron(III) compounds in their gamma-ray irradiation.[51] The large contribution of wings of magnetic hyperfine relaxation component in the non-irradiated Fe(dbm)$_3$ and Fe(dpm)$_3$ is attributed to the larger molecular sizes which lessen the intermolecular spin-spin interaction.  The fading of the wing component after irradiation is, however, not remarkable in Fe(dbm)$_3$ compared with the cases found in other compounds, indicating that aromatic π-conjugating system involved in the ligands of Fe(dbm)$_3$ suppress the radical formation as mentioned in the preceding section for the after-effects in EC decay.

The advantages of non-destructive analysis in solid samples can be demonstrated as well in the studies of local radiation damages caused by $^6$Li(n,α)t reaction in the samples, which include $^6$Li atoms.  The Mössbauer absorption spectra of lithium trisoxalatoferrate(III), Li$_3$[Fe(C$_2$O$_4$)$_3$]4.5H$_2$O, unirradiated and irradiated in a neutron reactor are shown in Fig. 10.  For the sake of comparison, absorption spectra of gamma-ray irradiated Li$_3$[Fe(C$_2$O$_4$)$_3$]4.5H$_2$O and of neutron-irradiated Na$_3$[Fe(C$_2$O$_4$)$_3$]5H$_2$O are also included in Fig. 10.  The total gamma-ray dose used for the spectrum (b) in Fig. 10 is about 7 times larger than the total gamma-ray dose for the irradiation site used for the spectra (e).  The Mössbauer absorption spectrum of Na$_3$[Fe(C$_2$O$_4$)$_3$]5H$_2$O irradiated with neutrons in the same condition as in the case of (e) in Fig. 10 shows hardly a remarkable component of Fe(II) species which consists of a main component in the spectrum of Li$_3$[Fe(C$_2$O$_4$)$_3$]4.5H$_2$O irradiated with thermal neutrons (e).[52]

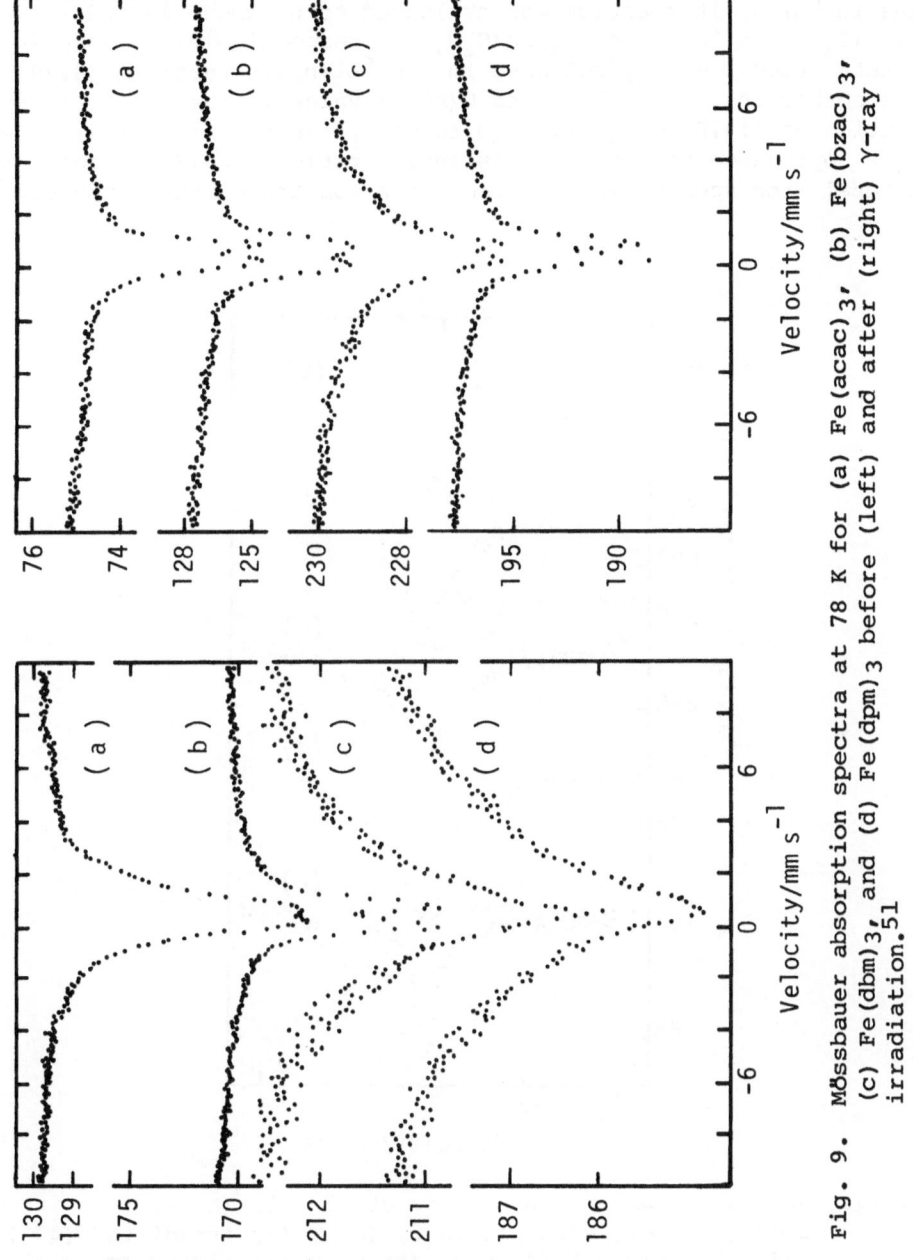

Fig. 9. Mössbauer absorption spectra at 78 K for (a) Fe(acac)₃, (b) Fe(bzac)₃, (c) Fe(dbm)₃, and (d) Fe(dpm)₃ before (left) and after (right) γ-ray irradiation.[51]

Based on the dependence of the areal intensity ratio of Fe(II) species to the total iron species in the Mössbauer absorption spectra in the different irradiation times or neutron fluxes, the $G(Fe^{2+})$ value in $^6Li(n,\alpha)t$ reaction was evaluated to be about 19 and 10 for $Li_3[Fe(C_2O_4)_3]4.5H_2O$ and $Li_3[Fe(C_2O_4)_3]$, respectively, by assuming the total recoil energy released in the $^6Li(n,\alpha)t$ reaction (4.78 MeV) was absorbed in the sample. The $G(Fe^{2+})$ value in the gamma-ray irradiation of $Li_3[Fe(C_2O_4)_3]4.5H_2O$ to be 8.4 in a similar way, based on the dependence of the areal intensity ratio of Fe(II) species to the total iron species in the absorption spactra on the gamma-ray

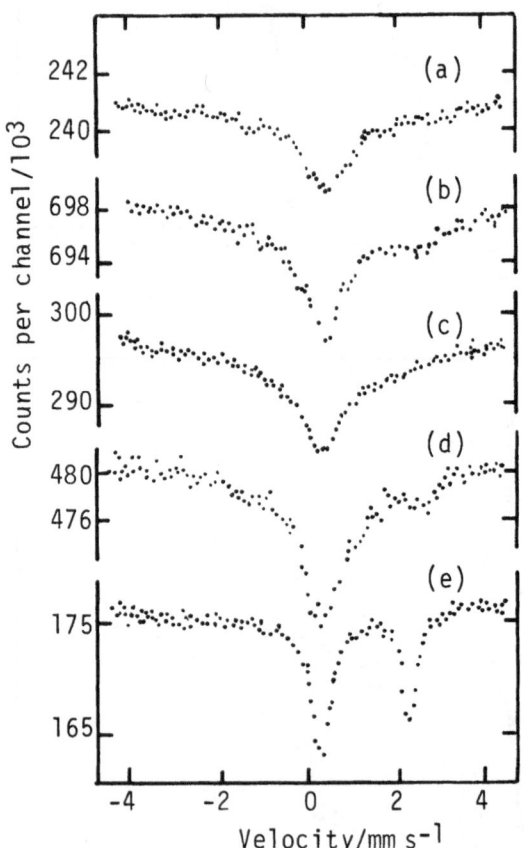

Fig. 10.  Mössbauer absorption spectra at 78 K of (a) neutron-irradiated (4 h) $Na_3[Fe(C_2O_4)_3]5H_2O$, (b) $\gamma$-ray irradiated ($4 \times 10^7$ rad) $Li_3[Fe(C_2O_4)_3]4.5H_2O$, (c) unirradiated $Li_3[Fe(C_2O_4)_3]$-$4.5H_2O$, (d) neutron-irradiated (4 h) $Li_3[Fe(C_2O_4)_3]4.5H_2O$ kept in a Cd capsule, and (e) neutron-irradiated (4 h) $Li_3$-$[Fe(C_2O_4)_3]4.5H_2O$.[52]  (Thermal neutron flux: $5 \times 10^{11}$/s $cm^2$; fast neutron flux: $4.5 \times 10^{10}$/s $cm^2$).

dose.  The results reveal that a higher density of the excited species, such as radicals or ions, produced by the high energy particles is more effective to cause the reduction of the Fe(III) to
Fe(II) species in this compound and that the hydrated water molecules play an important role in the reduction process.


APPLICATIONS TO THE STUDIES OF BLANKET MATERIALS

It is realized that the knowledge, such as radiation resist-
ance, radiolytic products, on the blanket materials used in a fusion
reactor, where $^6Li(n,\alpha)t$ reactions cause significant radiation damages as described in the preceding paragraphs, is becoming more and

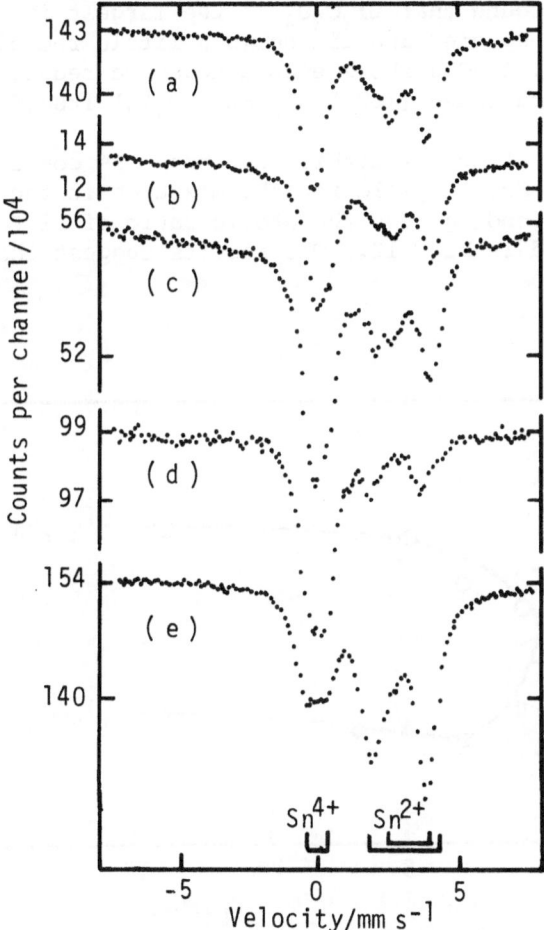

Fig. 11.  Mössbauer absorption spectra at 78 K of neutron irradiated
(a) $[(CH_3)_4N]_2Sn(C_2O_4)_3$, (b) $[(CH_3)_4N]_2Sn(C_2O_4)_3$ mixed with
LiF, (c) $[(CH_3)_4N]_2Sn(C_2O_4)_3$ mixed with LiH, (d) $[(CH_3)_4N]_2Sn$-
$(C_2O_4)_3$ mixed with $Li_2CO_3$, and (e) $[(CH_3)_4N]_2Sn(C_2O_4)_3$ mixed
with $Li_2C_2O_4$.[53] Neutron fluence: $5 \times 10^{11}$/s cm2 × 4 hours.

more indispensable in relation to future energy problems. Mössbauer elements can be introduced into the blenket materials as a probe in order to get in situ information about the local radiolytic products. Because the alpha and tritium particles produced in the $^6$Li$(n,\alpha)$t reaction have long recoil-ranges, a fine powdered compound which includes Mössbauer atoms can be used for this purpose as a mixture with powdered blanket materials.[53]

A typical example is demonstrated in Fig. 11, where the $^{119}$Sn-Mössbauer absorption spectra of neutron-irradiated powdered tetramethylammonium trisoxalatostannate(IV), $[(CH_3)_4N]_2[Sn(C_2O_4)_3]$, and its mixtures with various kinds of powdered lithium compounds, LiF, LiH, Li$_2$CO$_3$ and Li$_2$C$_2$O$_4$, in the same atomic ratio of Li to Sn are compared. It is found that Li$_2$C$_2$O$_4$ is the largest in reducing Sn(IV) species to Sn(II) species and LiH shows a little reducing effect. On the other hand, Li$_2$CO$_3$ shows even a negative reducing effect compared with the irradiated $[(CH_3)_4N]_2[Sn(C_2O_4)_3]$ itself.

The peculiar effect of lithium carbonate is confirmed in the comparison of the Sn(II) yield for the mixtures in the same irradiation condition depending upon the atomic ratio of Li to Sn in the mixtures, as found in Fig. 12. The results suggest that carbonate

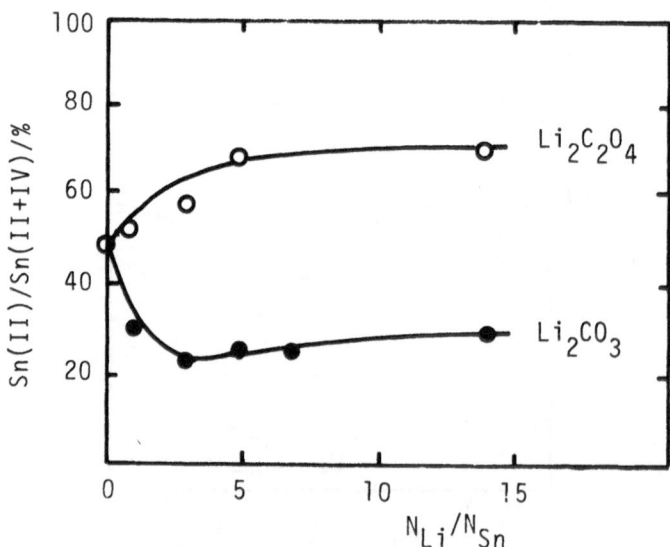

Fig. 12. Dependence of the yield of Sn(II) species on the atomic ratio of lithium to tin in the mixtures of $[(CH_3)_4N]_2$Sn-$(C_2O_4)_3$ with Li$_2$C$_2$O$_4$ and Li$_2$CO$_3$, respectively. Neutron irradiation was carried out in the same condition (neutron fluence: $5 \times 10^{11}$/s cm2 $\times$ 4 h).

anions should produce some oxidizing species, e. g., $O^-$, through the following hypothetical process:

$$CO_3^{2-} \longrightarrow CO + 2\ O^- \xrightarrow{2\ e^-} CO + 2\ O^{2-}$$

It was mentioned that the yield of $^{57}Fe(II)$ species is much lower in the EC decay of $^{57}Co$-labelled $K_3[Fe(CO_3)_3]3H_2O$ compared with that observed in the EC decay in $K_3[Fe(C_2O_4)_3]4.5H_2O$ in Fig. 1. Those data lead us into temptation to conclude that the accumulated studies of Mössbauer emission spectroscopy will provide us with unique in situ information about the trapped species and the microscopic radiolytic processes locally caused by nuclear events, although further efforts must be carried out to systematize the data and to apply the technique to various kinds of promising fields.

REFERENCES

1. R. L. Mössbauer, Naturwiss. 45:538 (1958); Z. Phys. 151: 124 (1958).
2. G. K. Wertheim, Phys. Rev. 124:764 (1961).
3. J. G. Mullen and H. N. Ok, Phys. Rev. Lett. 17:287 (1966).
4. H. N. Ok and J. G. Mullen, Phys. Rev. 168:550, 563 (1963).
5. W. Triftshäuser and P. P. Craig, Phys. Rev. 162:274 (1967).
6. D. Schrofer and W. Triftshäuser, Phys. Rev. Lett. 20:1242 (1968).
7. W. Trousdale and P. P. Craig, Phys. Lett. 27A:552 (1968).
8. G. K. Wertheim, W. R. Kingston, and R. H. Herber, J. Chem. Phys. 37:687 (1962).
9. H. Sano and M. Kanno, Chem. Commun. 1969:601 (1969).
10. M. Pasternak and T. Sonnino, Phys. Rev. 164:384 (1967).
11. G. Kaindl, W. Potzel, F. Wagner, U. Zahn, and R. L. Mössbauer, Z. Phys. 226:103 (1969).
12. F. Ambe and S. Ambe, Bull. Chem. Soc. Jpn. 47:2875 (1974); J. Chem. Phys. 63:4077 (1975).
13. F. Ambe, H. Shoji, S. Ambe, M. Takeda, and N. Saito, Chem. Phys. Lett. 14:522 (1972).
14. F. Ambe, S. Ambe, H. Shoji, and N. Saito, J. Chem. Phys. 60:3774 (1974).
15. F. Ambe and S. Ambe, Radiochim. Acta 20:141 (1973); Phys. Lett. 43A:399 (1973).
16. P. Jung and W. Triftshäuser, Phys. Rev. 175:512 (1968).
17. P. Glentworth, A. L. Nichols, N. R. Large, and R. J. Bullock, Chem. Commun. 1971:206 (1971).
18. B. Khurgin, S. Ofer, and M. Rakavy, Phys. Lett. 33A:219 (1970).
19. P. Rother, F. Wagner, and U. Zahn, Radiochim. Acta 11:203 (1963).

20. U. Zahn, "Perspectives in Mössbauer Spectroscopy", S. G. Cohen and M. Pasternak ed., Plenum Press, New York-London (1973).

21. M. O. Faltens and D. A. Shirley, J. Chem. Phys. 53:4249 (1970).

22. J. A. Stone and W. L. Pillinger, Phys. Rev. Lett. 13:200 (1964).

23. J. Gal, Z. Hadari, and E. Yanir, J. Inorg. Nucl. Chem. 32:2509 (1970).

24. H. Sano, Radioisotopes 24:357 (1975).

25. J. M. Friedt and J. Danon, Atomic Energy Rev. 18:4 (1980).

26. J. P. Adloff, "Effects Chimiques des Transformations Nucléaires", Bibliographie 1973, Rep. No. CRN/CNPA 74-28 and subsequent annual reports, Univ. de Strasbourg, 67037 Strasbourg-Cedex, France.

27. V. I. Gol'danskii and R. H. Herber ed., "Chemical Applications of Mössbauer Spectroscopy", Academic Press, New York (1968).

28. N. N. Greenwood and T. C. Gibb, "Mössbauer Spectroscopy", Chapman and Hall, London (1971).

29. H. Sano, "Mössbauer Spectroscopy-The Chemical Applications", Kodansha, Tokyo (1972).

30. P. Gütlich, R. Link, and A. Trautwein, "Mössbauer Spectroscopy and Transition Metal Chemistry", Inorg. Chem. Concepts Ser. No. 3, Springer-Verlag, Berlin-Heiderberg, New York (1978).

31. T. Tominaga and E. Tachikawa, "Modern Hot-atom Chemistry and Its Applications", Inorg. Chem. Concepts Ser. No. 5, Springer-Verlag, Berlin-Heiderberg, New York (1981).

32. H. Sano, "Recent Advances in the Applications of Mössbauer Spectroscopy in Chemistry", Tamkang Univ. Press, Taipei, Taiwan (1982).

33. H. Sano and Gütlich, Hot Atom Chemistry in Relation to Mössbauer Emission Spectroscopy, in: "Hot Atom Chemistry, Recent Trends and Applications in the Physical and Life Sciences and Technology", T. Matsuura, ed., Kodansha, Tokyo (1984).

34. V. P. Alekseev, V. I. Gol'danskii, V. E. Pruksakow, A. V. Neved'ev, and R. S. Stukan, JETP Lett. 16:43 (1972).

35. R. Grimm, P. Gütlich, E. Kankeleit, and R. Link, J. Chem. Phys. 67:5491 (1977).

36. G. R. Hoy, D. W. Hamill, and P. P. Wintersteiner, "Mössbauer Effect Methodology", Vol. 6, I. J. Gruverman, ed., Plenum Press, New York-London (1971).

37. G. R. Hoy and P. P. Wintersteiner, Phys. Rev. Lett. 28: 877 (1972).

38. H. Sano and F. Hashimoto, Bull. Chem. Soc. Jpn. 38:684 (1965).

39. R. Ingalls and G. De Pasquali, Phys. Lett. 15:262 (1965); R. Ingalls, C. J. Coston, G. De Pasquali, H. G.

Drickermer, and J. J. Pinajian, <u>J</u>. <u>Chem</u>. <u>Phys</u>. 45:1057 (1966).

40. J. M. Friedt, E. Baggio-Saitovitch, and J. Danon, <u>Chem</u>. <u>Phys</u>. <u>Lett</u>. 7:603 (1970); E. Baggio-Saitovitch, J. M. Friedt, and J. Danon, <u>J</u>. <u>Chem</u>. <u>Phys</u>. 56:1269 (1972).

41. H. Sano and H. Iwagami, <u>Chem</u>. <u>Commun</u>. 1971:1637 (1971).

42. H. Sano, K. Sato, and H. Iwagami, <u>Bull</u>. <u>Chem</u>. <u>Soc</u>. <u>Jpn</u>. 44:2570 (1971).

43. Y. Sakai, Ph. D. Thesis, Tokyo Metropolitan Univ. (1980).

44. H. Sano and T. Ohnuma, <u>Chem</u>. <u>Phys</u>. <u>Lett</u>. 26:348 (1974); <u>Chem</u>. <u>Lett</u>. 1974:589 (1974); <u>Bull</u>. <u>Chem</u>. <u>Soc</u>. <u>Jpn</u>. 48: 266 (1975).

45. H. Sano, N. Matsubara, and N. Saito, <u>Bull</u>. <u>Chem</u>. <u>Soc</u>. <u>Jpn</u>. 38:333 (1965).

46. H. Sano, <u>J</u>. <u>Radioanal</u>. <u>Chem</u>. 36:105 (1977).

47. Y. Sakai, K. Endo, and H. Sano, <u>Bull</u>. <u>Chem</u>. <u>Soc</u>. <u>Jpn</u>. 53: 1317 (1980).

48. K. Endo, H. Sano, and H.-H. Wei, <u>Radiochem</u>. <u>Radioanal</u>. <u>Lett</u>. 50:29 (1981).

49. Y. Sakai, K. Endo, and H. Sano, <u>Bull</u>. <u>Chem</u>. <u>Soc</u>. <u>Jpn</u>. 54: 3587 (1981).

50. H. Sano, M. Harada, and K. Endo, <u>Bull</u>. <u>Chem</u>. <u>Soc</u>. <u>Jpn</u>. 51: 2583 (1978).

51. Y. Sakai, K. Endo, and H. Sano, <u>Intern</u>. <u>J</u>. <u>Appl</u>. <u>Rad</u>. <u>Isotopes</u> 32:435 (1981).

52. J. Akashi, M. Katada, and H. Sano, <u>Chem</u>. <u>Lett</u>. 1982:817 (1982); J. Akashi, Y. Uchida, T. Kojima, M. Katada, and H. Sano, <u>Bull</u>. <u>Chem</u>. <u>Soc</u>. <u>Jpn</u>. 57:No.5 (1984).

53. J. Akashi, T. Kojima, M. Katada, and H. Sano, <u>Radiochem</u>. <u>Radioanal</u>. <u>Lett</u>. 58:237 (1983).

# STRUCTURE, BONDING, AND THE MOSSBAUER LATTICE TEMPERATURE

Rolfe H. Herber

Department of Chemistry
Rutgers University
New Brunswick, NJ 08903

## I. INTRODUCTION

Since the earliest days of Mossbauer Effect Spectroscopy and
its application to chemical problems, it has been well understood
that the various parameters which could be extracted from a
resonance spectrum all have characteristic dependencies on
temperature. Moverover, it soon became apparent that a good deal
of useful chemical information could be extracted from a
determination of these temperature coefficients, and that the
resulting systematics provided an additional means for
understanding the structure and bonding in chemical systems.

Of particular interest has been the temperature dependence of
the recoil-free fraction, and the relationship of this parameter to
the strength of <u>inter</u>- and <u>intra</u>-molecular forces. In the present
discussion is presented a brief review of the fundamental equations
describing the temperature dependence of the recoil-free fraction
and the isomer shift in Section II, followed in Section III by some
applications to inorganic and metallic systems, in Section IV by
some applications to organometallic compounds and in Section V by a
summary of the present status of such studies. Not intended to be
an exhaustive summary of the large amount of data which are
available in the literature, the present review is meant to be
indicative of the kind of information which can be extracted from
temperature dependent Mossbauer effect studies, especially those
involving the two nuclides $^{57}$Fe and $^{119}$Sn.

## II. FUNDAMENTAL EQUATIONS

The central feature related to the ability to observe resonance scattering of nuclear gamma rays in an experimental arrangement is the emission (in the source) and the absorption (in the absorber) of radiation without loss of appreciable energy by recoil. The microscopic details of the emission and absorption of radiation can be best treated by a quantum mechanical formalism[1,2] which will ultimately yield the probability that a given event will involve a zero-phonon process. This probability can be expressed by the recoil-free fraction, f, which is that fraction of all emission or absorption events in which (in a solid), no vibrational phonon of the lattice is involved. It has become customary, through the obvious relationship between f and the Debye-Waller factor of X-ray diffraction experiments, to relate the recoil-free fraction to the mean square vibrational amplitude of the Mossbauer atom. This relationship is specifically expressed by

$$f = \exp(-K^2 \langle x^2 \rangle) \qquad [1]$$

where K is the wave vector of the gamma radiation, equal to $\lambdabar^{-1}$ with

$$\lambdabar = hc/2\pi E_\gamma \qquad [2]$$

and $\langle x^2 \rangle$ is the mean square amplitude of vibration parallel to the gamma ray propagation axis. Since this vibrational amplitude is temperature dependent, the recoil-free fraction itself is temperature dependent, and it is through the temperature dependence of $\langle x^2 \rangle$ that chemical information can be extracted from f(T) measurements. In this context it is appropriate to note that the formalism which relates f(T) to the Debye-Waller factor of crystallographic studies, must take account of the fact that in the latter, it is the temperature dependence of the X-ray scattering from electrons which is involved, whereas in a Mossbauer experiment, it is the nuclear motion (and its termperature dependence) which is of crucial interest. In all chemical systems under conditions where the Mossbauer effect can be observed, the Born-Oppenheimer approximation appears to be valid, and thus the nuclear motion in condensed matter closely reflects the time averaged electronic motion. It is for this reason that the Debye-Waller factor is colosely identified with f.

The Debye model of solids leads[3] to an expression for f of the form

$$\ln f = \left\{ -\frac{3}{4} \frac{E_\gamma^2}{Mc^2 k_B \theta_D} \left[ 1 + 4 \left(\frac{T}{\theta_D}\right)^2 \int_0^{\theta_D/T} \frac{x dx}{e^x - 1} \right] \right\} \qquad [3]$$

in which $\theta_D$ is the Debye temperature of the solid.

Equation [3] can be expressed in a more useful form by differentiating with respect to temperature and solving [3] in the low temperature (T → 0) and high temperature (T > $\theta_D/2$) limit. In the low temperature limit

$$\ln f = - \frac{3E_\gamma^2}{Mk_B\theta_D} \qquad [4]$$

and hecomes temperature independent, with all atoms (nuclei) in their zeroeth vibrational state. Under these conditions the recoil-free fraction as given in [1] reflects the zero point motion of the atom in its (harmonic oscillator) potential energy well. For a given Mossbauer transition, the smaller the zero point motion, the larger will be the recoil free fraction, and such considerations play an important part in the choice of a matrix for the Mossbauer atom, especially in the design of sources which are usuable at experimentally convenient temperatures.

In the high temperature limit, [3] can be written in the form

$$\ln f = - \frac{3E_\gamma^2}{Mc^2 k_B \theta_D^2} T \qquad [5]$$

where all but the linear term in T of [3] have been ignored. As will be discussed below, the left side of Eq. [5] is experimentally accessible, but the right side contains two parameters, M and $\theta$, which in general are not independently known, especially in systems of chemical interest.

When a solid is comprised of individual Mossbauer active atoms not bonded by appreciable covalent forces to nearest neighbor atoms, the mass M in [5] can be replaced by the atomic mass of the Mossbauer atom, so that

$$\frac{d\ln f}{dT} = - \frac{3E_\gamma^2}{Mc^2 k_B \theta^2} \qquad [6]$$

from which the lattice temperature, $\theta$, can be extracted. Since all but the simplest solids do not exactly fulfill the Debye criterion that the solid be composed of a monoatomic isotropic array of atoms, it is preferable to characterize $\theta$ in [6] as a Mossbauer lattice temperature (rather than a Debye temperature), and in the subsequent discussion, this parameter will be represented by the symbol $\theta_M$.

In general, the determination of an absolute recoil-free

fraction in a Mossbauer experiment requires a detailed knowledge of the ratio of the Mossbauer transition gamma rays to all of the events detected by the nuclear radiation detector, the internal conversion coefficient of the gamma ray, the line width and recoil-free fraction of the source and the details of the geometry of the scattering system. Frequently such information is not readily determinable, and absolute recoil-free fractions are rarely reported in the literature. However, for the purposes of elucidating the Mossbauer lattice temperature from [6], it is only necessary to determine the temperature dependence of f. For a thin absorber for which saturation effects can be neglected, the temperature dependence of f is well approximated by the temperature dependence of the area under the resonance curve, provided that this area is calculated from data extending to at least ± 5 line widths of the resonance maximum.

The area under the resonance curve is proprotional to the product of the recoil-free fraction of source and absorber, as well as the number of absorber atoms per unit area of the sample measured over a temperature range, that is

$$A(T) = kf_s(T) \ f_a(T) \ n_a\sigma_o \qquad [7]$$

where k is a proportionality constant and $\sigma_o$ is the cross section for resonance scattering. It follows from [7] that if the source is kept at constant temperature and the same absorber is measured at various temperatures

$$d[A(T)]/dT = k' \ d \ f_a(T) \qquad [8]$$

and hence [7] can be re-written as

$$d \ \ln A(T)/dT = - \ \frac{3E_\gamma^2}{Mc^2k_B\theta_M^2} \qquad [9]$$

from which

$$\theta_M = \left[ \frac{-3E_\gamma^2}{Mck_B[_d\ln \ A(T)/dT]} \right]^{1/2} \qquad [10]$$

For $^{57}$Fe and $^{119}$Sn Mossbauer spectroscopy, using the appropriate atomic masses, this relationship results in

$$\theta_M(^{57}Fe) = -1.1659x10^1 [d\ln A/dT]^{-1/2} \qquad [11]$$

and

$$\theta_M(^{119}Sn) = -1.3381x10^1 [d\ln A/Dt]^{-1/2} \qquad [11']$$

202

As already noted, these equations assume that the appropriate mass, M, in [5] can be replaced by the appropriate atomic masses. This assumption is, however, of questionable validtiy when the Mossbauer active atom constitutes part of a covalently bonded moiety within the solid. Under such conditions the effective mass, $M_{eff}$, must be estimated from other experimental data. As will be discussed below, for many (simple) organic molecules, in which the inter-molecular bonding forces are much weaker than the intra-molecular forces, $M_{eff}$ can frequently be replaced by the molecular weight of the monomeric units making up the solid.

A more difficult problem arises in analyzing temperature dependent recoil-free fraction data for solids in which the Mossbauer atom forms part of an extended one-, two- or three-dimensional structure with covalent bonding forces to nearest neighbor atoms. Under such conditions, an effective vibrating mass can frequently be calculated from the temperature dependent second order doppler shift of the Mossbauer resonance. This shift is derived from the mean square velocity of an atom in a harmonic lattice. As shown in 1966 by Housley and Hess[4], this mean square velocity is given in the high temperature limit by

$$\langle v^2 \rangle = \frac{3k_BT}{M} \left\{ 1 + \frac{1}{20}\left(\frac{h}{k_BT}\right)^2 \omega_2{}^2 - \frac{1}{1680}\left(\frac{h}{k_BT}\right)^4 \omega_4{}^4 + \ldots \right. \quad [12]$$

where $\omega_n$ is the $n^{th}$ moment of the frequency distribution of the normal modes of the solid. In the low temperature limit

$$\langle v^2 \rangle = \frac{9h}{8M}\omega_1 \quad [13]$$

Wegener et al.[5] have shown that for a Debye Solid

$$\langle v^2 \rangle = \frac{9}{8}\frac{k\theta_D}{M} + \frac{3k_BT}{M} f\left(\frac{T}{\theta_D}\right) \quad [14]$$

where

$$f\left(\frac{T}{\theta_D}\right) = 3\left(\frac{T}{\theta_D}\right)^3 \int_o^{\theta_D/T} \frac{x^3 dx}{e^{x}-1} \quad [15]$$

At high temperatures, the difference in the isomer shift between a sample at different temperatures is given by

$$\Delta IS = \frac{-3k_B}{M_{eff}c} \Delta T \quad [16]$$

which on rearrangement yields

$$M_{eff} = \left[ -\frac{3k}{c} \frac{d\ IS}{dT} \right]^{-1} \qquad [16']$$

$$= -4.1602 \times 10^{-2} \left[ \frac{d\ IS}{dT} \right]^{-1} \qquad [16'']$$

independent of Mossbauer nuclide, so that an effective mass, $M_{eff}$ can be calculated from the temperature dependence of the isomer shift. Including this value of $M_{eff}$ in [5] yields, on rearranging

$$\theta'_M = \left[ \frac{E_\gamma}{k_B c} \frac{d\ IS/dT}{d\ln A/dT} \right]^{1/2} \qquad [17]$$

which is equivalent to

$$\theta'_M(^{57}Fe) = 4.3203 \times 10^2 \left[ \frac{d\ IS/dT}{d\ln A/dT} \right]^{1/2} \qquad [18]$$

and

$$\theta'_M(^{119}Sn) = 7.1564 \times 10^2 \left[ \frac{d\ IS/dT}{d\ln A/dT} \right]^{1/2} \qquad [18']$$

for the two common Mossbauer nuclides. It is important to recognize that these two Mossbauer lattice temperatures should be considered phenomenological lattice temperatures, with their primary utility lying in a comparison between structurally closely related samples. The use of this lattice dynamical parameter in characterizing the effect of radiation damage in solids, as well as the potential annealing of such effects, has been previously pointed out.[6]

Finally, in this context, it is worth noting that Eq. [6] can be solved for f, provided $\theta$ has been determined from other experimental data, such as neutron diffraction or low temperature heat capacity measurements, which can be used to determine the phonon spectrum of the sample in question. Solving [6] for the recoil-free fraction leads to

$$f = \exp\left[ -\frac{3E_\gamma^2 T}{k_B M c^2 \theta^2} \right] \qquad [19]$$

which is equivalent to

$$\ln f(^{57}Fe, 295 \text{ K}) = -4.0213 \times 10^4 \theta^{-2} \qquad [20]$$

and

$$\ln f(^{119}Sn, 295 \text{ K}) = -5.2839 \times 10^4 \theta^{-2} \qquad [20']$$

making the atomic mass assumptions already referred to. For other temperatures and effective masses, $M_{eff}$ (e.g.: as extracted from application of Eq. [16]) the modification of [19] is obvious.


## III. SOME APPLICATIONS TO INORGANIC SYSTEMS

### Lattice Dynamics of FeTi

The lattice dynamics of the 1:1 intermetallic compound FeTi have been studied in detail by Welter et al.[7], using inelastic neutron scattering of 4.78A neutrons over a scattering angle from 20 to 160°. The compound has the CsCl structure with a = 2.977 at room temperature, and is isoelectronic with chromium. The low temperature specific heat determined in these neutron scattering experiments was $\theta_D(0K) = 480$ K assuming a temperature independent phonon density of states. This value is significantly lower than the Debye temperature extracted from specific heat measurement[8,9], but the application of corrections for elastic constants at 0 K and the use of more realistic shear modulus values leads to a corrected value of 508 K in good agreement with the low temperature specific heat data.

The $^{57}Fe$ Mossbauer data for FeTi are summarized in Table I, from which it is seen that the lattice temperature calculation using Eq. [11] and the "bare atom" mass assumption, yields a value of 529 K in reasonable agreement with the specific heat data. More importantly, using either $\theta = 529$ K and M = 55.847 gm mol$^{-1}$ or $\theta' = 441$ K and $M_{eff} = 82.1$ gm mol$^{-1}$ leads to a calculated value of f(295 K) = 0.865 and f(78 K) = 0.963. These values can be compared to those calculated from the mean square amplitude vibration data extracted from Fig. 6 of Welter et al.[7] and the application of Eq. [1], which leads to f(295 K, neutron diff.) = 0.803 and f(78 K, neutron diff.) = 0.915. The temperature dependence of the recoil-free fraction calculated from the neutron diffraction data (assuming a linear temperature dependence of $\langle x^2 \rangle$, which is only strictly true in the range ~180 < T < 500 K) is ~ $-5.9 \times 10^{-4}$ K$^{-1}$, compared to the Mossbauer data value of $-4.9 \times 10^{-4}$ K$^{-1}$. It has been pointed out by Welter[10] that a somewhat better fit to the recoil-free fraction data is obtained by using the low temperature approximation for f(78 K) since the high temperature limit assumption that T > $\theta/2$ is clearly not valid at 78 K. The low temperature approximation is of the form

$$\ln f = - \frac{E_\gamma^2}{2Mc^2 k_B \theta} \left\{ 1.5 + \left(\frac{\pi T}{\theta}\right)^2 \right\} \qquad [21]$$

which reduces in the case of [57]Fe data to

$$\ln f(78 \text{ K}) = - \frac{23.170}{\theta} \left\{ 1.5 + \left(\frac{2.4504 \times 10^2}{\theta}\right)^2 \right\} \qquad [22]$$

and leads in the present case to f(78 K) = 0.909 in good agreement with the value extracted from the neutron diffraction results.

Clearly, the Mossbauer data for the lattice dynamical properties of FeTi are consistent with the results obtained by complementary techniques (neutron diffraction and specific heat measurements) and give some confidence that temperature dependent recoil-free fraction data can provide reliable values for such parameters for chemical systems which are less conveniently studied by other methods.

Table I.  Summary of Lattice Dynamical [57]Fe Mossbauer Data for FeTi

| | |
|---|---|
| IS(78), mm sec$^{-1}$ | $-0.020 \pm 0.006$ |
| | (wrt metallic iron at 295 K) |
| IS(T) = $0.02560 - 5.06 \times 10^{-4}$T, | $s^a$ = 0.987 (9 points) |
| lnA(T) = $0.0752 - 4.856 \times 10^{-4}$T, | $s^a$ = 0.961 (8 points)[b] |
| $M_{eff}$ = 82.1 gm mol$^{-1}$ | from Eq. [6] |
| $\theta_M$ = 529 K | from Eq. [11] |
| $\theta_M'$ = 441 K | from Eq. [18] |
| f(78) = 0.909 | from Eq. [22] |

[a] S = square of the linear correlation coefficient.

[b] The 78 K datum has been excluded because of low temperature curvature in the data.

## Lattice Dynamics of LiFeSnO$_4$

Lithium ferristannate can be prepared from Li$_2$CO$_3$, Fe$_2$O$_3$ and SnO$_2$ by the usual evacuated tube high temperature methods and the synthetic procedure can yield two different crystallographic forms, depending on the thermal conditions.[11] The high temperature form (ramsdellite) is a brown powder having orthorhombic symmetry, corresponding to space group P$_{mcn}$. The unit cell parameters are a = 3.066(1), b = 5.066(1), and c = 9.87(2) Å. Both iron and tin atoms are randomly distributed over octahedral sites, while the lithium atoms can occupy either four tetrahedral or one octahedral site. These lithium sites are located in tunnels which permit the light atoms considerable freedom of motion and presumably account for the high electrical conductivity of this material.

The X-ray powder pattern of the low temperature form of $LiFeSnO_4$ can be indexed in a hexagonal cell and is consistent with a $P6_3$ mc space group. In this structure tin is located on octahedral sites while the lithium and iron atoms are distributed on both octahedral and tetrahedral sites. The details of this structure have been discussed by Choisnet et al.[11]

Temperature dependent Mossbauer effect studies have been carried out using both $^{57}Fe$ and $^{119}Sn$ reporter atoms in both the high temperature (HT) and low temperature (LT) forms of $LiFeSnO_4$. The results are summarized numerically in Table II.

The major interest in the two forms of $LiFeSnO_4$ derives from the possibilities which exist to intercalate both of these crystallographic modifications with excess (small) cations, especially lithium, and detailed studies of the changes in the $^{57}Fe$ and $^{119}Sn$ hyperfine interactions of samples having the stoichiometry $Li_{1+x}FeSnO_4$ (0 < x < 1) have been carried out. In the context of the present discussion, however, attention will be focussed only on the lattice dynamical data derived from Mossbauer experiments on the x = 0 structures, since the consequences of the intercalation and de-intercalation of excess lithium will be discussed elsewhere.[12]

The data summarized in Table II readily show that the tin atom is in the + 4 state, with small isomer shifts (+ 0.10 and + 0.09 mm $sec^{-1}$ at 295 K for the LT and HT form, respectively. The two crystallographic structures have been discussed in detail by Choisnet et al.[11] and show that in the LT form the six nearest neighbor oxygen atoms are nearly equidistant from the metal center (2.03 ± 0.03 Å), while in the HT form the SnO distances vary from 2.00 to 2.15 Å. It is interesting to note that the room temperature quadrupole splitting of the $^{119}Sn$ resonance of the HT form (0.446 ± 0.006 mm $sec^{-1}$) is almost identical to that reported[13,14] for $SnO_2$ (0.497 ± 0.006 mm $sec^{-1}$), although the recoil-free fraction in the two compounds differs by more than a factor of 2.7 (vide infra). The iron atom isomer shift and quadrupole hyperfine interactions are characteristic of high-spin $Fe^{+3}$, and low temperature measurements[12] have shown that these metal centers order magnetically with an internal field of ~ 55 T at ~ 20° K.

The effective mass calculation based on the temperature dependence of the isomer shift, based on Eq. [16'], for both iron and tin atoms in both crystallographic modifications reveal a significant covalency in the bonding of the metal atoms to their nearest neighbors. In this context it is interesting to note that the Mossbauer lattice temperature, $\theta_M'$, calculated from Eq. [16'] is lower in the LT form as compared to the HT forms as probed by the iron reporter atom, whereas the opposite situation obtains in the case of the $^{119}Sn$ resonance.

Finally, in terms of the lattice dynamical data which can be extracted from the spectroscopic information, it is informative to consider the recoil-free fractions for the two reporter atoms in $LiFeSnO_4$. These values are essentially identical for Fe in the two crystallographic forms, and for Sn in the two crystallographic forms, but differ from each other in what appears to be a consistent manner. These results suggest that, even correcting for the differences in the effective masses of the two Mossbauer active

Table II. $^{57}Fe$ and $^{119}Sn$ Mossbauer Data for Two Crystallographic Forms of $LiFeSnO_4$

| Resonance | LT | | HT | |
|---|---|---|---|---|
| | Fe | Sn | Fe | Sn |
| IS(78) mm sec$^{-1}$ | 0.443±0.005 | 0.130±0.020 | 0.476±0.010 | 0.129±0.020 |
| dIS/dT mm sec$^{-1}$ K$^{-1}$x10$^4$ | 5.26 | 2.40 | 5.01 | 2.25 |
| M$_{eff}$ gm mol$^{-1}$ | 79.2 | 173 | 83.2 | 185 |
| QS(295) mm sec$^{-1}$ | 0.756±0.005 | 0.639±0.020 | 0.433±0.010 | 0.446±0.006 |
| dlnA/dT K$^{-1}$x10$^3$ | 0.786 | 1.292 | 0.610 | 1.400 |
| $\theta_M$, K | 416 | 371 | 472 | 357 |
| $\theta_M'$, K | 353 | 308 | 392 | 286 |
| $\Gamma$ mm sec$^{-1}$ | ~0.5 | 0.99±0.07 | ~0.4 | 0.84±0.04 |
| f(295 K) ± 0.05 | 0.85 | 0.77 | 0.88 | 0.77 |

atoms, the phonon spectrum sampled at the iron site is somewhat different than that sampled at the tin site. The differences in f follow qualitatively the differences in the two Mossbauer temperatures and are consistent with a stronger binding of the iron atoms in the structure than of the tin atoms. The recoil-free fraction for the $^{119}$Sn resonance is especially large at room temperature, exceeding not only the value in $SnO_2$[0.28 ± 0.03] but also that in the standard $^{119}$Sn source matrix, $BaSnO_3$[0.58 to 0.70][14,15] as will be discussed more fully, below.

## Lattice Dynamics of BaSnO$_3$

Because barium stannate is widely used as a source matrix for $^{119}$Sn Mossbauer effect spectroscopy, a number of studies have been reported with the object of determing the room temperature recoil-free fraction. Hembree and Price[15], using both SnNi and $SnO_2$ standard absorbers, determined values of 0.66 and 0.70 in modest agreement with the value of 0.58±0.02 reported earlier by Sano and Herber.[16] Sitek[17] who used a value for the resonance scattering cross section of $1.402 \times 10^{-18}$ cm$^2$ found a value of 0.52±0.02, similar to values of 0.51 and 0.49±0.02 reported by Hannaford and Wignall[18] in a comparison study of several matrices. In a later investigation, Hannaford and Horn[19], using selective modulation of the recoilless gamma rays, found f(295 K) = 0.57±0.02. A measurement by Puri[20], making use of Li drifted silicon detectors and a resonance cross section of $(1.31 \pm 0.01) \times 10^{-18}$ cm$^2$ found f = 0.65±0.01, but this value is based on resonance effect intensity rather than area measurements.

Temperature dependent Mossbauer experiments have been carried out in the author's laboratory, using a sample of BaSnO$_3$ obtained[21] by the controlled thermal decomposition of barium tin oxalate. The results extracted from experiments covering the temperature range 82 < T < 295 K are summarized in Table III and yield a recoil-free fraction value in good agreement with the results of the area determination reported previously.[16] Comparable data for CaSnO$_3$ are included in Table III for comparison.

Precision measurements have shown that the isomer shifts for BaSnO$_3$ and $SnO_2$ at 295 K are identical within an estimated experimental error of ±0.005 mm sec$^{-1}$. From the second order Doppler shift data cited in Table this experimental error is comparable to a temperature difference of about ‡ 20 K. Thus, the practice of reporting isomer shifts based on the centroid of the (wide) $SnO_2$ resonance spectrum interchangably with that based on a zero relative velocity experiment in which both source and absorber are identical (BaSnO$_3$), appears fully justified, provided that the experimental temperature is kept close to 295 K.

Table III. Summary of $^{119}$Sn Mossbauer Data for $BaSnO_3$ and $CaSnO_3$

| | | |
|---|---|---|
| IS(78) mm sec$^{-1}$ | 0.048±0.006 | 0.059±0.006 |
| IS(295) | ≡0 | 0.009±0.006 |
| dIS/dT mm sec$^{-1}$ K$^{-1}$ | 2.254x10$^{-4}$ | 2.22x10$^{-4}$ |
| $M_{eff}$ gm mol$^{-1}$ | 182 | 187 |
| dlnf/dT K$^{-1}$ | 1.814x10$^{-3}$ | 1.251x10$^{-3}$ |
| $\theta_M$, K | 314 | 378 |
| $\theta_M'$, K | 252 | 301 |
| f(295) | 0.58 | 0.69 |

## IV. LATTICE DYNAMICS OF ORGANOMETALLICS

In contrast to the extraction of lattice dynamical information from temperature dependent Mossbauer studies on inorganic and metallic (alloy) systems discussed above, the application of such studies to organometallic compounds has proven — on the whole — to be less informative. The reasons for this are to be found in the fact that (a) the temperature dependence of the recoil-free fraction is frequently very large, so that only a limited temperature range is accessible to the investigator; and (b) the temperature dependence of the isomer shift is small, and frequently not accurately determinable.

Nonetheless, studies of the temperature dependence of the recoil-free fraction have been related to details of the bonding of organometallic solids[22], especially to the formation of dimers and oligomers[23], and to the distinction between monomeric units and one-, two-, and three-dimensional polymers.[24, 25]

Because in non-polymeric organometallics the strength of the bonding between monomeric units is much weaker than the intra-molecular bonding forces, the appropriate effective mass in Eq. [5] or [6] can frequently be replaced by the molecular weight of the monomeric unit. In effect, such an assumption is equivalent to

210

substituting the atomic mass of the Mossbauer active atom by a point mass having the numerical value of the formula weight. The lattice dynamical equivalence of this procedure is to envision the solid as being composed of point masses oscillating about their equilibrium position without coupling of the acoustic (low frequency) modes with the optical (high frequency) modes characteristic of the <u>intra</u> molecular motions. It has been observed that many organometallic compounds, especially those of iron and tin being discussed in the present context, form solids in which the unit cell contains either two or four monomeric units. Such a unit cell will involve an <u>intra</u> unit cell vibration which can be probed by appropriate spectroscopic techniques which are sensitive to lattice modes in the solid, principally Raman spectroscopy.[26]

As has been shown in a number of detailed studies[27-30], it is frequently possible to identify a Raman active mode at low frequencies which can be correlated with the <u>inter</u>-molecular <u>intra</u>-unit cell vibration referred to above. If such a mode dominates the low frequency vibrational spectrum of the solid, its frequency can be related to the lattice temperature by the Debye relation

$$\theta_D = \frac{h\omega}{k_B} \qquad [23]$$

Replacing $\theta$ in [9] by this frequency leads to

$$dlnA/dT = -\left[ \frac{3E_\gamma^2 k_B}{Mc^2} \frac{}{h^2\omega^2} \right] \qquad [24]$$

which, on rearrangement leads to

$$M_{eff}\ \omega^2 = -3.7511 \times 10^3\ [dlnA/dT]^{-1} \qquad [25]$$

for $^{57}$Fe reporter atoms
and

$$M_{eff}\ \omega^2 = -1.0238 \times 10^4\ [dlnA/dT]^{-1} \qquad [25']$$

for $^{119}$Sn Mossbauer studies.

Several tests of this equation have been reported in the literature and studies on structurally and functionally similar compounds appear to fit the model well enough so that scattering maxima in the lattice mode region Raman spectra can be uniquely identified with the <u>inter</u>-molecular, <u>intra</u>-unit cell vibration. Rein and Herber[31] in a study of 11 organotin compounds were able to show that a plot of $M\omega^2$ <u>versus</u> $(dlnA/dT)^{-1}$ could be fitted to a linear regression passing through the origin, with a slope of $9.99 \times 10^3$ in reasonable agreement with the expected value of $1.024 \times 10^4$ (equal to $-3E_\gamma^2$ $kc^{-2}$ $h^{-2}$ from Eq. [24]).

Since the central assumption of the above model is that inter molecular vibrations are only mass (and not structure) dependent, it is informative to compare pairs of molecular compounds with equal (or nearly equal) molecular weights, but different chemical identities. Such a comparison is effected in Table IV. The major difference in the temperature dependence of the recoil-free fraction is noted for the triphenyl tin chloride- triphenyl tin cyanide pair in which this factor differs by ~23% although the formula weights differ by no more than 2.5%. This observation is presumably related to the presence of the polar halogen substituents, since it has been repeatedly noted[32] that the large quadrupole hyperfine interaction in $(C_6H_5)_{4-x}SnCl_x$ compounds (X = 1 to 3) is sensitive to the condensed state of the sample. In general, the QS interaction of the neat solid and that observed for a frozen solution sample in a glass forming solvent, at the same temperature, differ significantly. This difference must originate in the differences in the bonding forces between nearest neighbor molecules in the two matrices (neat solid and frozen solution) and the presence of these interactions weakens the validity of the basic assumptions of the effective vibrating mass model described above. Notwithstanding this <u>caveat</u>, it is clear that these results suggest that a good deal of information, especially that relating to the strength of intermolecular forces in covalent solids, can be extracted from temperature dependent Mossbauer effect data.

An additional parameter, related to the vibrational anisotropy in covalent compounds - the Gol'danskii-Karyagin effect - can also be extracted from temperature dependent Mossbauer spectra, and this parameter has proven to be of considerable utility in structure and bonding elucidation in organometallics.[33-36] A detailed discussion of this effect is beyond the scope of the present review, and the reader is referred to the extensive literature for a more complete treatment of such studies.


V.   SUMMARY AND ACKNOWLEDGMENTS

A fundamental assumption of our understanding of the solid state is that the motion of atoms and molecules as a result of thermal excitation reflects the nature of the forces which bind

Table IV. $^{119}$Sn Mossbauer Data for Pairs of Organotin Compounds Having (nearly) Identical Formula Weights.

| Compound | IS(78) mm sec$^{-1}$ | QS(78) mm sec$^{-1}$ | $-d\ln A/dT$ K$^{-1}$, x10$^2$ |
|---|---|---|---|
| $(C_6H_5)_3Sn(C_5H_4N)$ | 1.28 | - | 1.66 |
| $(C_6H_5)_4Sn$ | 1.300 | - | 1.63 |
| $(\underline{m}CF_3C_6H_4)_4Sn$ | 1.283 | - | 2.12 |
| $(p\ CF_3C_6H_4)_4Sn$ | 1.285 | - | 2.20 |
| $(CH_3)_3Sn(C_6H_5)$ | 1.238 | - | 3.11 |
| $(CH_3)_3Sn(C_5H_4N)$ | 1.294 | 0.614 | 3.40 |
| $(C_6H_5)_3SnCl$ | 1.355 | 2.558 | 2.17 |
| $(C_6H_5)_3SnCN$ | 1.316 | 3.182 | 1.73 |

atoms and molecules to each other. Some of the details of this motional behavior can be elucidated most readily by the application of nuclear gamma ray resonance techniques to the study of condensed matter systems.

In the present discussion, an attempt has been made to summarize some of the important aspects of the temperature dependence of Mossbauer parameters and to develop a set of relationships which relate these thermal factors to the properties of solids. A number of applications to chemical systems have been used to illustrate the kind of information which can be obtained. Although the present discussion has been limited to studies involving only two nuclides $^{57}$Fe and $^{119}$Sn, extension to other Mossbauer active atoms is completely straightforward.

It should be noted that there are a number of other parameters related to the motional behavior of atoms in solids, which have not been touched on in the present discussion. Thus, for example, no detailed mention has been made of the vibrational anisotropy in solids (the Gol'danskii-Kanyagrin effect) which can be extracted from the temperature dependence of the intensity ratio of a quadrupole doublet. Nor has the large amount of information which can be obtained relative to the presence of defects in solids - such as those, for example, which result from radiation damage, ion implantation, chemical effects of nuclear transformations, optical irradiation effects, etc. - been touched on in the present review. The wealth of data which has been accumulated in these areas, and the interpretation of such results, is beyond the scope of the present discussion, despite the close relationship between the study of the details of such phenomena and the lattice dynamical behavior of condensed matter systems.

The author is greatly indebted to many of his colleagues and former students for making available some of the data described in this chapter, as well as for numerous fruitful discussions and helpful comments. In particular the assistance of Dr. H. Eckert and Mr. J.E. Phillips with some of the measurements herein reported is gratefully acknowledged. The sample of FeTi, as well as an insightful analysis of the data, was provided by Dr. J.M. Welter. The several modifications of $LiFeSnO_4$ were provided by E. Wong and Prof. M. Greenblatt; the sample of $BaSnO_3$ by Dr. P.K. Gallagher and a number of the organotin compounds by Prof. Baldwin King.

Some of the research herein described has been supported by the National Science Foundation under grant DMR-S102940 and by a grant from the Center for Computer and Information Services of Rutgers University. This support is herewith ratefully acknowledged.

REFERENCES

1. R.L. Mossbauer, Z. Physik $\underline{151}$, 124 (1958).
2. See for example N.N. Greenwood and T.C. Gibb in "Mossbauer Spectroscopy", Chapman Hall, London, 1971; "Applications of Mossbauer Spectroscopy", V.I. Gol'danskii and R.H. Herber, Eds. in Chemical Academic Press, New York 1968; and references therein.
3. See for example W.M. Visscher Annals of Physics $\underline{1960}$, $\underline{9}$, 194; H. Frauenfelder, "The Mossbauer Effect", W.A. Benjamin Co., New York, 1962; G.K. Wertheim, "The Mossbauer Effect: Principles and Applications", Academic Press, new York 1964.
4. R.M. Housley and F. Hess, Phys. Rev. $\underline{1977}$, $\underline{146}$, 517.
5. H. Wegener "Der Mossbauer Effekt und Seine Anwendung in der Physik Und Chemie", Bibliogr. Inst. Mannheim, 1966.

6.  R.H. Herber and R. Kalish, Phys. Rev. 1977, B16, 1789; T.K. McGuire and R.H. Herber, "Proc. Symp. on Nuclear and Electron Resonance Spectroscopies" G. Kaufman and G. Shenoy Eds. North Holland Pub. Co., Amsterdam 1981; R.H. Herber and R. Kalish, J. Phys. Chem. Solids, 1984, 45, 89.

7.  U. Buchenau, H.R. Schober, J.-M. Welter, G. Arnold and R. Wagner, Physical Review 1983, B27, 955.

8.  E.A. Starke, Jr., C.H. Cheng and P.A. Beck, Physical Review 1962, 126, 1746.

9.  K. Ikeda, T. Nakamichi and M. Yamamoto, J. Phys. Soc. Japan 1974, 37, 652.

10. J.M. Welter, private communication.

11. J. Choisnet, M. Hervieu, B. Raeau and P. Tarte, J. Solid State Chem. 1981, 40, 344.

12. E. Wong, M. Greenblatt, H. Eckert and R.H. Herber, to be published; see Abstracts for the September 1984 Meeting, American Chemical Society, Philadelphia PA.

13. K.P. Mitrofanov, M.V. Plotnikova and V.S. Shpinel, J. Exp. Theoret. Physics (USSR) 1965, 48, 791.

14. H.A. Stockler, H. Sano and R.H. Herber, J. Chem. Phys. 1966, 45, 1182, see also Ref. 15.

15. G. Hembree and D.C. Price, Nuclear Instruments and Methods 1973, 108, 99.

16. H. Sano and R.H. Herber, J. Inorg. Nuclear Chem. 1968, 30, 409.

17. J. Sitek, Nuclear Instrum. Methods 1974, 114, 163.

18. P. Hannaford and J.W.G. Wignall, Phys. Stat. Solidi 1969, 35, 809.

19. P. Hannaford and R.G. Horn, J. Phys. C.: Solid State Physics, 1973, 6, 2223.

20. R.K. Puri, Nuclear Instrum. Methods 1974, 117, 381.

21. P.K. Gallagher and D.W. Johnson, Jr., Thermochim. Acta 4, 283 (1972); P.K. Gallagher and F. Schrey, Therm. Anal. Proc. Int. Conf., 3rd, 2, 623 (1972) (C.A. 78, 131560r); G. von Wagner and H. Binder, Z. Anorg. Allg. Chem. 297, 334 (1958).

22. R.H. Herber, J. Fischer and Y. Hazony, J. Chem. Phys. 1973, 58, 5185 and references therein; R.H. Herber, M.F. Leahy and Y. Hazony, J. Chem. Phys. 1974, 60, 5070; Y. Hazony and R.H. Herber, Journal de Physique 1974, 66, 131. R.D. Ernst, D.R. Wilson and R.H. Herber, J. Amer. Chem. Soc. 106, 1646 (1984).

23. R.H. Herber and M.F. Leahy in "Organotin Compounds: New Chemistry and Applications" J.J. Zuckerman, Ed., Advances in Chemistry Series No. 157, Amer. Chem. Soc. 1976.

24. K.C. Molloy, M.P. Bigwood, R.H. Herber and J.J. Zuckerman, Inorg. Chem. 1982, 21, 3709.

25. R. Barbieri, L. Pellerito, A. Silvestri, G. Ruisi and J.G. Noltes, J. Organomet Chem. 1981, 210, 43 and references therein.

26. R.H. Herber, Journal de Physique 1979, C2, 386 and references therein.
27. R.H. Herber and J. Fischer, Can. J. Spectroscopy, 1974, 19, 21.
28. R.H. Herber and M.F. Leahy, J. Chem. Phys. 1977, 67, 2718.
29. Y. Hazony and R.H. Herber, in "Mossbauer Effect Methodology", Vol. 8., I. Gruverman, Ed., Plenum Press, New York, 1973.
30. A.J. Rein and R.H. Herber in Proc. 5th Int. Conf. Raman Spectroscopy, E.D. Schmidt et al, Editors, H.F. Schulz Verlag, Freiburg (i. Br.) West Germany, 1976.
31. A.J. Rein and R.H. Herber, J. Chem. Phys. 63, 1021 (1975).
32. R.H. Herber, J. Inorg. Nuclear Chem. 1973, 35, 67.
33. V.I. Gol'danskii, E.F. Makarov and V.V. Khrapov, Physics Letters 3, 344 (1963); S.V. Karyagin, Soviet Physics (Solid State) 5, 1552 (1964); V.A. Bryukhanov, V.I. Gol'danskii, N.N. Delyagin, L.A. Korytko, E.F. Makarov, I.P. Susdalev and V.S. Shpinel, Zh. Eksperim. Teor. Fiz. 43, 448 (1963).
34. R.B. King, R.H. Herber and G.K. Wertheim, Inorg. Chem. 3, 101 (1964).
35. R.H. Herber, S.C. Chandra and Y. Hazony, J. Chem. Phys. 1970, 53, 3330.
36. R.H. Herber and S.C. Chandra, J. Chem. Phys. 1971, 54, 1847.

# MÖSSBAUER SPECTROSCOPY OF SOILS AND SEDIMENTS

Lawrence H. Bowen and Sterling B. Weed

Departments of Chemistry and Soil Science
North Carolina State University
Raleigh, N. C. 27695

## INTRODUCTION

Soils are the most wide-spread form of iron-containing material on the land portion of the earth's surface. Together with aquatic sediments they provide a varied and important group of materials for which Mössbauer spectroscopy is an appropriate technique of study. They have been, however, less studied by this technique than many more-specialized geological samples. Part of this is certainly due to the complicated iron mineralogy of soils. Another important reason is the requirement of scientific expertise in two diverse areas in order to have a reasonable expectation of obtaining significant results. This review will concentrate on papers of the last five years, during which time there has been significant increase in both quantity and quality of reports in this applied field of Mössbauer spectroscopy. As far as we are aware, all papers dealing with this subject refer to the $^{57}$Fe resonance.

In applying Mössbauer spectroscopy to such complex mixtures as soils and sediments, much reliance must be placed on spectra of simpler, well-characterized minerals. Thus a portion of this chapter is devoted to reviewing the results on synthetic samples which relate to soil materials. These include especially studies on aluminum-substituted iron oxides and oxyhydroxides. Also important in characterizing soils and sediments are results from the much more extensive literature on the layered alumino-silicate minerals. These are common components of soils and are found in varying degrees of weathering in especially the clay fraction (<2 μm) of soils.

Our own interest has centered on weathered, well-oxidized tropical and temperate soil clays. The major iron-containing components of such clays are not layer silicates, but aluminum-substituted goethite ($\alpha$-FeOOH) and hematite ($\alpha$-Fe$_2$O$_3$). A review of the literature on both synthetic and naturally-occurring iron oxides and oxyhydroxides was made by one of us in 1979.[1] We have concentrated in this chapter on papers appearing since that review, but have attempted to provide a reasonably complete survey of reported results on soils, soil clays, and sediments from this period. As discussed above, we also survey papers dealing with synthetic samples which are in our opinion pertinent to the subject.

REVIEWS AND GENERAL PAPERS

Goodman[2] has written an extensive chapter on the use of Mössbauer spectroscopy in soils and clay minerals research, which includes a basic description of the experimental technique and theory. He has also presented a review[3] of some of the important applications to clay minerals and hydroxides/-oxides of iron. Coey[4] has surveyed the field of clay minerals and their transformations up to ~1979. One major section of this review is devoted to studies of weathering, soils, and sediments. For an extensive survey of the Mössbauer literature on layer silicates, the review of Heller-Kallai and Rozenson[5] is recommended. Although this review does not discuss soils as such, it is certainly a good reference for finding what is known about the Mössbauer spectra of specific layer silicates. It also provides a discussion of the correlation between structure and Mössbauer parameters for these systems.

In a more specialized review, Dolnicar[6] has compiled results of a coordinated IAEA sponsored research program to apply Mössbauer spectroscopy in mineralogy, soil science, and ceramics over the period 1977-1980. The advantages of this technique for developing nations are emphasized - in particular, low cost, rapid new results, and the geographic specificity and economic importance of the applications. Tominaga[7] has discussed some environmental and geochemical applications including amorphous minerals and sediments, as well as air particulates and archaeology. Among other examples, several studies of alteration and weathering of minerals are presented.

SYNTHETIC AND NATURAL SINGLE PHASES

Goethite, $\alpha$-FeOOH

The mineral goethite is well-characterized in the Mössbauer literature. It is antiferromagnetically ordered below its Néel temperature of about 400K, has an extrapolated

218

internal field of about 504 kOe at low temperature, a field
gradient perpendicular to the magnetic axis which gives a
quadrupole interaction $\Delta = (e^2qQ/4)(3\cos^2\theta-1)$ of about $-0.3$
mm/s, and isomer shift typical of high-spin, octahedral $Fe^{3+}$.
However, both particle size effects and aluminum substitution
can cause collapse of the sextet to a doublet at RT and even
below. The earlier work is reviewed by Bowen[1]. There have
been a number of recent studies on substituted goethites relevant to interpretation of soils. The paper by Golden et al.
from our laboratory in 1978[8] was not the first to discuss the
effect of aluminum substitution, but was the most complete
study at that time, and included both synthetic and soil clay
samples. It was shown that for synthetic samples the RT spectrum collapsed to a doublet between 9 and 13 mole percent
aluminum, with strong evidence of superparamagnetic line broadening. The 77K spectra remained magnetically ordered to 25%
Al, although considerable line broadening was observed. Correlation equations were given for these spectra between the
average magnetic field and aluminum content, and because of the
decrease in particle size with Al substitution, with both Al
and surface area. Table 1 gives these equations, as well as
subsequent correlations developed by other groups. The substitution predicted from the magnetic fields was compared to
chemical analysis by citrate-dithionite extraction for a series
of soil fine clays ($<0.2\ \mu m$). The comparisons gave good agreement, even when hematite was present, except when chemical
analysis gave results above 35% Al. Mössbauer values were
always lower in these cases, presumably due to extraction of
other forms of Al in the chemical treatment.

Table 1.  Relation between magnetic field H (kOe) and
          aluminum substitution c (mole %) for goethite.

| T | Equation* | Reference |
|---|---|---|
| 125K | $H = 485.3 - 1.40\ c$ | 15 |
| 77K | $H = 500 - 1.77\ c$ | 8 |
| | $H = 498 - 1.36\ c - 0.11\ S$ | 8, 11 |
| 4K | $H = 505 - 0.52\ c$ | 13 |
| | $H = 505.4 - 0.33\ c - 0.036\ S$ | 15 |
| | $H = 506.5 - 0.42\ c - 87/D$ | 15 |

* S is surface area in $m^2/g$, D is crystal diameter
  (in nm) determined from the (111) X-ray line width.

Considering subsequent papers in approximate chronology, Murad[9] examined natural goethites of very low aluminum content and found considerable line-broadening asymmetry in the RT sextets. These deviations were ascribed to the presence of small, but significant amounts of silicon in the samples, which would disrupt the magnetic structure. Fleisch et al.[10] emphasized a different approach to the determination of aluminum content in synthetic goethites. From a temperature variation of the minimum count rate at zero velocity, an approximate temperature for the collapse of magnetic order can be obtained. They showed that a 21 mole percent sample studied in detail obeyed the Brillouin function, so that the observed collapse temperature should be related to the Néel temperature and thus to aluminum content. The observed decrease is approximately linear up to about 15% Al with a slope of about -7 K per percent Al. It should be noted that the observed decrease is much greater than predicted by a simple model and the slope apparently becomes less negative at higher aluminum content. Goodman and Lewis[11] prepared synthetic goethites with aluminum substitution up to ~30%. Varied particle sizes were obtained also, using different preparation methods. They compared their results on magnetic field at 77K with both aluminum substitution and surface area, and found general agreement with the equation of Golden et al.[8] (Table 1), but noted that the sextets were asymmetric and thus it is incorrect to fit such spectra to a single sextet. In some cases appreciable doublet was present even at 77K. They pointed out the difficulty of distinguishing spectra of substituted goethite from akaganéite ($\beta$-FeOOH).

Johnston and Norrish[12] examined nine goethites of natural origin, using RT Mössbauer spectroscopy as well as other techniques. They report a wide scatter in the RT magnetic field variation with either particle size (from X-ray line broadening) or aluminum content, and indicate that caution should be applied in interpreting such spectra as due to a particular cause. Fysh and Clark[13] report RT, 77K, and 4K spectra of aluminous goethites prepared by both hydrothermal and low-temperature synthesis. As observed by Golden[8], they report an almost-complete collapse of the RT spectra to a doublet by 15% substitution, although vestige of magnetic relaxation is seen in the material with larger particle size at 15% Al. They indicate that little is gained from fitting broad, asymmetric sextets to a distribution of magnetic fields, and do not report detailed results for the higher temperatures. As shown below, we feel such empirical fits to a distribution of fields have much merit. The 4K spectra in the above paper were well characterized by one-sextet fits and the equation for H vs. Al is given in Table 1. In addition, this paper reports recoilless fraction measurements which increase from f = 0.69 for pure goethite to 0.89 for 19% Al at 4K. At RT f = 0.65 for pure goethite according to these authors, but decreases to 0.50 for 19% Al.

Murad[14] illustrates the usefulness of fitting RT Mössbauer spectra of goethite to a distribution of fields. He points out that the hyperfine field of maximum probability and the upper and lower (generally broader) half-widths of such fitted distributions are useful in characterizing the crystallinity and stoichiometry of the goethite. Murad and Schwertmann[15] have provided a detailed study of the relation between both aluminum substitution and crystallinity and the Mössbauer spectra of goethite at 4K. In contrast to Fysh and Clark,[13] they find a distinct effect of particle size even at 4K (Table 1). They point out that because two variables affect the measured magnetic field, even at very low temperature, Mössbauer results cannot give unambiguous determination of either variable without additional information.

Both RT and 77K Mössbauer spectra of Yugoslav bauxites are reported by Music et al.[16] Both goethite and hematite were found. The RT spectra consisted of doublet plus sextet (the sextet being hematite) and the 78K spectra two sextets plus doublet. These authors did not attempt to relate the observed low field for goethite at 78K to aluminum content, but did indicate that small particle size affects the spectra. Fysh and Clark[17] have used Mössbauer techniques to study the acid leaching of bauxite. Measurements at both RT and 4K were taken. Prior to leaching these samples contained mostly Al-substituted hematite. The acid leaching increased the goethite component, and no evidence of akaganéite formation was observed, presumably due to the nucleating effect of goethite present in the bauxites.

## Hematite, $\alpha$-$Fe_2O_3$

Another common component of soils is hematite, which is generally responsible for the red color of many clays. As with goethite, this mineral has been well-characterized by Mössbauer spectroscopy in its pure state.[1] Hematite has a much higher Néel temperature than goethite, about 956K, and at low temperature is antiferromagnetic also, with H(OK) ~ 544 kOe. The field gradient axis in this state is parallel with the magnetic axis, giving a positive $\Delta$ ~ 0.4 mm/s. At the Morin transition, about 260K, the spins flip by ~90°, producing a negative $\Delta$ ~ -0.2 mm/s and lowering the magnetic field by about 8 kOe. This state is weakly ferromagnetic. Both small particle size and high aluminum content will lower the temperature of the Morin transition, and in soil samples no evidence of the low-temperature antiferromagnetic state has been reported. There continue to be reports in the literature of studies of the Morin transition under extreme conditions, such as high pressure. We will consider only a few studies on this transition concerning the effect of aluminum substitution.

Jonas et al.[18] report a RT Mössbauer study of synthetic aluminum-substituted hematites. As for goethite at 77K, the field for hematite decreases linearly with Al content up to about 15% Al (Table 2). In the notes to this table, preparation temperatures are listed, as the properties of the hematites appear to depend on preparation method. Jonas et al. observed no decrease in X-ray lattice spacing above about 15% Al, but the observed magnetic field continued to decrease up to

Table 2. Relation between magnetic field H(kOe) and aluminum substitution c (mole %) for hematite (WF phase).

| T | Equation | Reference |
|---|---|---|
| 295K | $H = 509 - 1.4 \, c$ | 18[*] |
| | $H = 513 - 1.3 \, c$ | 19[+#] |
| | $H = 517.9 - 0.82 \, c$ | 21[**] |
| | $H = 515.9 - 0.86 \, c$ | 21[***] |
| 77K | $H = 532 - 0.39 \, c$ | 19[#] |
| 4K | $H = 536 - 0.42 \, c$ | 21[***] |

   * estimated from graph; hematites prepared from goethite by firing at 550°C.
   + approximation valid for c < 15.
   # hematites prepared from goethite by firing at 500°C.
  ** hematites fired at 950°C.
*** hematites prepared from goethite by firing at 590°C.

the experimental limit of somewhat greater than 20% Al. In 1982 appeared three independent articles on aluminous hematites.[19] Considering first the report from our group, both RT and 77K spectra are discussed. Unlike goethite, all RT spectra are sextets even up to the limit of ~32% Al. As discussed in this article, the relatively low preparation temperature (500° C) allows the formation of highly-substituted hematites. Higher firing temperatures of 900°C improve crystallinity but also expel aluminum from the hematite lattice, leaving a maximum substitution of ~15% Al. The asymmetry of the RT spectra was ascribed to the substitution rather than particle size. Both RT and 77K spectra were fit by a molecular field model assuming random substitution and a super-transferred field of about 0.39 kOe per percent Al. The variation of magnetic field with Al was linear at 77K, but had marked curvature at RT, particularly for higher Al substitutions. An estimate is made in Table 2 for the linear portion of the RT curve, and the reported 77K line is given. Considerably higher Al contents were observed here than in previous reports. Using a much higher firing temperature, Tomov et al.[20] found solubilities of $Al_2O_3$ in $Fe_2O_3$ only up to 10% Al for mixtures of the two oxides. In addition to the higher temperature, the starting materials were different, previous workers having used already-substituted goethites. No general relation between H and aluminum content was given, although the observed field did decrease with increased Al. Fysh and Clark have reported results on aluminous hematites prepared by three methods.[21] The high temperature preparation, at 950°C, had a solubility limit of ~15% Al, whereas 590°C produced up to 19% substitution. In Table 2 we quote their equations for variation of H with Al. These differ somewhat from the equations of other groups reported at RT. However, it should be noted that the 590°C preparations gave RT spectra which were arbitrarily fit to two sextets and the equation quoted in Table 2 refers to the high-field sextet, which should have a higher intercept and smaller slope than the average field from a one-sextet fit. The quadrupole interaction at RT varied with Al for the 950°C series whereas it was constant at about $\Delta = -0.22$ mm/s for the 590°C series. At 4K even for the 950°C series, about 15% of the non-substituted sample was above the Morin transition. This series showed no effect of Al on the magnetic field at 4K, whereas the 590°C series had a linear decrease (Table 2). Reported recoilless fractions for hematite at 4K are f = 0.70 (unsubstituted) and 0.82 (14% Al).

Studies of the Morin transition in aluminous hematite have been reported by DeGrave et al. In a short paper[22] it was shown that the fraction of the weakly-ferromagnetic (WF) state, which is above the Morin transition, varies at 80K from 90% for 7.7% Al to 29% for 2.9% Al. In addition, the transition temperature region, difference in magnetic field between the two states and quadrupole interaction vary with Al. The spectra, however, consist of two distinct sextets for the two spin phases. A more detailed study of the sample with 4.8% Al[23] has been reported by the same group. In this case the dependence of $\Delta$ and H for the two phases with temperature show rather complex spin reorientations occurring during the transition, maintaining however two distinct phases. If a Morin transition were observed in a soil hematite it would certainly indicate an unusually well-crystallized, low-substituted species.

## Akaganéite, β-FeOOH

This crystalline form of oxyhydroxide has large holes or channels in the structure and is prepared synthetically by hydrolysis of $FeCl_3$ solutions. The chloride ion is found in the crystal and is apparently required to stabilize the channel structure. There has been controversy in the literature over the occurrence of β-FeOOH in soil (see Bowen[1]) and over the location of halide in the structure. Several studies of synthetic materials have clarified the Mössbauer spectra of β-FeOOH. Chambaere et al.[24] report the RT spectrum to consist of two distinct doublets with quadrupole splittings of about 0.55 and 0.95 mm/s. Both $Cl^-$ and $F^-$ in the lattice appeared to affect the relative proportion of these doublets, the one with Q.S. = 0.55 mm/s increasing in intensity with increasing halogen. Johnston and Logan[25] also report two distinct doublets with about the same values of Q.S., although the intensity ratio of the doublets did not show a monotonic change with chloride content. These authors present a 77K magnetically-split spectrum which clearly shows more than one sextet, but was not analyzed. Murad[26] has measured Mössbauer spectra down to 4K. At 4K at least three sextets were required to fit the spectrum, with magnetic fields of 473, 479, and 486 kOe and widely-varying Δ. Murad points out that higher temperatures (between 80K and RT) are more useful for distinguishing β-FeOOH from goethite in natural samples because of the greater separation in magnetic fields. Ohyabu and Ujihira[27] report RT Mössbauer spectra of a series of chloride and fluoride substituted samples. They found a different dependence of the doublet intensity ratio with halide than reported by Chambaere et al.[24] and suggested that $F^-$ could occupy either of two open sites in the structure whereas $Cl^-$ was restricted to one. Childs et al.[28] studied a series of synthetic akaganéites containing chloride and fluoride. At 77K a three-site model was required for the magnetic spectra. In one case a 4K spectrum also required three components. These authors propose that two distinct octahedra, $FeO_3(OH)_3$ and $FeO_2(OH)_4$ are responsible for the observed spectra, with only secondary effects from the halide. In two recent papers, Chambaere and De Grave report their detailed Mössbauer studies of synthetic akaganéites. The first paper concerns the Néel temperature[29] and its variations. From measurements with external field it is shown that superparamagnetic relaxation is absent and the transition is a true magnetic ordering. From studies with varying hydrolysis time, it is shown that the variation of Néel temperature is primarily a function of interstitial water ($T_N$ = 303.3 − 23.5 $X_{H_2O}$), and does not depend on halide content. This variation is interpreted as due to a decrease in average spin of $Fe^{3+}$. In the second paper,[30] these authors show that the three sextets required for fitting low temperature spectra are from only two primary sites for iron, with the third sextet being due to a thermally induced magnetic state. The two sites are presumed to be, in agreement with Childs et al.,[28] $FeO_3(OH)_3$ and $FeO_2(OH)_4$. They point out that these sites exhibit distributions of quadrupole interaction and magnetic field.

## Lepidocrocite, $\gamma$-FeOOH

This mineral is the only well-crystallized iron oxyhydroxide which has a magnetic ordering temperature below liquid nitrogen temperature. At 4K it has a distinctly lower magnetic field (about 460 kOe) than goethite and also smaller quadrupole interaction, although at RT it gives a doublet essentially identical to goethite of small particle size. Considering the effect of particle size on the Mössbauer spectra of goethite and hematite, it might be expected that lepidocrocite also show effects of crystallinity. A very recent paper by Murad and Schwertmann[31] addresses this subject. Two synthetic lepidocrocites of markedly different crystallinity were studied. The more crystalline sample showed onset of magnetic order at ~77K with the doublet, however, still present at 60K with reduced intensity. The less crystalline sample, in contrast, had only a doublet at 50K. Both samples had magnetic spectra at 4K with maximum absorption for a field of about 460 kOe. However, a field distribution was necessary to properly describe the experimental spectra, with a much broader range required for the poorly crystalline sample.

## Clay Minerals

In this section we discuss two very different papers which have relevance to Mössbauer spectroscopy of soils. Gendler et al.[32] have studied the precipitation and crystallization of ferric hydroxide in the presence of kaolinite (and as a control, without). Precipitation was carried out at pH 5 and both RT and 77K spectra were obtained immediately after washing, after heat treatment, and after aging for periods up to two years. Magnetic sextets were not observed at 80K for any samples except for the kaolinite sample heated to 120 C. Variations were noted in quadrupole splitting, line width, and especially intensity ratio of the 80K to 300K spectra. This ratio was higher for the precipitate on kaolinite than for that without the clay. Although these authors postulate the formation of ultrafine hematite particles in some cases, it seems difficult to determine from the Mössbauer doublets to what crystalline form these iron species correspond. The authors conclude from their data that the kaolinite catalyzes the crystallization process.

Diamant et al.[33] report characterization studies on the
iron adsorbed on the interlayer surfaces of montmorillonite.
The unique aspect of their work is to show clearly the marked
effect of interlayer water on the Mössbauer spectra of adsorbed
$Fe^{2+}$. At about 210K the peak intensity of this species is
drastically reduced, an effect not seen for adsorbed $Fe^{3+}$ or
structural iron of either valence. The variation of intensity
below this temperature is similar for all species and corresponds to a Debye temperature of ~189K. They interpret these
results as indicating the adsorbed $Fe^{2+}$ is a solvated free ion
whereas adsorbed $Fe^{3+}$ hydrolyzes and polymerizes on the clay
surface.

SEDIMENTS

Petersen and Rasmussen[34] have determined the mineralogy of
the clay fraction of several glacial sediments from Greenland.
Mössbauer spectroscopy was used together with X-ray diffraction
and chemical analysis. The unique character of these clays is
their close relation to the rocks from which they formed.
Mössbauer data primarily gave the $Fe^{2+}/Fe^{3+}$ ratios in the
material subsequent to extraction of a small fraction of oxide
component by dithionite. About 40% oxidation had taken place.
These clays were highly crystalline alteration products of
biotite. Manning and coworkers have reported a series of
papers on Great Lakes sediments. Cores from Hamilton Harbor,
Ontario were examined[35] as to the complex iron mineralogy due
in part to natural processes and in part to the surrounding
steel industry. In this paper the role of iron for phosphate
fixation in sediments was proposed. Subsequent papers emphasize
this feature. Sediment cores from both oxic and anoxic waters
of the Great Lakes[36] were examined by RT Mössbauer spectroscopy. The two profiles differed markedly in their $Fe^{2+}/Fe^{3+}$
ratios, with the oxic core having a high $Fe^{3+}$ concentration
near the surface and about constant, lower values below, while
the $Fe^{3+}$ in the anoxic core decreased exponentially. The
ferric hydroxide concentration was correlated with estimated
adsorbed phosphorus. This relation between ferric hydroxide
and non-apatite inorganic phosphorus has been further explored
by Mössbauer studies of sediment cores from Lakes Erie and
Ontario.[37,38] In the first of these it was suggested that the
sediments have been saturated with phosphorus for some time.
The amorphous ferric hydroxide coats the clay particles and
provides the substrate for adsorption of bio-available phosphorus. In the second,[38] distinctly-colored layers from a deep
core were analyzed by Mössbauer spectroscopy and chemical techniques. Varying amounts, 6-15%, of magnetic component were
present at RT with field 493-514 kOe, ascribed to hematite. A
very black layer at about 5.5 cm depth was ascribed to high
$MnO_2$, whereas a nearby red layer was considered to have a

226

color due to amorphous ferric oxide gel, which gave at 4K a broad magnetic field sextet at about 458 kOe. It seems surprising that (a) an ultrafine gel would give a red color (most gels and poorly crystalline oxyhydroxides are rather brown) and (b) that the 14% hematite component at room temperature was not seen at 4K. This amount of hematite is more than enough to account for a distinctly red color. Although evidence is presented in this paper as well as the previous ones that phosphorus is bonded as phosphate to iron, the Mössbauer spectra unfortunately provide no direct evidence of such bonding.

Minai et al.[39] report a study of sediments from deep sea locations in the Pacific. Both red and blue clays were found, and the relation between composition and location was explored. Mössbauer spectra were obtained at RT and 78K. They consisted of high-spin ferrous and ferric doublets with in some cases a sextet due to hematite. The paramagnetic ferric component was highest near the equator, ferrous iron showing a reverse dependence on latitude. The ferrous component increased in abundance towards the west. The above results were found for red clays. The blue clays had larger ferrous component and also iron sulfides. The latter may have contained low-spin $Fe^{2+}$, such as in pyrite $FeS_2$, which overlaps the high spin $Fe^{3+}$ doublet. Greek lake sediments were examined by Papamarinopoulos et al.,[40] both by magnetic techniques and by RT Mössbauer spectroscopy of the magnetic concentrates. In these the predominant component was non-stoichiometric magnetite, with smaller amounts of hematite and ferrous and ferric doublets. Suttill et al.[41] studied the geochemistry of iron in tidal flat sediments from East Anglia by Mössbauer spectroscopy, X-ray diffraction, and magnetic measurements. Both RT and 77K Mössbauer measurements were taken. Depth profiles of the distribution of iron among the various minerals were obtained from analysis of the complex Mössbauer spectra. Contributions from chlorite, pyrite, and greigite ($Fe_3S_4$) were distinguished, with pyrite increasing with depth. A magnetic concentrate was shown to give a Mössbauer spectrum similar to magnetite.

Johnston and Glasby[42] used RT Mössbauer spectroscopy and X-ray diffraction to characterize a series of Pacific Ocean sediments between New Zealand and Rarotonga. With distance from New Zealand the $Fe^{3+}/Fe^{2+}$ ratio increased, with corresponding darkening of color from yellow to brown. The main variation in mineralogy was deduced to be the increased contribution of ferrihydrite to the $Fe^{3+}$ Mössbauer peaks at further distance from New Zealand. Surendranath and Bansal[43] report Mössbauer results on a polymetallic nodule from the Indian Ocean. Both at RT and 77K the spectra were ferric doublets with quadrupole splitting Q.S. = 0.66 mm/s, which was ascribed to fine-particle FeOOH. Mørup and Lindgreen[44] have examined by

Mössbauer spectroscopy core samples from the oil fields of the North Sea. Most of the iron was present as $Fe^{2+}$ in clay minerals, with minor amounts of $Fe^{3+}$ and pyrite $FeS_2$. Some ferrous carbonates were also present in some samples. At lower depth, magnetic components such as hematite and goethite were observed. The presence of the mixed iron-magnesium-calcium carbonate ankerite (isomer shift with respect to Fe 1.23-1.27 mm/s, Q.S. = 1.40-1.55 mm/s) appeared to relate to the age of the source rock. Murad[45] used Mössbauer spectroscopy and X-ray diffraction to study the iron oxide bands found in a hydrothermal assemblage from the caldera wall of the active volcano on Santorini Island. The presence of hematite and the absence of ferrihydrite in this assemblage were contrasted with results on recent nearby submarine deposits.

SOILS

In two 1978 papers from our laboratory, Bigham et al.[46,47] report on the iron oxide mineralogy of the fine clay fractions from five North Carolina ultisols and two oxisols from Brazil. Mössbauer spectra at RT, 78K and 4K were obtained and used in conjunction with X-ray diffraction and chemical extraction to characterize the mineralogy. The RT spectra exhibited most prominently an $Fe^{3+}$ doublet with in some cases a sextet due to hematite. The low temperature spectra had much-reduced intensity from the doublet and gave general agreement with the X-ray results indicating varying proportions of aluminous goethite and hematite, although in one case hematite was detected in the 4K Mössbauer spectra but not in the X-ray pattern. These authors quote[46] a relative ratio of recoilless fraction aluminous hematite to goethite at 4K of 1.25 and used this value to obtain quantitative estimates of the two materials. In the second paper[47] emphasis is on the relation of color, surface area and phosphate retention to the iron mineralogy. Soils were grouped into pairs of similar layer silicate mineralogy, but differing in color. The redder clays correlated with higher proportion of hematite. The yellow, goethite-rich clays had generally higher surface area and better phosphate adsorption. Bigham et al.[48] also studied the variation in iron mineralogy within a profile from an Appling soil from North Carolina to compare with the obvious color variation from brown to red as one goes deeper in the profile. The presence of both hematite and goethite in the lower C horizon, with only goethite found in the higher B horizon was interpreted as evidence for a pedogenic transformation of hematite to goethite.

Longworth et al.[49] used both Mössbauer and magnetic techniques to investigate the nature of secondary ferrimagnetic iron oxides formed on sedimentary rocks. Samples from north England, Wales and eastern France were studied, the first two

sites having been subjected to recent and extensive fires.
Mössbauer measurements were made on both the original samples
and on magnetic extracts. Measurements were made with an
applied magnetic field at 4K in order to separate the various
sextets. Both Mössbauer and magnetic results were consistent
with the presence of magnetite, $Fe_3O_4$ rather than maghemite, $\gamma$-
$Fe_2O_3$ as the carrier of the ferrimagnetism, in contrast to
earlier studies of these types of soils. Mössbauer spectra
indicated a degree of non-stoichiometry in the magnetite, as
well as the additional presence of varying amounts of hematite
and other iron components.

Childs et al.[50] studied a series of New Zealand soils of
both red and yellow/brown color grouped in pairs like those of
Bigham et al.[47] Their RT and 77K spectra also clearly show the
relation between the red color and presence of hematite. The
red samples gave always a sextet due to hematite as well as a
doublet at RT. Most of the red and all of the yellow soils
produced a new sextet at 77K of lower field, attributed either
to small particle-size goethite or akaganéite. These authors
indicate that X-ray diffraction, IR, and electron diffraction
proved of little help in identifying the iron components.
Silver et al.[51] report studies of clay samples from the Caribbean Islands. These clays contained montmorillonite, kaolinite
and/or mica as major components, and a ferrous doublet was seen
in a number of samples in addition to the ferric doublet.
Goethite, when observed at 80K, was collapsed to a doublet at
RT, as generally observed in the previous papers discussed. We
have reported[52] Mössbauer studies of the effect of chemical
concentration of the iron oxides in soil clays by boiling 5M
NaOH. Although this treatment is commonly used to remove
silicates and aluminum oxides from soils in order to study the
concentrated iron oxide fraction, it is rather drastic, and we
considered Mössbauer spectroscopy the best way to observe what,
if any, changes occurred in the nature of the concentrated
fraction. Samples of soil clays were studied with both hematite and goethite present, with no hematite but appreciable
goethite, and with poorly crystalline material, as evidenced by
highly-broadened magnetic peaks at 78K and appreciable doublet.
The iron content was appreciably raised by the treatment.
Estimates of aluminum substitution in the original and concentrated samples were made from the observed magnetic fields for
goethite at 78K and for hematite at RT. For goethite-rich
samples the estimates were in good agreement with chemical
analysis of the concentrated samples, while the agreement was
not as good for samples containing both hematite and goethite.
We now ascribe some of the discrepancies to the neglect of
particle size effects. In most samples the doublet component
at 78K decreased appreciably on treatment. This appears due to
an increase in crystallinity on treatment with base. There was

no evidence in this work that the aluminum substitution changed markedly on treatment, however.

Schwertmann et al.[53] used Mössbauer spectroscopy and differential X-ray diffraction to study the possible occurrence of hematite in soils of axeric temperate areas, specifically in the alfisols of the Alpine forelands. It is well-established that tropical climates promote the formation of hematite. The work reported by these authors shows that hematite can form also in temperate zones under suitable geologic conditions. The 4K Mössbauer spectra of the red samples had a component with magnetic field ~531 kOe, ascribed to hematite of reduced crystallinity and/or high aluminum substitution. A minimum of about 3% hematite in the presence of goethite could be observed. Both climatic conditions (increased temperature, decreased rainfall) and nature of the parent material are apparently important for the production of hematite.

Maheshwari et al.[54] give a brief report of their study of two soil clays from India. Both ferric and ferrous doublets were found at RT. At 20K, magnetic sextets were observed due to hematite (H ~ 513-520 kOe) and FeOOH (fit to two sextets with H ~ 450-495 kOe). The lower field observed for both $Fe_2O_3$ and FeOOH in one of the soils was attributed to less oxidizing conditions (this soil had more $Fe^{2+}$ and less hematite as well). Labenski et al.[55] used Mössbauer techniques to study sandstone uranium deposits of Argentina. The higher concentrations of uranium were found in samples with higher $Fe^{3+}/Fe^{2+}$ and in some cases a magnetic component ascribed to either maghemite or hematite (although the quoted field values seem to indicate hematite is more likely). De Geyter et al.[56] report on the mineralogy of the ferriferous soils in the region of the springs of Blanchimont in Belgium, with particular emphasis on the iron components as identified by Mössbauer techniques among others. A distinctive feature is the occurrence of ferrihydrite in a sample from 50 cm depth and its transformation at the surface to goethite. Das et al.[57] studied ferrous to ferric ratios in soils and rocks of the foothills of the Himalayas. They compare their results to those of other workers, but it is unclear from their data what the correlation is between these ratios and the geological nature of the samples.

More recently, our laboratory has reported[58] a study of iron oxides in petroferric materials - the term petroferric referring to a contact between soil and indurated material of high iron oxide/oxyhydroxide content. Again, soil color appears related to the iron components, with hematite causing reddening. Mössbauer spectroscopy was used to estimate the relative goethite/hematite concentrations and also their aluminum content. Most samples contained >10% Al substitution.

Alkali treatment did not consistently change the values observed for magnetic fields (and thus estimated Al substitution), but did appear to increase the observed fields for some samples. Estimates of solubility limits for Al in $\alpha$-$Fe_2O_3$ appeared in some cases to exceed the expected value of 15-18%. However, alkali treatment indicated the highest estimated aluminum substitution might be a particle size effect, as the concentrated samples showed no greater than 19% Al in hematite. A comparison was made between oxalate extractable iron (so-called active iron) and observed hematite content. These generally varied inversely, but it was pointed out that lepidocrocite, detected in some samples by X-ray diffraction, would increase the oxalate extractable fraction. One unusual effect was noted, that for samples with little or no hematite present, a high active iron ratio appeared correlated with better crystallinity of the goethite, indicated by the presence of broadened sextets at RT. This is the opposite of what might be expected if oxalate extraction removes poorly-crystalline goethite.

A final paper to be discussed in this section is the study by Fysh and Ostwald of Australian iron ores,[59] some hematite-rich and some goethite-rich. Spectra both at RT and 4K were reported. The possible presence of ferrihydrite in the goethite-rich samples was excluded by spectra determined after oxalate extraction, which indicated removal only of finely-divided goethite. The 4K spectra indicated, from observed fields, that low or no aluminum substitution was present. At 4K some of the hematite-rich samples indicated two magnetic phases, one below and one above the Morin transition. As discussed earlier in this review, the observation of a Morin transition in hematite (not observed in any reported soils studies) indicates high crystallinity and low substitution.

HUMIC ACID COMPLEXES AND FERRIHYDRITE

In this section we review papers which discuss the organic complexes of iron in soil and the least crystalline of the iron oxide/hydroxide minerals, ferrihydrite, whose stabilization in natural environments is presumed to be due to organic material.

Considering organic complexes with humic acid, the Mössbauer data seem uncertain at present. Goodman and Cheshire[60] studied the pH variation of the reaction. The initial slurry at pH 3 had two doublets with Q.S. = 0.59 and 0.97 mm/s. Reduction of pH below 2 caused the iron to be reduced to a complex ferrous doublet. Increasing pH caused the precipitation of a magnetically-ordered species at 77K, which was attributed to $\beta$-FeOOH. Dickson et al.[61] report RT and 82K Mössbauer spectra of humic and fulvic acid mixtures. The RT spectra of the humic acid samples were $Fe^{3+}$ doublets with Q.S.

= 0.51 mm/s. The 82K spectrum of the fulvic acid samples were doublets of ferrous iron with Mössbauer parameters similar to that of $Fe^{2+}$ in frozen aqueous solution. Neither of the above two papers present any evidence of iron-organic bonding. Senesi[62] has criticized previous reports and indicates that much spectroscopic evidence favors the presence of iron-organic bonding in humic acid, pointing out that Mössbauer data alone are insufficient to answer the question of bonding.

Ferrihydrite is a poorly-crystalline mineral with bulk composition $5Fe_2O_3.9H_2O$ and characteristic broad X-ray lines at ~1.5A and 2.5A. The Mössbauer spectra of synthetic ferrihydrites prepared by hydrolysis at pH 7-8 of $Fe(NO_3)_3$ and $Fe/Al(NO_3)_3$ were studied at RT and 77K by Childs and Johnston[63] with the objective of comparing to spectra of natural soils. The observed doublets were broad and could be fitted with two sets of lines with Q.S. ~ 0.60 and 0.95 mm/s. They make the point that their earlier assignment of similar doublets in soils to akaganéite is in error and ferrihydrite is more likely the component causing the large observed quadrupole splittings in some New Zealand soils. Murad and Schwertmann[64] report RT, 130K and 4K spectra of a number of synthetic and natural ferrihydrite samples with varying degrees of crystallinity. All had the basic two X-ray lines characteristic of planar $Fe-[O(H)]_6$ octahedra, but some better-crystallized samples had 4 other characteristic X-ray peaks. The RT and 130K spectra required two doublets for fitting, as in the above, with Q.S. ~ 0.54, 0.89 mm/s. The 4K spectra were magnetically-split and broad, requiring at least three sextets for a reasonable fit, with field values ranging from 445 to 510 kOe. Some of these were fit by up to 12 sextets, producing a distribution that peaked somewhat below 500 kOe. This is lower than the field of hematite at this temperature and higher than that of lepidocrocite, but of similar value to aluminum-substituted goethite. One distinctive feature of the ferrihydrites is their low, almost zero, quadrupole interaction at 4K (about -0.03 with the sign convention of this review). These authors point out the similarities in the spectra of ferrihydrite to the iron-storage protein ferritin and to microcrystalline hematite. Schwertmann et al.[65] have published criteria for identification of ferrihydrite in soils. Techniques required are X-ray diffraction, Mössbauer spectroscopy, and chemical extraction. The most useful X-ray technique was a differential one involving subtraction of the oxalate-extracted diffraction pattern from the original. Mössbauer spectra at 4K were also considered important in identification, in particular as affected by oxalate treatment. Murad has applied Mössbauer techniques to the study of ferrihydrite deposits on an artesian well in Bavaria.[66] These were very poorly-crystalline, giving only two line X-ray patterns and broad Mössbauer field distributions at 4K with maximum absorption for ~470 kOe. This latter maximum, coupled with a half-width of the field distribution of ~80 kOe agrees with results from other poorly-crystalline ferrihydrites. As crystallinity increases the maximum of the field distribution also increases to about 500 kOe with corresponding decrease in halfwidth to about 45 kOe. The contributions of silica and of bacterial activity to the production and stabilization of the ferrihydrite were also discussed.

Finally in this section we mention a recent paper by Murad and Taylor[67] on hydroxycarbonate green rusts. These are not poorly-crystalline iron oxides, but are likely unstable precursors of at least some of the iron oxides in soils. Both $Fe^{2+}$, $Fe^{3+}$ and $Fe^{2+}$, $Al^{3+}$ synthetic green rusts were studied. Mössbauer spectra were obtained at 120K and 4K. The 4K spectra were complex due to magnetic ordering of several components. The 120K spectra indicated two $Fe^{2+}$ and two $Fe^{3+}$ doublets for the $Fe^{2+}$, $Fe^{3+}$ sample, with only one low intensity $Fe^{3+}$ doublet for the $Fe^{2+}$, $Al^{3+}$ sample. These spectra changed rapidly on exposure of the samples to air, ferric iron of course increasing. The observation of two $Fe^{2+}$ doublets was explained on the basis of an ordered cation structure with excess $Fe^{3+}$ affecting the Mössbauer parameters of neighboring $Fe^{2+}$.

## EXAMPLE SOIL CLAYS AND CONCLUSIONS

A wide variety of papers have been surveyed. Many of these emphasize difficulties or ambiguities encountered in this field. However, it is clear that the application of Mössbauer spectroscopy to soils and sediments is actively being pursued and can be made with much greater assurance than was possible five years ago. Although the developments surveyed in preceding sections indicate Mössbauer spectroscopy is important as a technique for soil mineralogical examination, it must be used in conjunction with other techniques such as, especially, X-ray diffraction and chemical extraction. The simplest Mössbauer experiments at RT are unfortunately of limited usefulness. Spectra at 80K and even lower are required in order to characterize magnetically-ordered species which often are seen only at lower temperatures for soil samples than for pure materials, due to poor crystallinity and/or impurity substitution. Several papers have reported fitting the broad and asymmetric magnetic peaks with a distribution of fields. The method of Wivel and Mørup[68] is one we have found convenient to use.

We have examined recently two soil clays down to 12K and give some results here to show why low temperature measurements are important. We also illustrate the use of a field distribution in fitting spectra at intermediate temperature. The two soils are Fannin, a typic Hapludult from the North Carolina mountains, and Djibelor $R_p$, a Sols lessives from Senegal. Kaolinite and vermiculite are the major components of Fannin, while Djibelor $R_p$ is predominantly kaolinite. We chose these samples as having low, but measureable citrate-bicarbonate-dithionite-(CBD) extractable iron, and indicating hematite, goethite and a significant doublet in their 77K Mössbauer spectra[*]. Spectra at 12-13K are shown in Figure 1 and at 100K in Figure 2. These were fitted by two sextets plus doublet. Within the precision of measurement these fits are acceptable, although the detailed shape of the absorption is clearly more complex. There is appreciable growth of the doublet component between 12K and 100K, moreso for Fannin (12% to 40% of the absorption area) than for Djibelor $R_p$ (18% to 28%).

After subtraction of the doublet, the remaining sextets at 100K exhibit magnetic field distributions of quite different character (Figures 3 and 4). In both, a relatively sharp maximum above 510 kOe corresponds to the hematite component. The Djibelor distribution between 300 and 500 kOe has distinct structure which we attribute to aluminum substitution in goethite. Fannin has a much broader, structureless distribution extending to lower fields. The fit was arbitrarily cut off below 150 kOe. It seems clear that relaxation processes due to small particle size occur to much greater extent in Fannin than in Djibelor $R_p$. One advantage of the field distribution is that the relative contributions of hematite and goethite to the absorption can be better ascertained than from a two sextet fit, which over-emphasizes the goethite contribution. The hematite contribution is clearly separated for Djibelor (Figure 3). Overlap between hematite and goethite for Fannin (Figure 4) makes the determination of relative contributions less certain, but still more realistic than from the simpler fit.

Further information is shown in the plot of average (two sextet) magnetic fields versus temperature in Figure 5. Both hematite and goethite fields decrease more rapidly with temperature than a Brillouin function, again indicating fine particle size effects, more noticeable for Fannin. From the empirical equations of Tables 1 and 2, estimates can be made for aluminum substitution. For Djibelor $R_p$, both hematite and

---

[*] Fannin has 8% Fe, 1.3% Al extractable by CBD, Djibelor $R_p$ 4% Fe, 0.4% Al. Details of the characterization of these clays are given in the M.S. thesis of Aminata Niane, N. C. State University (1983), which also reports RT and 77K Mössbauer spectra. According to the 77K Mössbauer analysis, Fannin has 48% of its iron as goethite, 22% as hematite. Djibelor has 70% goethite, 7% hematite.

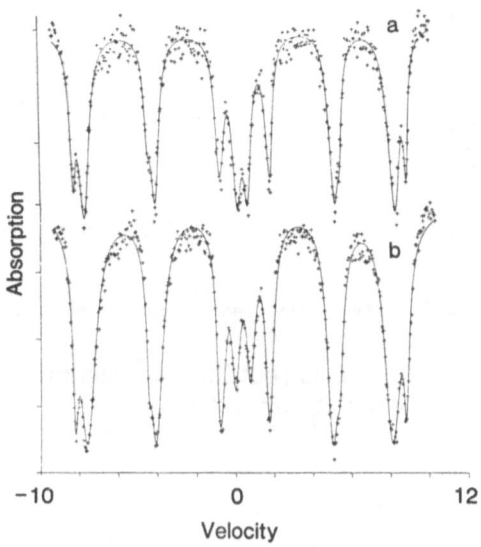

Fig. 1. Low temperature Mössbauer spectra. (a) Djibelor $R_p$ at 12K. (b) Fannin at 13K.

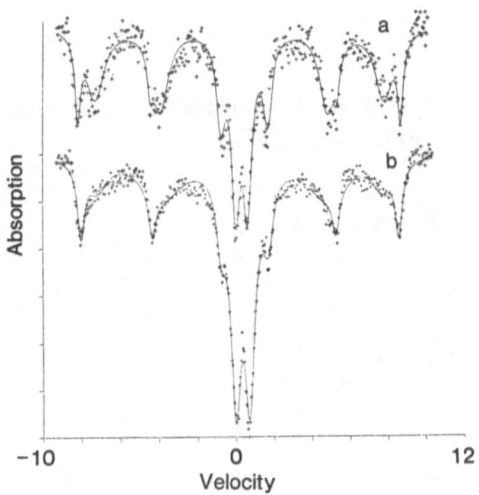

Fig. 2. Mössbauer spectra at 1ØØK. (a) Djibelor $R_p$. (b) Fannin.

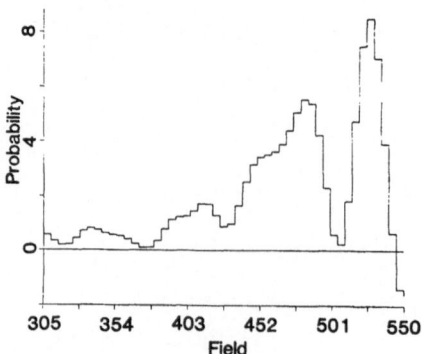

Fig. 3. Probability (%) distribution of magnetic field (in kOe) for Djibelor R$_p$ at 100K.

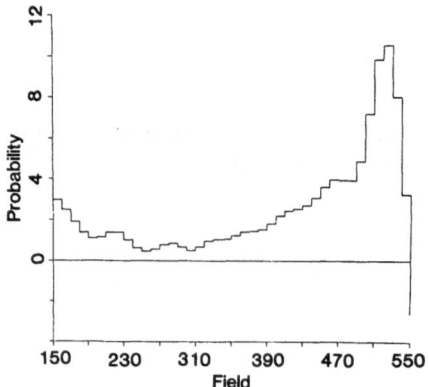

Fig. 4. Probability (%) distribution of magnetic field (in kOe) for Fannin at 100K.

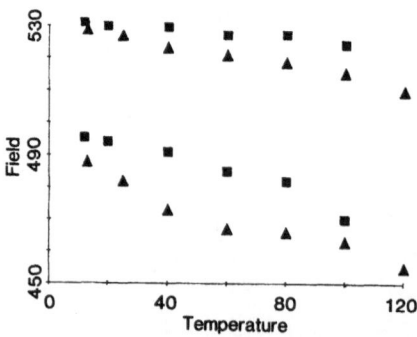

Fig. 5. Average magnetic fields (in kOe) vs. T (in K). Hematite fields are the higher pairs of points. ■ Djibelor R$_p$, ▲ Fannin.

goethite have about 10-15% Al according to these estimates, which is in agreement with bulk CBD extraction results. For Fannin at 4K the estimate gives 12% Al in hematite and 23% in goethite, compared to a CBD value of 25% Al. However, other sources of aluminum such as the mica and gibbsite observed by X-ray diffraction could contribute to the latter. The field distributions and large decreases in average magnetic field with temperature indicate fine particle effects over-estimate the goethite substitution. An approximate surface area correction reduces the 23% to 11% Al in goethite, more in line with hematite. Estimated hematite substitution in Fannin from higher temperature data is unreasonably large (>30%), again presumably due to fine particles. Ferrihydrite or other oxalate-extractable species whose magnetic structure would collapse below 80K are present at less than 5% of the CBD iron. Thus the apparent change in curvature between 40-80K is attributed to the simplified fitting procedure rather than a significant physical effect. The field of maximum probability for hematite in both clays behaves similarly with temperature as the average field. For the goethite component of Djibelor the field of maximum probability coincides with the average field at 20K but is higher by about 12 kOe at 100K. As shown in Figure 4, it was not possible to observe a clear maximum in the goethite distribution for Fannin.

It is seen that low temperature data add much to the information on the nature of the iron components. However, such a temperature study is time consuming, and would be unnecessary if the 77K spectrum gives well-defined magnetic peaks with little doublet component. Measurement of spectra at 77K following chemical treatment such as CBD or oxalate extraction or alkali concentration will give some of the same information. However, the low temperature measurements are required to utilize the full potential of the Mössbauer technique.

ACKNOWLEDGMENTS: The financial assistance of the National Science Foundation under grant EAR-7904834-A01 is gratefully acknowledged. Bibliographical assistance was provided by John G. Stevens and the Mössbauer Effect Data Center. The program for analyzing field distributions was kindly provided by S. Mørup. The chemical and physical characterization of the two clays discussed in the last section was performed by Aminata Niane, who also supplied the Djibelor clay from Senegal. Joyce Weatherspoon provided expert typing of this manuscript.

REFERENCES

1.   L. H. Bowen, Mössbauer Spectroscopy of Ferric Oxides and Hydroxides, Mössbauer Effect Data J. 2:76-94(1979).

2.  B. A. Goodman, Mössbauer Spectroscopy, in "Advanced Chemical Methods for Soil and Clay Minerals Research", J. W. Stucki and W. L. Banwart, ed., p.1-92, D. Reidel, Dordrecht (1980).

3.  B. A. Goodman, Mössbauer Spectroscopy, in "Advanced Techniques for Clay Mineral Analysis", J. J. Fripiat, ed. p. 113-137, Elsevier Scientific Publishing Co., Amsterdam (1981).

4.  J.M.D. Coey, Clay Minerals and Their Transformations Studied with Nuclear Techniques: The Contribution of Mössbauer Spectroscopy, At. Energy Rev., 18:73-124(1980).

5.  L. Heller-Kallai and I. Rozenson, The Use of Mössbauer Spectroscopy of Iron in Clay Mineralogy, Phys. Chem. Min., 7:223-238(1981).

6.  J. Dolnicar, Application of Mössbauer Spectroscopy in Mineralogy, Soil Science and Ceramics: Co-ordinated Research Programme, 1977-1980, At. Energy Rev. Suppl., 2:257-268(1981).

7.  T. Tominaga, Environmental and Geochemical Applications of Mössbauer Spectroscopy, in "Short-Lived Radionuclides in Chemistry and Biology", J. W. Root and K. A. Krohn, ed., p. 495-509, American Chemical Society, Washington, D.C. (1981).

8.  D. C. Golden, L. H. Bowen, S. B. Weed and J. M. Bigham, Mössbauer Studies of Synthetic and Soil-Occurring Aluminum-substituted Goethites, Soil Sci. Soc. Amer. J., 43:802-808(1979).

9.  E. Murad, Mössbauer Spectra of Goethite: evidence for structural imperfections, Min. Mag., 43:355-361(1979).

10. J. Fleisch, R. Grimm, J. Grübler and P. Gütlich, Determination of the Aluminum Content of Natural and Synthetic Alumogoethites using Mössbauer Spectroscopy, J. de Phys., 41-C1:169-170(1980).

11. B. A. Goodman and D. G. Lewis, Mössbauer Spectra of Aluminous Goethites ($\alpha$-FeOOH), J. Soil Sci., 32:351-363(1981).

12. J. H. Johnston and K. Norrish, A $^{57}$Fe Mössbauer Spectroscopic Study of a Selection of Australian and Other Goethites, Aust. J. Soil Res., 19:231-237(1981).

13. S. A. Fysh and P. E. Clark, Aluminous Goethite: A Mössbauer Study, Phys. Chem. Min., 8:180-187(1982).

14. E. Murad, The Characterization of Goethite by Mössbauer Spectroscopy, Amer. Min., 67:1007-1011(1982).

15. E. Murad and U. Schwertmann, The Influence of Aluminum Substitution and Crystallinity on the Mössbauer Spectra of Goethite, Clay Min., 18:301-312(1983).

16. S. Music, Z. Dragcevic, O. Lahodny-Sarc, I. Nagy-Czako, and A. Vertés, Mössbauer Effect Study of Some Yugoslav Bauxites, J. de Phys., 41-C1:305-306(1980).

17. S. A. Fysh and P. E. Clark, A. Mössbauer Study of the Iron Mineralogy of Acid-Leached Bauxites, Hydrometallurgy, 10:285-303(1983).

18. K. Jonas, K. Solymar and J. Zöldi, Some Applications of Mössbauer Spectroscopy for the Quantitative Analysis of Minerals and Mineral Mixtures, J. Mol. Struct., 60:449-452(1980).

19. E. DeGrave, L. H. Bowen and S. B. Weed, Mössbauer Study of Aluminum-Substituted Hematites, J. Mag. Mag. Mat., 27:98-108(1982).

20. T. Tomov, D. Klissurski and I. Mitov, Mössbauer Study of the Formation of Solid Solutions in the $\alpha$-$Fe_2O_3$-$Al_2O_3$ System, Phys. Stat. Sol., 73:249-254(1982).

21. S. A. Fysh and P. E. Clark, Aluminous Hematite: A Mössbauer Study, Phys. Chem. Min., 8:257-267(1982).

22. E. DeGrave, L. H. Bowen and G. G. Robbrecht, $^{57}$Fe Mössbauer Effect Study of the Morin Transition in Some Aluminum Substituted Hematites, in "Studies in Inorganic Chemistry Vol. 3: Solid State Chemistry 1982", R. Metselaar, H. J. M. Heijligers and J. Schoonman, ed., p. 571-575, Elsevier Scientific Publishing Co., Amsterdam (1983).

23. E. DeGrave, D. Chambaere, and L. H. Bowen, Nature of the Morin Transition in Al-substituted Hematite, J. Mag. Mag. Mat., 30:349-354(1983).

24. D. Chambaere, A. Govaert, J. De Sitter and E. DeGrave, A Mössbauer Investigation of the Quadrupole Splitting in $\beta$-FeOOH, Sol. St. Comm., 26:657-659(1978).

25. J. H. Johnston and N. E. Logan, A Precise Iron-57 Mössbauer Spectroscopic Study of Iron(III) in the Octahedral and Channel Sites of Akaganéite ($\beta$-iron hydroxide oxide), J. Chem. Soc., Dalton, 13-16(1979).

26. E. Murad, Mössbauer and X-ray Data on $\beta$-FeOOH (Akaganéite), Clay Min., 14:273-283(1979).

27. M. Ohyabu and Y. Ujihira, Study of the Chemical States of Chlorine and Fluorine in Akaganéite, J. Inorg. Nucl. Chem., 43: 3125-3129(1981).

28. C. W. Childs, B. A. Goodman, E. Paterson, and F.W.D. Woodhams, The Nature of Iron in Akaganéite ($\beta$-FeOOH), Aust. J. Chem., 33:15-26(1980).

29. D. Chambaere and E. De Grave, On the Néel Temperature of $\beta$FeOOH: Structural Dependance and Its Implications, J. Mag. Mag. Mat., in press (1984).

30. D. G. Chambaere and E. DeGrave, On the Influence of the Dual Iron Co-ordination on the Hyperfine Field in $\beta$FeOOH, J. Mag. Mag. Mat., in press (1984).

31. E. Murad and U. Schwertmann, The Influence of Crystallinity on the Mössbauer Spectrum of Lepidocrocite, Min. Mag., in press (1984).

32. T. S. Gendler, L. S. Yershova, L. O. Karpachevskiy, and R. N. Kuz'min, Nuclear Gamma Resonance Study of Iron Oxides and Hydroxides on Kaolinite, Doklady Akad. Nauk SSSR, 258:1205-1208(1981). Eng. Transl.: Sov. Soil Sci., 13:87-90(1981).

33. A. Diamant, M. Pasternak, and A. Banin, Characterization of Adsorbed Iron in Montmorillonite by Mössbauer Spectroscopy, Clays Clay Min., 30:63-66(1982).

34. L. Petersen and K. Rasmussen, Mineralogical Composition of the Clay Fraction of Two Fluvio-Glacial Sediments from East Greenland, Clay Min., 15:135-145(1980).

35. P. G. Manning, W. Jones and T. Birchall, Mössbauer Spectral Studies of Iron-Enriched Sediments from Hamilton Harbor, Ontario, Can. Min., 18:291-299(1980).

36. P. G. Manning, T. Birchall, and W. Jones, Ferric Hydroxides in Surficial Sediments of the Great Lakes and their Role in Phosphorus Availability: A Mössbauer Spectral Study, Can. Min., 19:525-530(1981).

37. P. G. Manning and W. Jones, The Binding Capacity of Ferric Hydroxides for Non-Apatite Inorganic Phosphorus in Sediments of the Depositional Basins of Lakes Erie and Ontario, Can. Min., 20:169-176(1982).

38. P. G. Manning, K. R. Lum, and T. Birchall, Forms of Iron, Phosphorus and Trace-Metal Ions in a Layered Sediment Core from Lake Ontario, Can. Min., 21:121-128(1983).

39. Y. Minai, T. Furuta, K. Kobayashi, and T. Tominaga, A Mössbauer Study of Deep Sea Sediments, Radiochem. Radioanal. Lett., 48:165-174(1981).

40. S. Papamarinopoulos, P. W. Readman, Y. Maniatis, and A. Simopoulos, Magnetic Characterization and Mössbauer Spectroscopy of Magnetic Concentrates from Greek Lake Sediments, Earth Planet. Sci. Lett., 57:173-181(1982).

41. R. J. Suttill, P. Turner, and D. J. Vaughan, The Geochemistry of Iron in Recent Tidal-Flat Sediments of the Wash Area, England: a mineralogical, Mössbauer, and magnetic study, Geochim. Cosmochim. Acta, 46:205-217(1982).

42. J. H. Johnston and G. P. Glasby, A Mössbauer Spectroscopic and X-ray Diffraction Study of the Iron Mineralogy of Some Sediments from the Southwestern Pacific Basin, Marine Chem., 11:437-448(1982).

43. K. Surendranath and C. Bansal, Mössbauer Effect Study of Polymetallic Nodules from Indian Ocean Bed, Phys. Stat. Sol., 73:K133-136(1982).

44. S. Mørup and H. Lindgreen, Applications of Mössbauer Spectroscopy in Oil Prospecting, "Proc. Internat. Conf. on the Applications of the Mössbauer Effects", p.290-292, Indian Nat. Sci. Academy, New Delhi (1982).

45. E. Murad, Iron Oxide Mineralogy of a Hydrothermal Assemblage on Santorini Island, Aegean Sea, Min. Mag., 46:89-93(1982).

46. J. M. Bigham, D. C. Golden, L. H. Bowen, S. W. Buol and S. B. Weed, Iron Oxide Mineralogy of Well-drained Ultisols and Oxisols: I. Characterization of Iron Oxides in Soil Clays by Mössbauer Spectroscopy, X-ray Diffractometry, and Selected Chemical Techniques, Soil Sci. Soc. Amer. J., 42:816-825(1978).

47. J. M. Bigham, D. C. Golden, S. W. Buol, S. B. Weed and L. H. Bowen, Iron Oxide Mineralogy of Well-drained Ultisols and Oxisols: II. Influence on Color, Surface Area, and Phosphate Retention, Soil Sci. Soc. Amer. J., 42:825-830(1978).

48. J. M. Bigham, D. C. Golden, L. H. Bowen, S. W. Buol, and S. B. Weed, Mössbauer and X-ray Evidence for the Pedogenic Transformation of Hematite to Goethite, Soil Sci. Soc. Amer. J., 42:979-981(1978).

49. G. Longworth, L. W. Becker, R. Thompson, F. Oldfield, J. A. Dearing and T. A. Rummery, Mössbauer Effect and Magnetic Studies of Secondary Iron Oxides in Soils, J. Soil Sci., 30:93-110(1979).

50. C. W. Childs and B. A. Goodman, Application of Mössbauer Spectroscopy to the Study of Iron Oxides in Some Red and Yellow/Brown Soil Samples from New Zealand, in "International Clay Conference 1978", M. M. Mortland and V. C. Farmer, ed., p.555-565, Elsevier Scientific Publishing Co., Amsterdam (1979).

51. J. Silver, M. Sweeney, and I.E.G. Morrison, A Mössbauer Spectroscopic Study of Some Clay Minerals of the Eastern Caribbean West Indies. Part I: Spectra from 80 to 300K, Thermochim. Acta, 35:153-167(1980).

52. L. H. Bowen and S. B. Weed, Mössbauer Spectroscopic Analysis of Iron Oxides in Soil, in "Mössbauer Spectroscopy and its Chemical Applications", J. G. Stevens and G. K. Shenoy, ed., p.247-261, American Chemical Society, Washington (1981).

53. U. Schwertmann, E. Murad, and D. G. Schulze, Is There Holocene Reddening (Hematite Formation) in Soils of Axeric Temperate Areas?, Geoderma, 27:209-223(1982).

54. V. K. Maheshwari, J. S. Samra, A. K. Singh, and K. Chandra, Mössbauer Studies of Two Indian Soil Clays, in "Proc. Internat. Conf. on the Applications of the Mossbauer Effect", p.866-868, Indian Nat. Sci. Academy, New Delhi (1982).

55. F. Labenski, H. B. Nicolli, and C. Saragovi-Badler, Genesis of Sandstone-Type Uranium Deposits in the Sierra Pintada District, Mendoza, Argentina: A Mössbauer Study, Uranium, 1:1-18(1982).

56. G. De Geyter, S. Hoste, G. Stoops, R. E. Vandenberghe, and L. Verdonck, Mineralogy of the Ferriferous Soil Materials in the Source Area of Blanchimont (Province of Liege, Belgium), Pedologie, 32:349-366(1982).

57. S. C. Das, S. K. Sengupta, N. C. Paul, N. Bhattacharya, J. B. Basu and N. Chaudhuri, Mössbauer Spectroscopic Analysis of Iron in Soils and Rocks in the Eastern Himalayan Foothill Region, Indian J. Pure Appl. Phys., 21:376-378(1983).

58. I. J. Ibanga, S. W. Buol, S. B. Weed, and L. H. Bowen, Iron Oxides in Petroferric Materials, Soil Sci. Soc. Amer. J., 47:1240-1246(1983).

59. S. A. Fysh and J. Ostwald, A Mössbauer Study of Some Australian Iron Ore Minerals, Min. Mag., 47:209-217(1983).

60. B. A. Goodman and M. V. Cheshire, A Mössbauer Spectroscopic Study of the Effect of pH on the Reaction between Iron and Humic Acid in Aqueous Media, J. Soil Sci., 30:85-91(1979).

61. D.P.E. Dickson, L. Heller-Kallai, and I. Rozenson, Mössbauer Spectroscopic Studies of Humic Acid and Fulvic Acid Soil Fractions, J. de Phys., 41-C1:409-410(1980).

62. N. Senesi, Spectroscopic Evidence on Organically-bound Iron in Natural and Synthetic Complexes with Humic Substances, Geochim. Cosmochim. Acta, 45:269-272(1981).

63. C. W. Childs and J. H. Johnston, Mössbauer Spectra of Proto-ferrihydrite at 77K and 295K, and a Reappraisal of the Possible Presence of Akaganéite in New Zealand Soils, Aust. J. Soil Res., 18:245-250(1980).

64. E. Murad and U. Schwertmann, The Mössbauer Spectrum of Ferrihydrite and its Relation to Those of Other Iron Oxides, Amer. Min., 65:1044-1049(1980).

65. U. Schwertmann, D. G. Schulze, and E. Murad, Identification of Ferrihydrite in Soils by Dissolution Kinetics, X-ray Diffraction, and Mössbauer Spectroscopy, Soil Sci. Soc. Amer. J., 46:869-875(1982).

66. E. Murad, Ferrihydrite Deposits on an Artesian Fountain in Lower Bavaria, N. Jb. Miner. Mh., 45-56(1982).

67. E. Murad and R. M. Taylor, The Mössbauer Spectra of Hydroxycarbonate Green Rusts, Clay Min., 19:77-83(1984).

68. C. Wivel and S. Mørup, Improved Computational Procedure for Evaluation of Overlapping Hyperfine Parameter Distributions in Mössbauer Spectra, J. Phys. E, 14:605-610(1981).

# MÖSSBAUER STUDIES OF LIQUID CRYSTALS

D.L. Uhrich

Department of Physics and Liquid Crystal Institute
Kent State University
Kent, Ohio 44242

## ABSTRACT

The Mössbauer Effect has been used to study thermotropic
liquid crystalline systems since 1969 and has progressed from
studies of orientational ordering to studies of the molecular
vibrational anisotropy and rotational diffusion in supercooled
liquid crystals and liquid crystalline glasses. Liquid crystalline phases span the gap between the solid and isotropic liquid
phases via a stepwise introduction of orientational and spatial
ordering. For example, the nematic phase has only orientational
order and there are a variety of smectic layered arrangements.
By introducing Mössbauer probes into these liquid crystals, one
has the opportunity to study molecular relaxations in the vicinity
of the supercooled liquid crystal-liquid crystalline glass transition as a function of orientational and translational order in a
stepwise manner. These applications of the Mössbauer technique
will be discussed along with their potential use in the study of
lyotropic liquid crystalline glasses, membranes and other bilayer
systems whose layered arrangements are similar to those encountered
in thermotropic smectic liquid crystals.

## INTRODUCTION

The first use of the Mössbauer Effect to probe liquid crystalline systems occurred in 1969 and was published in 1970.[1] The
initial idea involved the use of the layered structure of the
smectic liquid crystalline phase to provide the rigidity needed
for a recoil-free event. In addition, since the liquid crystalline
molecules could easily be aligned in normal laboratory magnetic

243

fields, solute "monocrystals" could be formed. As a result, orientational Mössbauer studies could be carried out without the previously required single crystals.[1,2] The probability of a recoil-free event in true solute "monocrystals," however, could not be observed unless the liquid crystalline host was supercooled or in the glassy liquid crystalline state.[3,4] The observations in the normal liquid crystalline phases were for suspensions of microcrystallites rather than for solute molecules.[4]

The orientational order provided by aligned liquid crystalline systems has proved fruitful for Mössbauer investigations. Two very different kinds of information have been obtained. The first concerns the properties of the molecules bearing the Mössbauer nuclide. Determinations of the sign of the principal component of the electric field gradient tensor have been made for several iron- and tin-bearing molecules.[5-10] These studies have even included molecules for which the quadrupole splitting is not resolved.[6,7,9] The second area involves the properties of the liquid crystals themselves. For example, the molecular order of Fe-57 and Sn-119 solute molecules has depended on the size and shape of the solute molecules relative to the host liquid crystalline molecules.[5,6,9-13] Most fruitful, however, has been the study of molecular relaxations in the various liquid crystalline phases. These studies include measurements of the lattice contribution to the nuclear vibrational anisotropy,[6,9,13,14] translational diffusion,[15,16] Mössbauer-Debye temperatures in the glassy phases of various liquid crystalline systems,[3,10,12,17-19] supercooled liquid crystal-liquid crystalline glass transition temperatures and so far the only observations of rotational diffusion via the Mössbauer technique.[18,19] The latter observation was for the probe molecule 1,1'-diacetylferrocene in a supercooled nematic[18] and a supercooled smectic G liquid crystal.[19]

Fe-57, Sn-119, and I-129[20,21] bearing solute molecules have been used as probes in various liquid crystalline systems. There have been only a few studies of liquid crystalline systems in which the Mössbauer nuclide was actually incorporated as part of the liquid crystalline molecule. These molecules were designed molecules which contained Sn-119 or I-129. Most of the work has focused on thermotropic liquid crystalline materials. The Russian group has illustrated the potential of the Mössbauer Effect in lyotropic liquid crystals by studying lipid-Sn-119 complexes[22] and the dynamic properties of iron-bearing proteins bound in membranes.[23,24]

The following will include a brief description of liquid crystalline systems, a review of several observations of the Mössbauer Effect in liquid crystalline systems, and a discussion of potential applications of Mössbauer spectroscopy in the study of liquid crystalline materials. The review of Mössbauer studies of liquid crystals will touch on the highlights of what has been done and not be totally comprehensive because of the space limitations. For other reviews of Mössbauer spectroscopy in liquid crystals please consult References 25 and 26. The review by V.Ya. Rochev is especially recommended.

## LIQUID CRYSTALS

Liquid crystals have been described in several review articles and books.[27-31] Here, a brief description of these interesting materials will be given. Many of us are more aware of liquid crystals now than in the past because of their emergence in visual displays for items such as watches and calculators. These materials are distinguished from and intermediate to normal isotropic liquid and solid phases. Liquid crystal phases are separated from the solid and liquid states by first order phase transitions. The three liquid crystalline phases which are usually distinguished are the nematic, smectic and cholesteric phases. The latter is sometimes referred to as a twisted nematic, and it is not fundamentally different from the normal nematic phase.

Molecules which form liquid crystalline phases are relatively long and have two distinct regions: a central core region which is characterized by the presence of benzene rings and linkage groups with multiple bonds and a tail region which is comprised of floppy and flexible chains which are usually alkyl groups. Of importance to the formation of the liquid crystalline structure is the anisotropic diamagnetic and dielectric susceptibilities of the long liquid crystal molecules which provide the mechanisms for their characteristic molecular order. Figure 1 illustrates a schematic view of the nematic phase, a smectic A phase, and a tilted smectic phase.

The nematic is very fluid and has three degrees of translational mobility. It is characterized by a preferred molecular

Nematic          Smectic A          Smectic C

FIGURE 1. Structure of three liquid crystalline phases. The lines are projections of the molecules in the plane of the drawing. From Uhrich [11].

direction. The long axes of the molecules are distributed about the preferred direction in such a way that the molecules tend to be parallel. The smectic phase is a more highly ordered phase in that besides the orientational order of the molecules, their centers of gravity form layers. The molecules have less mobility perpendicular to the layers than within the layers. The more highly ordered smectic phase appears at a lower temperature than the nematic if a material exhibits both phases. The smectic A phase is characterized by the planar normal being parallel to the preferred molecular direction and the molecules being free to translate within the layers. The smectic C phase is similar to the smectic A except for the fact that the planar normals are tilted away from the preferred molecular direction of the molecules. Other smectic phases such as smectic B and G are comparable to the A and C phases, respectively. They differ only in the fact that they exhibit molecular order within the layers. These phases are more viscous than the A and C phases, and usually for a given system they will be the lowest temperature liquid crystalline phases.

Most non-Mössbauer Effect liquid crystal studies have been concerned with the normal temperature regions above room temperature. The study of the "cold" liquid crystal which includes the supercooled region and the liquid crystalline glass has been less ambitious. Differential scanning calorimetry and differential thermal analysis have been used to identify phase transitions and to locate the supercooled liquid crystal-liquid crystalline glass transition temperatures.[33-34] Recently, the group of Sorai and Seki at Osaka University, Japan, has been most active.[35] From the point of view of Mössbauer spectroscopy, the ability to form an aligned glass has afforded investigators numerous avenues of study. Before we turn to these areas, it is interesting to note that liquid crystalline material can be used to systematically augment the molecular order of the glass. One can change the molecular order in a stepwise fashion all the way from the isotropic glass to the anisotropic solid in discrete steps. The stepwise increase of orientational and translational order in cold liquid crystalline systems can be observed through the orientational and temperature dependencies of relaxation phenomena which are observable with the Mössbauer Effect.

MÖSSBAUER STUDIES IN LIQUID CRYSTALS

Initial Observations

The first observation of the Mössbauer Effect in liquid crystals was for the Fe-57 probe 1,1'-diacetylferrocene (DAF) in the liquid crystalline material 4,4'-bis(heptyloxy)azoxybenzene

(HOAOB) [solid $\xrightarrow[\longleftarrow]{74^{\circ}C}$ smectic C $\xrightarrow[\longleftarrow]{95^{\circ}C}$ nematic $\xrightarrow[\longleftarrow]{124^{\circ}C}$ isotropic

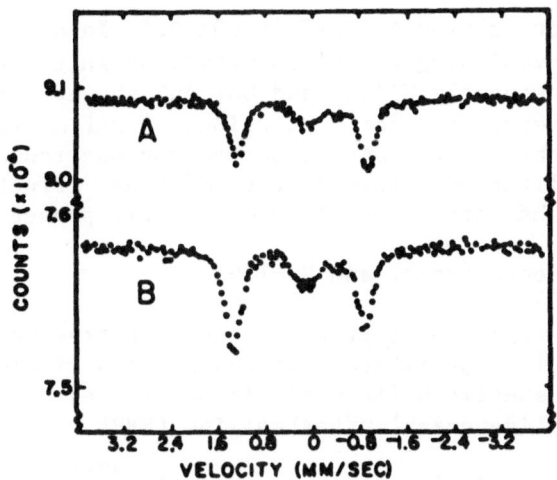

FIGURE 2.    Mössbauer spectra of a 7 wt% solution of DAF in HOAOB at
75°C.    For A. $\theta=0^\circ$ and B. $\theta=75^\circ$.    The outer lines are
the DAF absorption.    The small center line is due to Fe
in the Be discs of the sample holder.    From Uhrich [1].

liquid].    The Kent group found that the effect was observable in
the smectic C phase of HOAOB for a 7% (by weight) sample of DAF in
HOAOB.[1]    The quadrupole split doublet exhibited by the DAF gave an
orientational dependence as shown in Figures 2 and 3.    The spectra
showed a clear asymmetry which was a function of the angle ($\theta$)
between the $\gamma$-ray direction and the direction of an aligning 4kG
magnetic field.    For these experiments the $\gamma$-ray beam was directed
along the normal to a disc-shaped sample.    Goldanskii et al.  also
observed the orientation dependence of the intensity ratio of the

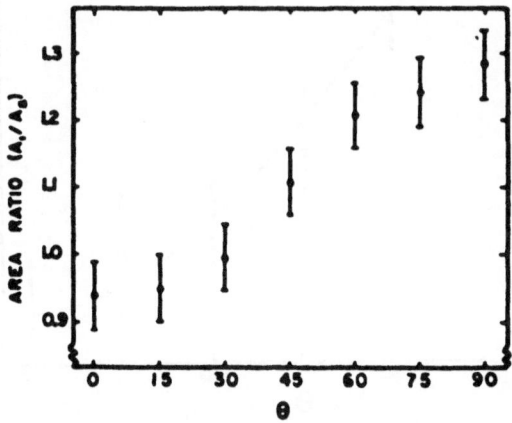

FIGURE 3.    Plot of the area ratio for the quadrupole split lines of
a 7 wt% solution of DAF in HOAOB as a function of $\theta$ at
75°C.    From Uhrich [1].

247

quadrupole split doublet for DAF in HOAOB.[2]  In addition, they
observed a non-vanishing recoilless-fraction in the nematic phase
of HOAOB.  Subsequently Wilson and Uhrich[4] showed that these
initial experiments were for suspensions of DAF microcrystallites
in HOAOB, and that if the DAF concentration was reduced so that
no DAF crystallites were formed, the Mössbauer Effect was not
observable in the normal smectic C or nematic phases.

## Orientation Effects (Fe-57 and Sn-119)

The first report of a true solution spectrum was for a solution of triethyltin palmitate (3EtSnPalm) in a liquid crystal was
for the "cold" smectic G (formerly identified as H) phase of
4-n-hexoxybenzylidene-4-n'-propylaniline (HBPA or 60,3).[5]

HBPA possesses the following phases:  Solid $\xrightarrow{23^{\circ}C}$ smectic G
$\xrightleftharpoons{66^{\circ}C}$ smectic A $\xrightleftharpoons{68^{\circ}C}$ nematic $\xrightleftharpoons{85^{\circ}C}$ isotropic liquid.  The

smectic G phase is easily supercooled and forms a glass.  The
3EtSnPalm molecules were aligned as evidenced by the asymmetry in
the intensity of the quadrupole split lines (see Figure 4).
Figure 5 shows that the total recoil-free fraction also depended
upon the alignment of the solute molecules.

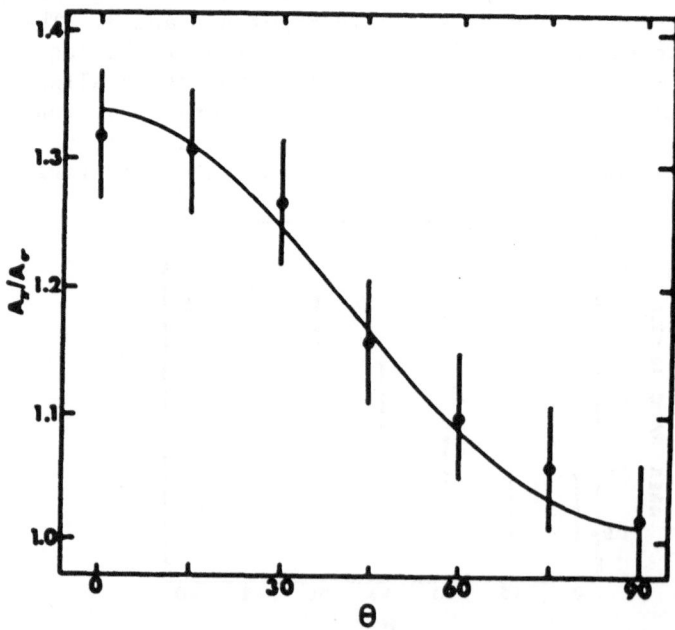

FIGURE 4.  Plot of the Sn-119 intensity ratio of the quadrupole
split lines vs $\theta$ for a 3EtSnPalm solution of HBPA at
77K.  Solid line is a fit of Eq. (4) to the data.  Here
$\varepsilon_M = -0.80$ and $S = 0.17$.  From Uhrich [5].

248

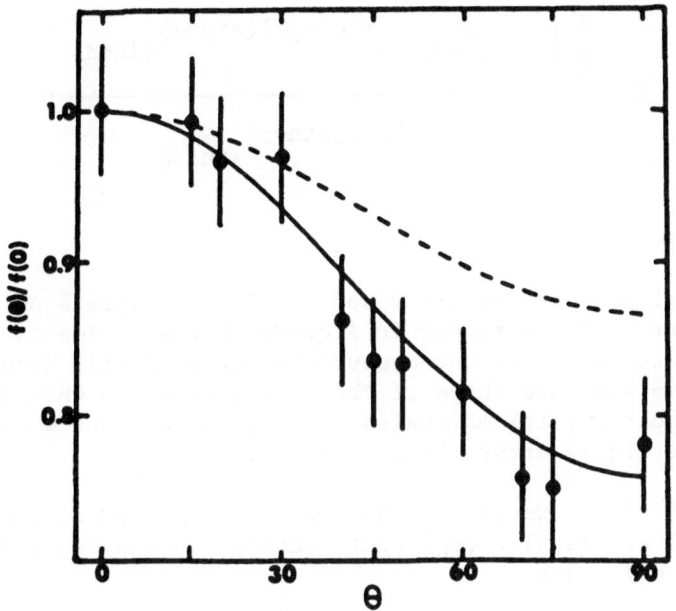

FIGURE 5.   A plot of the recoil-free fraction vs $\theta$ for 3EtSnPalm
            in HBPA at 77K.  Solid line is a best fit of the theory
            to the data without separating the inter- and intra-
            molecular contributions to the vibrational anisotropy.
            The dashed line is a fit of the theory to the data
            using $\varepsilon_M = -0.8$.  From Uhrich [5].

The theory which was developed to explain the above and
similar data is provided in the bibliography.[5,36-39]  It is based
on the mean-field approach to nematic order given by Saupe and
Maier[40] which yields the following distribution of the molecules
about the preferred direction.

$$F = Ce^{-(q/k_BT)\sin^2\delta} \qquad (1)$$

Where C is a constant, q is a measure of the orientational order,
T is the absolute temperature, $k_B$ is Boltzmann's constant and $\delta$
is the angle between the long molecular axis of a particular
molecule and the preferred direction.  Equation 1 may also be
used to describe the orientation order in smectic phases.  The
smectics, however, also have additional anisotropy associated with
the layered structures.  It has turned out, however, that Eq. (1)
has been adequate to explain the Mössbauer data to this point.
The molecular order parameter is given by:

$$S = \frac{\int\limits_{o}^{\pi} \frac{1}{2}(3\cos^2\delta-1) \; e^{-(q/k_BT)(\sin^2\delta)} \sin\delta d\delta}{\int\limits_{o}^{\pi} e^{-(q/k_BT)(\sin^2\delta)} \sin\delta d\delta} . \qquad (2)$$

Note that S = 1 for perfect order and S = 0 for complete orientational disorder as in an isotropic liquid. For experiments involving solute molecules the measured value of S will depend on the relative size and shape of the solute molecule when compared to the host crystal molecules and specifically on where it resides in the liquid crystalline structure.

The recoil-free fraction is approximated by a product of the molecular ($f_M$) and lattice ($f_L$) contributions in the usual way for molecular crystals:

$$f = f_M f_L . \qquad (3)$$

For the case of Fe-57 and Sn-119 where the nuclear transition is between the nuclear spin states 3/2 and 1/2 and for the case of an axial electric field gradient tensor (efg) where the main efg axis is parallel to the long molecular axis, the intensity ratio of the quadrupole split doublet becomes:[5]

$$\frac{A_\pi}{A_\sigma} = \frac{\int\limits_{o}^{2\pi} d\alpha \int\limits_{o}^{\pi} \sin\delta d\delta [1+(\frac{1}{4})(3\cos^2\gamma-1)] \; e^{-\varepsilon_M\cos^2\gamma} e^{-(q/k_BT)\sin^2\delta}}{\int\limits_{o}^{2\pi} d\alpha \int\limits_{o}^{\pi} \sin\delta d\delta [1-(\frac{1}{4})(3\cos^2\gamma-1)] \; e^{-\varepsilon_M\cos^2\gamma} e^{-(q/k_BT)\sin^2\delta}} . \qquad (4)$$

Here $\gamma$, $\alpha$, and $\delta$ are shown on Figure 6. This expression allows one to determine both the order parameter and the molecular contribution to the nuclear vibrational asymmetry [$\varepsilon_M = \frac{1}{\lambda^2} (\langle X_{||}^2\rangle_M - \langle X_{\perp}^2\rangle_M)$].

For vanishing $\varepsilon_M$, Eq. (4) becomes[6]

$$\frac{A_\pi}{A_\sigma} (\theta) = \frac{1 + \frac{1}{4}(3\cos^2\theta-1)S}{1 - \frac{1}{4}(3\cos^2\theta-1)S} . \qquad (5)$$

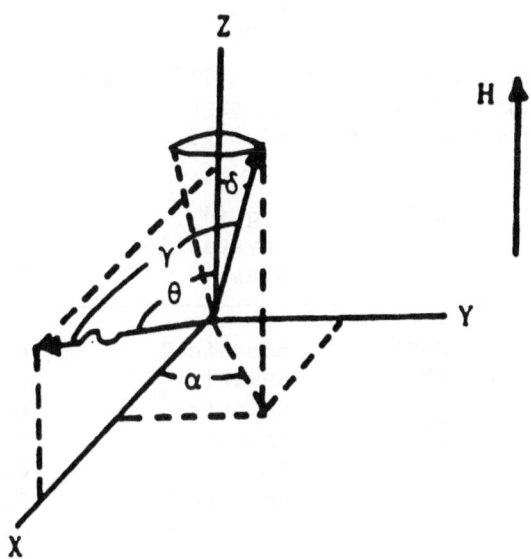

FIGURE 6. The polar and azimuthal angle of the molecular direction in the laboratory system are given by δ and α, respectively. θ is the experimental angle between the preferred molecular direction and the γ-ray direction. From Uhrich [11].

Since S must be positive if the efg axis is parallel to the long molecular axis, we note that for $\theta = 0$; $A_\pi/A_\sigma > 1$ and $\theta = 90°$; $A_\pi/A_\sigma < 1$. As a result, the sign of the electric field gradient ($V_{ZZ}$) can be determined. For the case where the efg axis is perpendicular to the long molecular axis the sense of $A_\pi/A_\sigma$ is reversed and $A_\pi/A_\sigma(\theta=0°) < A_\pi/A_\sigma(\theta=90°)$.

The total recoil-free fraction is also orientation dependent, and for the case of small $\varepsilon_M$, it is given by[6]

$$\frac{f(\theta)}{f(\theta=0)} = e^{\varepsilon_L \sin^2\theta}. \tag{6}$$

Table I lists the order parameters for several solute molecules in various types of liquid crystalline systems. The sign of the efg is also given.

Three Sn-bearing probes, SnBBA, SnUBBA, and SnUBOA listed in Table I are of interest because their Mössbauer spectra are unresolved doublets.[6,7,9] The apparent line centroid was observed to be a function of the direction between the aligning magnetic field and the γ-ray direction. SnUBBA and SnUBOA are two of three Sn-bearing materials which exhibit liquid crystalline phases for which Mössbauer spectra have been reported.[9]

251

TABLE I

Order parameter S for Fe-57 and Sn-119 bearing solutes in various liquid crystalline systems. The sign of the electric field gradient $V_{ZZ}$ for the probe molecule is also given.

| Solute[a] | Liquid Crystalline Host[b] | Liquid Crystalline Glass | S | Sign of $V_{ZZ}$ | Ref. |
|---|---|---|---|---|---|
| 3EtSnPalm | HBPA | Smectic G | 0.17 | + | 5 |
| 3EtSnPalm | MBBA | Nematic | 0.09 | + | 12 |
| SnBBA | BBOA | Smectic B | 0.48 | + | 7 |
| SnBBA | $\overline{7}$S5 | Smectic C | 0.29 | + | 12 |
| FMA | BBOA | Smectic B | 0.37 | + | 13 |
| FBA | BBOA | Smectic B | 0.27 | + | 13 |
| DOF | BBOA | Smectic B | 0.11 | + | 13 |
| DAF | MBBA,HBPA,BBOA | Nematic, Smectics G and B | 0 | | 18,19 |
| DAF | $\overline{7}$S5 | Smectic C | 0.12 | + | 10,12 |
| SnUBBA | Pure | Smectic B | | − | 9 |
| SnUBOA | Pure | Smectic B | | − | 9 |
| Sn-3BC | H-37 | Nematic | | + | 17 |
| Sn-3BL | H-109 | Nematic | | + | 17 |
| Sn-2M2Cl+Sal | H-109 | Nematic | | − | 17 |

[a] 3EtSnPalm = triethyltinpalmitate
SnBBA  = trimethyltin-benzylidene-4'-n-butylaniline
FMA  = ferrocynal-4'-methoxyaniline
FBA  = ferrocynal-4'-butylaniline
DOF  = 1,1'-di-n-octanoyl ferrocene
DAF  = 1,1'-diacetylferrocene
SnUBBA  = p-(11-trimethyltin)undecyloxy-benzylidene-n-butyl-aniline
SnUBOA  = p-(11-trimethyltin)undecyloxy-benzylidene-n-octyl-aniline
Sn-3BC  = tin-tributyl capronate
Sn-3BL  = tin-tributyl laurinate
Sn-2M2Cl  = tindimethyldichloride plus salicylidenaniline (1:2 composition)

[b] HBPA  = 4-n-hexoxybenzylidene-4'-n-propylaniline
BBOA  = 4-n-butoxybenzylidene-4'-n-octylaniline
MBBA  = 4-n-methoxybenzylidene-4'-n-butylaniline
$\overline{7}$S5  = 4-n-pentylphenyl-4'-n-heptyloxythiobenzoate
H-37  = commercially obtained liquid crystal
H-109  = commercially obtained liquid crystal

## Temperature Effects (Fe-57 and Sn-119)

The first observation of the temperature dependence of the Mössbauer spectra was for DAF in the smectic G (formerly called H) liquid crystal HBPA.[3] Here, it was shown that the "cold" liquid crystal was less rigid than the solid over the same temperature region. Further, the solid and liquid crystalline glass phases followed Debye-like behavior and the lnf vs T plot was linear with a negative slope as predicted:

$$\ln f = \frac{-3E\gamma^2 T}{Mc^2 k_B \theta_L^2} \qquad (T \geq \theta_L/2) \qquad (7)$$

where $E_\gamma$ is the energy of the Mössbauer $\gamma$-ray, M is the mass of the vibrating unit, $k_B$ is Boltzmann's constant, and $\theta_L$ is the Mössbauer-Debye temperature. To obtain $\theta_L$ for the host, one must correct for the difference in mass between the host and solute molecules.[41] Figure 7 compares the lnf vs T plots for DAF in MBBA for the crystal state and the nematic glass-supercooled nematic. The deviation from linear behavior is interpreted as the super-cooled nematic-nematic glass transition. The following expression[18] was used to fit the data

$$\ln f = AT + B(T-Tg)^2 + C \qquad (8)$$

where A, B and C are constants and A is determined from the linear portion of the data. Several probes have been used to obtain $\theta_L$ for many liquid crystalline systems. Table II illustrates the values of $\theta_L$ for several probes in MBBA, BBOA and HBPA. The spread of values clearly indicate that the size and shape of the probe with respect to the host is an important factor. In particular, the shape and size of the probe molecule affects its location, and the resulting distortions of the "cold" liquid crystalline arrangement. Figure 8 shows behavior similar to Fig. 7 for Sn-bearing probe molecules in MBBA. For these probes, however, the deviation from linear Debye behavior occurs even in the crystal phase of MBBA. Clearly, this deviation from linear behavior must be associated with relaxations other than those which occur at the supercooled liquid crystal-liquid crystalline glass transition temperature. Typically the Sn-119 Mössbauer Effect in the cold liquid crystals becomes vanishingly small at a temperature below the supercooled liquid crystal-liquid crystalline glass transition temperature.

Table II compares the measured Mössbauer-Debye temperatures for several probes in the nematic glass of MBBA, the smectic B glass of BBOA and the smectic G glass of HBPA. Not only do the probe molecules provide different Mössbauer-Debye temperatures but even the anisotropies differ. The results depend on the size

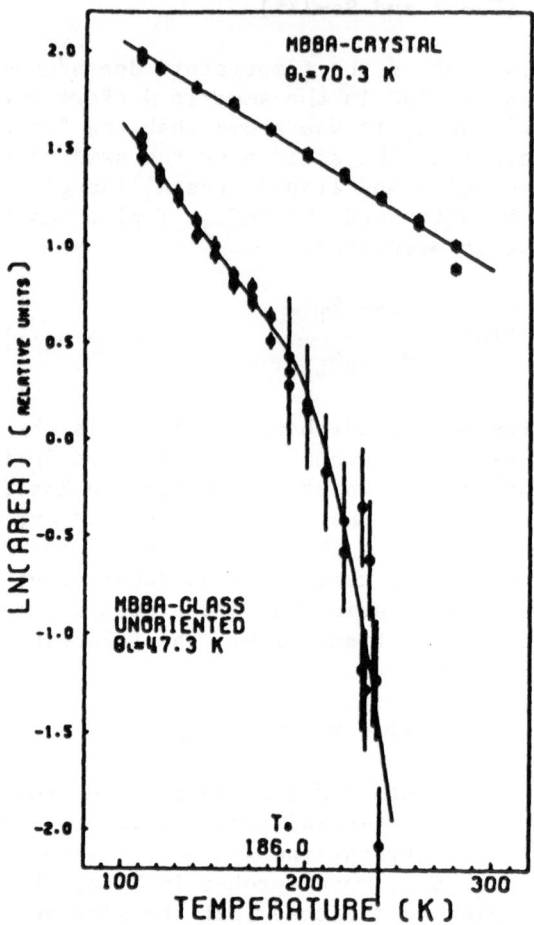

FIGURE 7.  Plots of lnf vs T for DAF in crystalline MBBA and for
DAF in the unoriented cold nematic phase of MBBA.  The
solid lines are a best fit of Eq. (7) to the data for
the crystal and Eq. (8) to the data for the cold nematic.  From LaPrice [18].

and shape of the solute molecules with respect to the host molecule and on their location in the liquid crystalline glass.
Measurements of the Mössbauer-Debye temperature for several
probes in smectic A and smectic C phases also have been reported
showing similar solute dependent results.[10,12,42]

     Temperature dependent diffusion has also been observed in
liquid crystalline systems.  The first observation was by the
Kent Group for DAF in the cold smectic G phase of HBPA.[15]  Line
broadening was observed in the temperature region 233-237K.
Initially, the data were explained in terms of translational

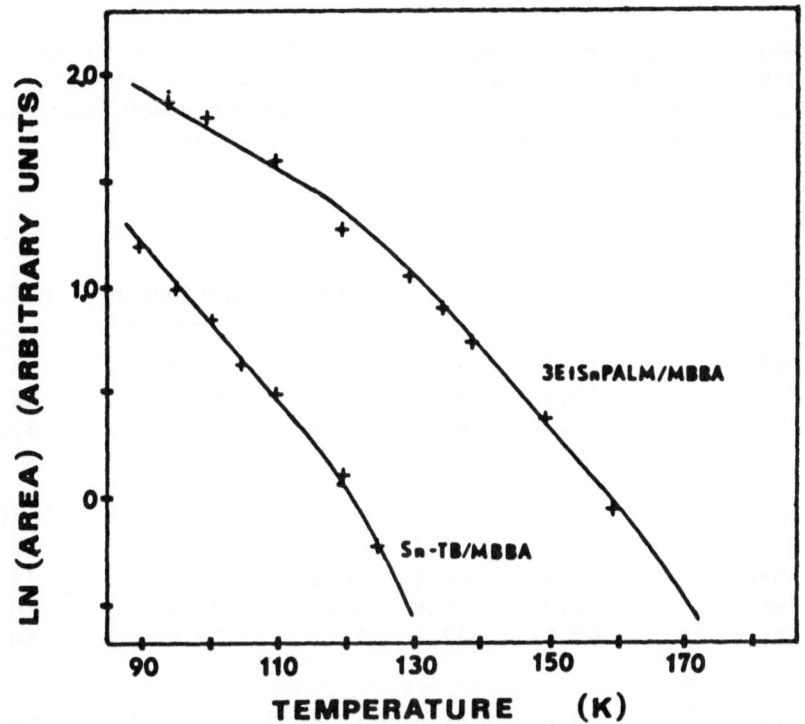

FIGURE 8.  Plots of lnf vs T for solutions of 3EtSnPalm and Sn-TB
in solid MBBA.  The solid lines are best fits of
Eq. (8) to the data.  From Todoroff [12].

diffusion of the DAF molecules in the supercooled smectic G
lattice.  For the case of continuous translational diffusion
which proceeds via a large number of small jumps the broadening
is given by:

$$\Delta\Gamma = 2\hbar\ k^2 D \tag{9}$$

where k is the γ-ray wave vector and D is the diffusion constant.
If on the other hand the diffusion proceeded via large uncor-
related jumps then the line broadening is given by

$$\Delta\Gamma = 2\hbar/t_j \tag{10}$$

where $t_j$ is the average time between jumps for the probe molecules.

## TABLE II

Mössbauer–Debye temperatures for several probe molecules in MBBA, BBOA and HBPA. The data include the crystal phase and the glass aligned via an external magnetic field at $\theta=0^\circ$ and $\theta=90^\circ$, and the glass not oriented by an external field (unoriented).

$$\theta_L \text{ (K)}$$

| Liquid Crystal | Probe Molecule | Crystal | $\theta=0^\circ$ | $\theta=90^\circ$ | Unoriented | Ref. |
|---|---|---|---|---|---|---|
| MBBA | DAF | 70.3 | 46.8 | 47.4 | 47.3 | 18 |
| | Sn–TB | 49.3 | 36.7 | 45.0 | – – | 12 |
| | 3EtSnPalm | 77.2 | 63.1 | 50.4 | – – | 12 |
| | | | | | | |
| BBOA | DAF[a] | 74.7 | 40.0 | 48.0 | 52.0 | 19,10 |
| | Sn–TB | 38.6 | 37.5 | 31.5 | – – | 12 |
| | Sn–BBA | 50.0 | 48.9 | 40.6 | – – | 12 |
| | FMA | 54.0 | 28.8 | 33.9 | 30.5 | 10 |
| | FBA | 62.0 | 27.2 | 34.7 | 32.0 | 10 |
| | | | | | | |
| HBPA | DAF | 69.0 | 46.3 | 45.3 | 46.5 | 19 |
| | 3EtSnPalm | 62.5 | 56.7 | 54.4 | 54.4 | 12 |

Sn–TB = Tin tetrabutyl

[a] Difference from LaPrice [19] according to Marande [10].

A subsequent study showed that coincident with the line broadening there was also a relaxation of the Fe-57 quadrupole split doublet.[19]  Similar phenomena were also observed by the Kent Group for DAF in MBBA.[18]  Figures 9 and 10 illustrate the data from 100K ⟵⟶ 240K for DAF in MBBA crystal, the cold MBBA nematic for $\theta=0$ and $\theta=90^\circ$ and for the unaligned system.

Dattagupta developed a theoretical model which incorporates both translational diffusion and rotational diffusion.[43,44]  The theoretical model accounted for the case of an atom or molecule which undergoes frequent small jumps (continuous diffusion limit) such that at each jump the efg redirects by $\pm \Delta\theta$.  Flinn et al. applied this model to motion of the ferrous ion in cold phosphoric acid and to motion of ferrocene in cold butylphthalate.[45,46]  They show that if the translational and rotational motions are connected by Stokes Law, then rotational diffusion will be difficult to see.  On the other hand, their suggestion that the best possibility

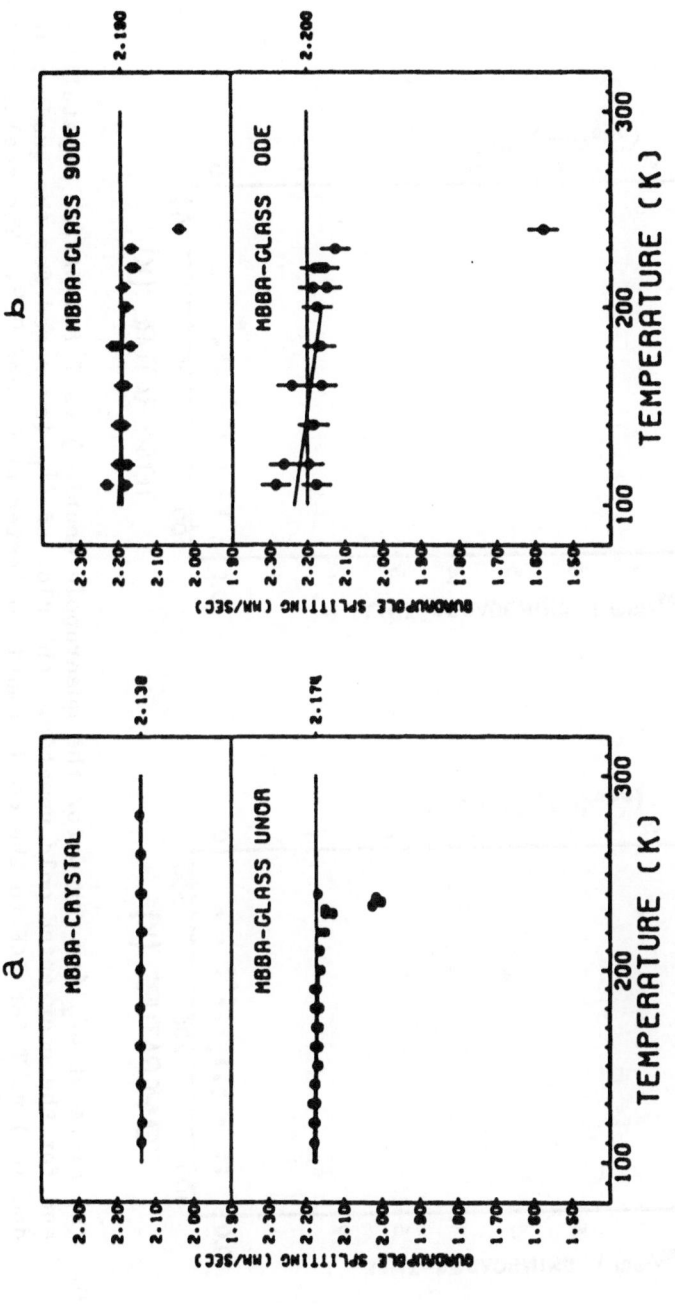

FIGURE 9. (a) Plots of $\Delta E_Q$ vs T for DAF in crystalline MBBA and for DAF in the unoriented cold nematic. (b) Plots of $\Delta E_Q$ vs T for DAF in the cold nematic oriented at 0° and 90°. The horizontal lines correspond to the average value of $\Delta E_Q$ in the range 100–200 K. The least squares best linear fit to these data are also shown. From LaPrice [18].

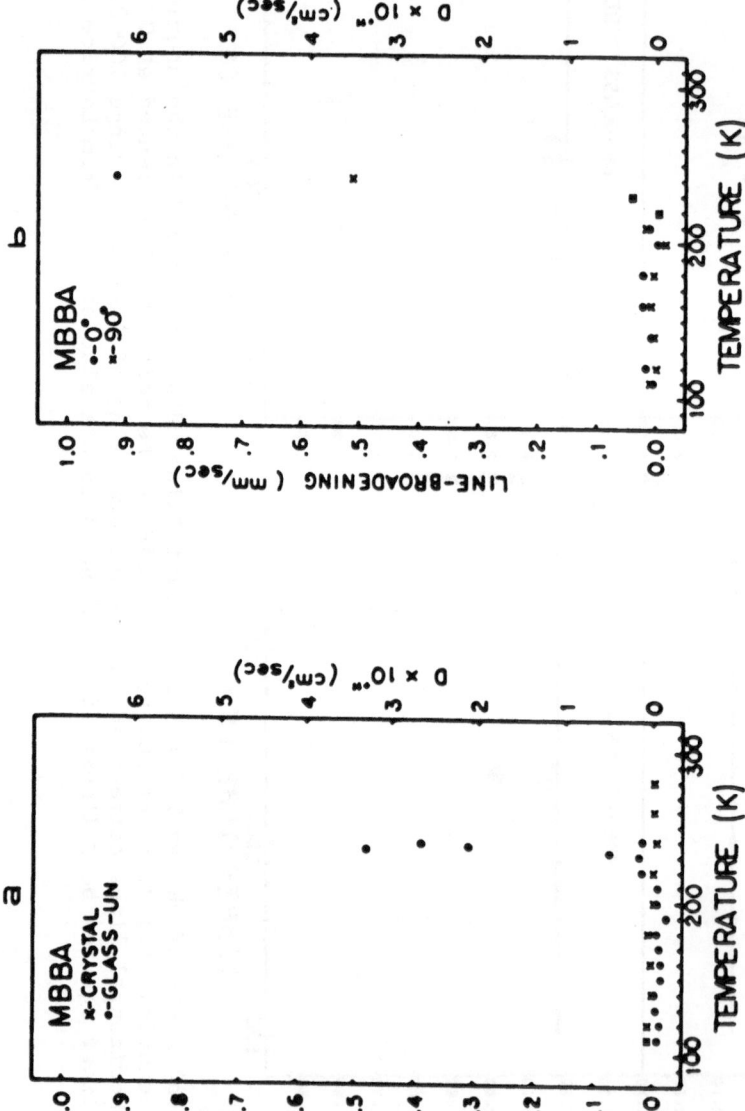

FIGURE 10. (a) Plots of $\Delta\Gamma_m$ (averaged for the quadrupole doublet) vs T for DAF in crystalline MBBA and for the unoriented cold nematic. (b) Plot of $\Delta\Gamma_m$ (averaged for the quadrupole doublet) vs T for DAF in the cold nematic oriented at 0° and 90°. The scales on the right of (a) and (b) show how the translational diffusion constant D would vary with T if the broadening could be attributed solely to translational diffusion. From LaPrice [18].

258

for seeing rotational diffusion via the Mössbauer Effect is where translational diffusion is vanishingly small applies to the data for DAF in cold MBBA and cold HBPA.  Rotational diffusion to date has not been observed via the Mössbauer technique in any other system.

Kandpal and Bhide[16] reported the observation of anisotropic diffusion of small particles of $FeSnO_3$ in the normal nematic phase of n-p-ethoxybenzylidene-p'-butylaniline (EBBA) and the normal nematic and monotropic smectic B phase of p-n-hexyloxy-benzylidene-p'-toluidine (HBT).   They find $D_{||}/D_{\perp} > 1$ for the nematic phase and $D_{||}/D_{\perp} < 1$ for the smectic B phase of HBT.

## Iodine 129 Mössbauer Spectroscopy

Potasek et al.  have observed the I-129 Mössbauer Effect in two liquid crystalline systems.[20,21]  They were the first group to observe the Mössbauer Effect in a liquid crystal system in which the Mössbauer nuclide was a part of the liquid crystal molecule. This material is 6-heptyloxy-5-iodo-2-naphthoic acid (HINA). In addition their 1972 report was the first Mössbauer alignment study of a system which had been quenched to the "cold" liquid crystalline phase.

The second system which they studied was a solution of 4-n-hexoxybenzylidene-4'-129-iodoaniline dissolved in the very similar HBPA.  Observation in both systems showed differences between the aligned system and a random powder sample.  The differences were attributed to the Goldanskii-Karyagin Effect, alignment by the external field and also surface alignment by the container walls.  They determined the sign of the quadrupole splitting to be negative and by observing the angular dependence of the average of the normalized ΔM=0 transition intensities divided by the average normalized ΔM=±1 transition intensities M (see Fig. 11) they were able to determine the tilt angle of the smectic G phase of HBPA.  Their result of 50° is in pretty good

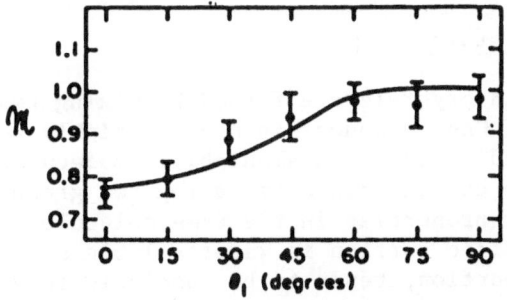

FIGURE 11.   A plot of M values versus θ.  The solid curve corresponds to a tilt angle of 50°.  From Potasek [21].

agreement with the later x-ray determination of ~40° by Deniz et al.[47]

## Liquid Crystal-Label Complexes

Goldanskii et al. dissolved ferrocene (F) into the liquid crystalline material cholesterylmyristate (CM).[48] They found two quadrupole split doublets in the Mössbauer spectrum of the sample which was cooled to room temperature from the isotropic phase. The wider doublet was characteristic of the ferrocene spectrum and it was present before heating. The second doublet which was produced by the heating cycle was characteristic of ferrocenium ($F^+$), $Fe^+(C_5H_5)_2$, compounds and salts. The authors attributed this effect to the formation of charge transfer complexes resulting from the chemical reaction between F and CM. Rochev et al. later observed similar phenomena in a solution of F in the liquid crystalline material ethyl-p-anisalaminocinnamate (EAAC).[49] They found that the relative amount of the $F^+$ form increased with the total heating time. Further, the asymmetry of the intensity ratio for each pair of the quadrupole split lines did not depend on their relative concentration. Of interest, only the intensity ratio of the $F^+$ form exhibited a dependence on the orientation provided by an external field (20-40kG). That is, the intensity ratio for the $F^+$ form depended on the angle between the external field and the γ-ray direction. This angular dependence was evidence that the $F^+$ form had complexed to the oriented liquid crystal molecule. The mechanism for the complex formation was suggested as the donor-acceptor interaction associated with the electron transfer from π-molecular orbitals of F to EAAC molecular orbitals.

Their experimental data yield S=0.7 and are consistent with the $F^+$ efg axis being perpendicular to the long molecular axis of the EAAC molecule. Both results suggest that the probable site of the $F^+$-fragment is near the benzene rings of the central core region of the liquid crystal molecules. The order parameter of 0.7 is the highest measured via the Mössbauer technique.

## Lyotropic Liquid Crystals

Lyotropic liquid crystals are formed by mixing two or more components.[27,29,50] One component is usually highly polar (e.g., water) and the second is often an amphiphilic molecule. The latter compounds are characterized by having two groups with very different solubility properties in the same molecule. One portion known as the hydrophilic portion is water soluble and the second portion, lyophilic portion, tends to be insoluble in water but soluble in hydrocarbons.

Amphiphilic systems exhibit several different structures as the

ratio of water to amphiphile is changed.  Schematically, for
example, one finds with the addition of water:


Solid
Amphi-  $\xrightleftharpoons[-H_2O]{+H_2O}$  Lamellar  $\xrightleftharpoons[-H_2O]{+H_2O}$  Hexagonal  $\xrightleftharpoons[-H_2O]{+H_2O}$  Micellar  $\xrightleftharpoons[-H_2O]{+H_2O}$  Solution
phile           Structure              Structure              System


The molecular packing in the lamellar structure is such that the
molecules pack in double layers with their long axes parallel to
the planar normal.  The ionic part of the molecule dissolves in
the water and the water insoluble tails dissolve in each other.

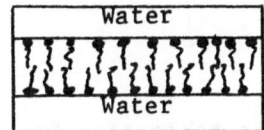

The addition of water to the system increases the surface area
per hydrophilic group and the lamellar structure becomes unstable
and the hexagonal structure appears.  Here groups of molecules
form cylinders with the tails  all pointing toward the center.  As
more water is added the micellar phase is initiated.

    The first report of Mössbauer spectroscopy in a lyotropic
system was reported by Kothekar and Mishra in 1973.[51]  They introduced Co-57 into egg-yolk lecithin with 30, 15 and 5% hydration.
With a decrease in hydration they found the absorption using a
stainless steel absorber to increase.  Further, they found the
Mössbauer linewidth increased with increased hydration which they
attributed to the increased diffusion of the Co-57 atoms in the
free water zone.  They also point out that the increased water
content makes the fatty acid chain more mobile which of course
occurs when the lamellar structure breaks up.

    These lipid-water systems are of interest because their
structure is similar to the structures of biological membranes.
Here, two studies involving lipids will be discussed.  The first
involved the Mössbauer Effect of Sn-119 in the cephalin + $SnCl_4$
system. Rochev et al.[22] point out that ethanolamino-phosphoglycide
(cephalin) and cholinophosphoglycide (lecithin) and the main
lipid components in living cells.  In contrast to the above
described lecithin study Rochev et al. introduced $SnCl_4$ into the
cephalin in order to form a complex between the Mössbauer probe
and the lipid molecule.[22]  As a result, information could be
obtained about the lipid itself.  They prepared two films for

Mössbauer study. The first was a lipid-free film (LFF) of $SnCl_4$
prepared from a solution of $SnCl_4$ in a chloroform and methanol
mixture (2:1). The second film was prepared from a solution of
$SnCl_4$ and cephalin (1:2) in the chloroform-methanol mixture. The
second film was identified as the lipid film (LF). The LF contained the $SnCl_4$-lipid complex. The suggested form for the Sn
configuration in each film is shown below

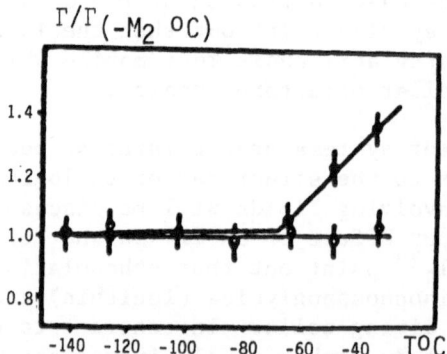

The Mössbauer data were taken as a function of temperature
for $-176 \longleftrightarrow -22^\circ C$. The temperature dependence of the recoil-
free fraction was identified for both samples. This result showed
that the $SnCl_4$ was complexed to the solvent in each film. The
temperature dependence of the linewidth for both samples differed
and is shown in Figure 12. The broadening above $-64^\circ C$ for the LF
was interpreted as being due to the "defreezing" of the lipid.
The lipid molecule conformational motion resulted in the diffusion
motion of the label. Even though the probe was not complexed to
the lipid in this case the research direction is sound particularly
if a probe exhibiting a quadrupole splitting can be attached
to the lipid.

FIGURE 12.   Temperature dependence of the reduced linewidths for
$SnCl_4$ in the lipid film ● and the lipid-free film ○ .
From Rochev [22].

262

The second study provides a correlation between a photoinduced electron transition and dynamic properties of the chromatophone membranes from Rhodospirillum (R) rubrum membranes.[24] Parak et al. investigated the recoil-fraction of Fe-57 incorporated in the membrane protein of chromatophores of the above named photosynthetic bacterium. They point out that a large number of membrane-bound proteins which are involved in electron transport contain iron.

Figure 13 shows the plot of -lnf vs T for the Fe-containing chromatophores of R rubrum. Note the similarity between it and the thermotropic liquid crystalline data, e.g. DAF in MBBA (see Fig. 7). Figure 13 shows a distinct deviation from linear behavior in the region >170K. This increase in the mean square vibrational amplitude was found to be coincident with an increase in the efficiency of photo-induced electron transfer in this system. Clearly, the electron-transfer efficiency is tied to the mobility of the system. While Parak et al. could not correlate their recoil-free fraction measurements with a specific iron-containing protein of the membrane, the results are significant.

## Conclusions and Projections

The Mössbauer Effect investigation in both thermotropic and lyotropic liquid crystalline systems have provided us with an insight into these ordered systems. Furthermore, liquid crystals

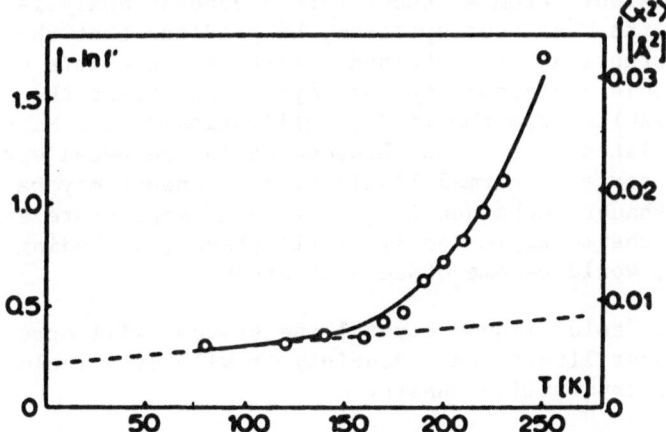

FIGURE 13.   -lnf vs T for Fe-containing parts of the chromatophores of R-rubrum. The values are normalized to
-lnf'=0.2231 at 4.2 K. The dashed line is the assumed
temperature dependence of the vibrational contribution
to the mean square displacement. The solid line is a
result of introducing a contribution from the fluctuations between conformational substates. From Parak
[24].

have provided Mössbauer spectroscopists with a method of observing anisotropic phenomena in ordered systems without resorting to single crystals. In "cold" thermotropic liquid crystalline systems the effects of the ordered arrangements on vibrational, translational and rotational diffusive motions have been observed. Clearly, the variety of available probe molecules coupled with the variety of liquid crystalline phases not only will allow investigators to learn more about the details of liquid crystalline systems and their relaxations but also to learn more about the glass-supercooled fluid transition itself.

Studies in lyotropic systems have not been as extensive as in the thermotropic phases but the same potential is there. Complexes of Fe-57 and Sn-119 probes with lipid molecules will allow one to study relaxation and the ordered structure of biological systems. In addition, the study of the dynamics of iron bearing proteins in membranes is a reality and it will yield important details in the future. As yet these studies have been limited to the "cold" lipid phases.

In 1975 P. de Gennes[52] proposed the theoretical possibility of observing the Mössbauer Effect in perfectly aligned smectic liquid crystals. In particular, for a Mössbauer γ-beam directed perpendicular to a set of perfectly stacked smectic planes he showed that the Mössbauer absorption will be a function of both the splay elastic constant and the coefficient of elasticity for layer compression. From a temperature dependent analysis of the lineshape of the Mössbauer spectrum, he predicts that the ratio of these constants may be obtained. Further, he points out how diffusion within and normal to the layers can affect the spectrum. The former results from the finite collimation of the Mössbauer beam and the latter will cause broadening in the usual way. If de Gennes is correct, normal liquid crystal phases may be studied with the Mössbauer technique (e.g., at room temperature and above). Both layered thermotropic and lyotropic phases, including proteins in membranes, would become objects of study.

Clearly, "cold" liquid crystalline systems will appear in the future Mössbauer literature. Possibly it will even include the normal liquid crystalline phases.

REFERENCES

1. D.L. Uhrich, J.M. Wilson, and W.A. Resch, Phys. Rev. Lett. 24:355 (1970).
2. V.I. Goldanskii, N.K. Kivrina, V.Ya. Rochev, R.A. Stukan, I.G. Chistyakov, and L.S. Shabishev, Mol. Cryst. Liq. Cryst. 24:239 (1973).
3. R.E. Detjen, D.L. Uhrich, and C.F. Sheley, Phys. Lett. 42A:522 (1973).

4. J.M. Wilson and D.L. Uhrich, Mol. Cryst. Liq. Cryst. 25:113 (1975).

5. D.L. Uhrich, Y.Y. Hsu, D.L. Fishel, and J.M. Wilson, Mol. Cryst. Liq. Cryst. 20:349 (1973).

6. D.L. Uhrich, V.O. Aimiuwu, P.I. Ktorides, and W.J. LaPrice, Phys. Rev. 12A:211 (1975).

7. P.I. Ktorides and D.L. Uhrich, Mol. Cryst. Liq. Cryst. 40:285 (1977).

8. V.Ya. Rochev, O.P. Kevdin, N.K. Kivrina, and E.F. Makarov, Acta Phys. Pol. A56:129 (1979).

9. P.I. Ktorides, D.L. Uhrich, R.M. D'Sidocky, and D.L. Fishel, J. Chem. Phys. 77:4188 (1982).

10. R. Marande, Ph.D. Dissertation, Kent State University, 1984 (unpublished).

11. D.L. Uhrich, R.E. Detjen, and J.M. Wilson, in "Mössbauer Effect Methodology," Vol. 8, I.J. Gruverman and C.W. Seidel, eds. Plenum Press, New York (1973), p. 175.

12. D. Todoroff, Ph.D. Dissertation, Kent State University, 1983 (unpublished).

13. V.O. Aimiuwu and D.L. Uhrich, Mol. Cryst. Liq. Cryst. 43:295 (1977).

14. D.L. Uhrich, J. Stroh, R. D'Sidocky, and D.L. Fishel, Chem. Phys. Lett. 24:539 (1974).

15. R.E. Detjen and D.L. Uhrich, in "Mössbauer Effect Methodology," Vol. 9, I.J. Gruverman, C.W. Seidel, and D.K. Dieterly, eds., Plenum Press, New York (1973), p. 113.

16. M.C. Kandpal and V.G. Bhide, Proc. Int. Liq. Cryst. Conf. Bangalore (1979), p. 421.

17. V.Ya. Rochev, N.K. Kivrina, and O.P. Kevdin, Acta Phys. Pol. A56:137 (1979).

18. W.J. LaPrice and D.L. Uhrich, J. Chem. Phys. 71:1498 (1979).

19. W.J. LaPrice and D.L. Uhrich, J. Chem. Phys. 72:678 (1980).

21. M.J. Potasek, E. Munck, J.L. Groves, and P.G. Debrunner, Chem. Phys. Lett. 15:55 (1972).

21. M.J. Potasek, P.G. Debrunner, and G. DePasquali, Phys. Rev. 13A:1605 (1976).

22. V.Ya. Rochev, O.P. Kevdin, N.K. Kivrina, E.A. Gilbukh, and L. May, Proc. Liq. Cryst. Conf., 3rd, Budapest, 2:777 (1980).

23. G.R. Kalamkarov, V.E. Prusakov, M.A. Ostrovsky, R.A. Stuken, and V.I. Goldanskii, Dokl. Akad. Nauk SSSR 219:1245 (1974).

24. F. Parak, E.N. Frolov, A.A. Kononenko, R.L. Mössbauer, V.I. Goldanskii, and A.B. Rubin, FEBS Lett. 117:368 (1980).

25. V.Ya. Rochev, in "Advances in Liquid Crystals," Vol. 5, G.H. Brown, ed., Academic Press, New York (1982), p. 79.

26. F.J. Litterst and G.M. Kalvius, Proc. Int. Conf. Mossbauer Spectrosc. Cracow 2:189 (1975).

27. G.H. Brown, Am. Scientist 60:64 (1972).

28. I.G. Chistyakov, Sov. Phys. Usp. 9:551 (1967).

29. G.H. Brown and J.W. Doane, Appl. Phys. 4:1 (1974).

30. P.G. de Gennes, "The Physics of Liquid Crystals," Oxford University Press, Oxford (1974).

31. "Advances in Liquid Crystals," Vols. 1,2,3,4,5, and 6, G.H. Brown, ed., Academic Press, New York (1975, 1976, 1978, 1979, 1982 and 1983).

32. S.E.B. Petrie, H.K. Bucher, R.T. Klingbiel, and R.I. Rose, Eastman Organic Chemical Bulletin 45:2 (1973).

33. M. Sorai, T. Nakamura, and S. Seki, Proc. Int. Conf. Liq. Cryst., Bangalore, India (1973).

34. M. Sorai and S. Seki, Mol. Cryst. Liq. Cryst. 22:299 (1973).

35. For example, M. Sorai, K. Tani, and H. Suga, Mol. Cryst. Liq. Cryst. 97:365 (1983).

36. J.M. Wilson and D.L. Uhrich, Mol. Cryst. Liq. Cryst. 13:85 (1971).

37. V.I. Goldanskii, O.P. Kevdin, N.K. Kivrina, E.F. Makarov, V.Ya. Rochev, and R.A. Stukan, Sov. Phys. JETP 36:1226 (1973).

38. J.I. Kaplan and M.L. Glasser, Mol. Cryst. Liq. Cryst. 11:103 (1970).

39. P.I. Ktorides and D.L. Uhrich, Mol. Cryst. Liq. Cryst. 87:69 (1982).

40. W. Maier and A. Saupe, Z. für Naturfors. 14A:882 (1959).

41. V.I. Goldanskii and E.F. Makarov, in "Chemical Applications of Mössbauer Spectroscopy," V.I. Goldanskii and R.H. Herber, eds., Academic Press, New York (1968).

42. D.L. Uhrich and R.P. Marande, Liq. Cryst. Symp., St. Louis ACS Meeting (April 1984).

43. S. Dattagupta, Phys. Rev. 12B:47 (1975).

44. S. Dattagupta, Phys. Rev. 14B:1329 (1976).

45. P.A. Flinn, B.J. Zabransky, and S.L. Ruby, J. Phys. (Paris) Colloq. 37:C6-739 (1976).

46. S.L. Ruby, B.J. Zabransky, and P.A. Flinn, J. Phys. (Paris) Colloq. 37:C6-745 (1976).

47. K.U. Deniz, U.R.K. Rao, A.I. Mehta, A.S. Paranjpe, and P.S. Paruathanathan, Mol. Cryst. Liq. Cryst. 42:127 (1977).

48. V.I. Goldanskii, O.P. Kevdin, N.K. Kivrina, E.F. Makarov, V.Ya. Rochev, R.A. Stukan, I.G. Chistyakov, and L.S. Shabyshev, Dokl. Akad. Nauk SSSR 209:139 (1973).

49. V.Ya. Rochev, O.P. Kevdin, F. Parak, Proc. Liq. Cryst. Conf., 3rd Budapest 1:363 (1980).

50. J. Charvolin, J. de Chim. Phys. 80:15 (1983).

51. V. Kothekar and R.K. Mishra, Ind. J. Med. Res. 61:216 (1973).

52. P.G. de Gennes, J. de Phys. 36:603 (1975).

# ORGANOTIN-119m MÖSSBAUER SPECTROSCOPY:    THE FIRST QUARTER CENTURY

J. J. Zuckerman

Department of Chemistry
University of Oklahoma
Norman, OK 73019

## INTRODUCTION

The first observation of the Mössbauer effect in tin[1] was published in 1960, and was soon followed by an application of tin-119m Mössbauer spectroscopy to a chemical problem.[2]  The Mössbauer effect stands as a lonely exception to the general rule in science history that phenomena never bear the names of their true discoverers.  Delivery to the scientific community is far more frequently honored, and in tin-Mössbauer spectroscopy the rules were first worked out and the news brought forward by V. I. Goldanskii in the USSR, N. N. Greenwood in the UK and in the USA by the editor of this volume.

My own introduction came at the Cornell University commencement exercises of 1963 when I found myself marching behind another young Assistant Professor whose gown bore the colors of the Johns Hopkins University.  He was Alan J. Bearden of the Physics Department, and he told me that he was interested in the Mössbauer effect. I responded that I had heard that the Mössbauer effect could be applied to chemistry (where I got this information, I cannot recall). The next day in his office we agreed to embark on a collaboration. By the end of the week we had moved one of Bearden's home-built, constant-acceleration, cam-drive spectrometers into my laboratory, I had copied all the papers reporting tin-119m Mössbauer data, and a physics graduate student (Howard S. Marsh) joined my group.  The following week our $^{119m}SnO_2$ source arrived from New England Nuclear. We were in the Mössbauer business!

This reminiscence will include an anthology of selected applications to organotin problems that have been made since the early

267

years. The limitations inherent in the use of the technique will be stressed, and the article will conclude with a consideration of what the next quarter century might hold in store.

## ORGANOTIN CHEMISTRY

Organotin chemistry is among the very strongest areas in organometallic chemistry. Basic studies in organotin chemistry are stimulated by the success with which a large number of modern physical techniques can be applied to organotin compounds, the availability of two oxidation states, Sn(II) and Sn(IV), with contrasting chemistries, and the wide variety of structural types known.

Equally important are the well-developed commercial applications of organotin chemicals.[3] Organotins became a leading commercial organometallic first through their use in polyvinylchloride (PVC) stabilization, and now as biocides, where their success is based upon their extremely high performance/unit weight ratio, and their degradation by chemical action into non-toxic inorganic tin compounds.[4] Industrial production of organotin chemicals now exceeds 25,000 tons annually.

Organotin compounds assume a wide variety of structural types,[5,6] which encompasses two-to eight-coordination in neutral, cationic and anionic species, with intra- and inter-molecular interactions to give bridged dimers and other oligomers in one-, two- and three-dimensional crystal lattices. The most interesting examples are those in which the structure depends upon the conditions. Organotin compounds are known which exhibit different geometries and coordination numbers in the gas phase and the solid state or in solution as a function of concentration. In these cases data from ancillary physical techniques can play an important role in establishing the true nature of these compounds in the solid state as well as in the various other phases.

Organotin compounds are particularly illustrative of the use of these ancillary spectroscopic methods because of the particular properties of the tin atom. For example, tin-bearing fragments are easy to identify in the mass spectrum since tin has the largest number of stable isotopes of any element (10). The tin-carbon stretching frequency lies in a convenient region and yields information concerning the configuration of alkyl groups about tin, and many cases of linear and planar organotin systems in the solid state have come to light through this means. The most important property of tin to the chemist interested in elucidating the structure of organotin derivatives by spectroscopy, however, has to do with the nuclear properties of the tin-119 isotope. This isotope, which is found in 8.6% natural abundance, possesses a unique combination of a spin of one-half nuclear ground state which makes it suitable for nuclear magnetic resonance[7] studies, along with an accessible isomeric state of suitable spin, lifetime and energy whose gamma decay makes it an ideal Mössbauer nuclide.

Measurements of tin-119 spin-spin couplings in proton, carbon-13, fluorine-19 and phosphorus-31 nmr spectra have proved exceedingly useful in organotin studies. The tin-119 chemical shift itself has developed into a powerful tool for investigating the structure and conformation of organotin molecules in solution. The correlation between shift and coordination number at tin can be used to investigate inter- and intramolecular association and the effect of dispersion forces originating from neighboring polarizable groups by nmr.

Since its first application to tin, the Mössbauer isomer shift (IS) and quadrupole splitting (QS) have contributed to the solution of many organotin problems.[8-10] Information from the IS value which reflects electronic situation and oxidation state, and from the magnitude of the QS which specifies the site symmetry at the tin atom is complementary to that received from the nmr experiment, since the Mössbauer studies are carried out in the solid state, while the nmr measurements are recorded on neat liquids or solutions. Observation of the Mössbauer effect at ambient temperatures in organotin compounds reflects large lattice forces, usually found in associated solids or polymers. Results from variable-temperature Mössbauer spectroscopy enable organotin monomers and polymers to be distinguished. The combination of the data from the nmr and Mössbauer techniques can provide a complete picture of the structural situation in an organotin compound as it passes from phase to phase.

Modern organometallic investigations utilize just such combinations of powerful techniques for the characterization of newly synthesized materials. Key to this effort to understand the intriguing organotin derivatives is of course single-crystal X-ray crystallography. The rate-limiting step, aside from crystal growth, is diffractometer time. There is also the requirement that the data can be made to converge to a single solution. Many unsolved data sets await renewed attention after initial failure in the world's crystallography laboratories. However, knowledge of the molecular parameters in the solid state (internuclear distances, angles and packing) would make much of chemical spectroscopy, which yields at best circumstantial structural evidence, redundant. The superior cost effectiveness has been amply demonstrated, and the impact of crystal structure results on chemistry will surely be much greater in the future.

On the other hand, organotin examples abound in which the detailed structural information provided by X-ray diffraction studies of the solid state is itself ambiguous.[5] Such cases arise, for instance, when neither the intra- or intermolecular internuclear distances nor the measured angular distortions from idealized geometries suffice to determine explicitly the coordination number at the tin atom, or even to suggest which of the ideal geometries has been deviated from. The problem is particularly vexing with regard to the distinction between molecular solids, into which individual molecules are held by van der Waals forces, and associated solids, where the individual molecules preserve their integrity but with additional components to the energy of the lattice. Where these additional lattice forces dominate the choice of crystalline arrangement and affect molecular geometry greatly, the existence of association in the solid is easily demonstrated. Where the forces produced by association in the solid are more feeble, the question often cannot be settled on the basis of structural information alone since the distortions wrought on the gas phase molecular structure by packing into the solid along with the various intermolecular juxtapositions of potentially bridging atoms may be only suggestive of association. In these cases, data from ancillary physical techniques can play an important role in establishing the true nature of the structure of these compounds in the solid state as well as the various other phases. Here Mössbauer spectroscopy's role will continue to expand as ever-increasing numbers of X-ray results present more and more intriguing questions concerning the nature of tin lattices and the details of their dynamics.

Reproducible chemical, physical and biological studies demand pure compounds of known formula and structure in the phases of interest. Lavoisier transformed chemistry into an exact science by applying separation, isolation, purification and identification techniques in a process whose further development continues to this day. These steps can be guided by the application of spectroscopic techniques, applied routinely at every stage. The ultimate goal of the organotin chemist is to determine constitution (elements present), composition (empirical formula), molecularity (formula multiple), connectivity (molecular formula), configuration (spatial arrangement), molecular dimensions (shape and size), packing and interactions between molecules in condensed phases and time-dependent properties which preserve molecular integrity (translation, vibration, rotation, pseudo-rotation, inversion, fluxionality) or disrupt it (dissociation, exchange, inter-conversion of isomers). Since biology is a watery province, knowledge of the aqueous situation is crucial, but usually least known.

The pure compounds sought arise uniquely from certain syntheses, or must be produced using volatility or solubility differences, or thermodynamic, mechanical, adsorptivity, electrical, magnetic, mass and charge, chromatographic, chemical or biological properties. Identity with a known, or inferences from analyses, molecular weight determinations, physical properties and resonance with radiation are used to assign formula and structure.

# TIN-MÖSSBAUER SPECTROSCOPY

Tin-119m Mössbauer spectroscopy has now established itself as one of the indispensable spectroscopic techniques in organotin chemistry. Unfortunately, while the range of velocities is about the same as for iron-57, the much greater natural line width of the tin-119m resonances reduces resolution and information content. Also, none of the compounds of primary interest in this article are magnetic, and the applied fields necessary to produce clean Zeeman-splitting range to 50 kG (5T) and are rarely applied. However, no known tin compound fails to yield its Mössbauer spectrum at 77K, which is easily attainable with conventional cryostats and modern $Ca^{119m}SnO_3$ sources give a satisfactory flux of the 23.8 keV gamma rays when held at ambient temperatures, allowing the relatively rapid collection of spectra. Thus can tin-119m Mössbauer spectroscopy contribute to the battery of results amassed from the other commonly applied physical methods (infrared, mass, nuclear magnetic resonance spectroscopy).[11]

The greatest impact on chemistry comes, however, in providing uniquely decisive results which can definitively rule out possible hypotheses. Here it is particularly imperative to eschew negative evidence.

The simplest and most straight forward application is finger-printing - based upon the principle that identical systems will behave in an identical way, and the next is analysis of a physical mixture - on the principle that physical mixing does not alter spectroscopic properties. However, these analytical applications are rarely made.[12] Some of the difficulties of applying Mössbauer spectroscopy in quantitative analysis have been elaborated by us elsewhere.[13]

Most of the information from tin-Mössbauer spectroscopy has been collected in order to gain increased understanding of molecular and solid-state structure, chemical bonding, lattice dynamics, etc. While these items would seem at first to offer more challenge, they leave wider scope for speculation, rationalization and prediction. It is earnestly hoped that the future will bring forth a greater proportion of predictions testable with recourse to experiment. Much of this will necessarily lie in the realm of solid-state structure testable by X-ray diffraction.

# APPLICATIONS OF MÖSSBAUER SPECTROSCOPY TO ORGANOTIN CHEMISTRY

## Determination of Oxidation State

It is the common practise of the chemist to distinguish the formal oxidation states of tin(II) and tin(IV). The electron population of the 5s-subshell should decrease in going from tin(II) to tin(IV), and the isomer-shift values seem to follow the same change. Near the midpoint of the now known spectral range of -0.61 to +4.69 mm s$^{-1}$ with respect to $SnO_2$ at zero (large quadrupole splittings may place spectral lines outside these quoted IS values) lie the singlet resonances of the two allotropes of tin, the α- or grey form with the diamond structure common to its lighter cogeners of the fourth group, and β- or white tin with a tetragonal metallic structure.[14,15] The former exhibits an IS of +2.10 and the latter +2.65 mm s$^{-1}$. The IS value of β-tin has generally been taken as the dividing line between the two oxidation states, tin(IV) compounds falling below, and tin(II) compounds above. That this is so[16] was first recognized in 1962.

In cases where the oxidation state and hence the structure cannot readily be determined from the formula, the IS information thus becomes crucial. Such a situation exists in the $R_2Sn$ compounds, known for over a century as examples of divalent species. Goldanskii[17] was able to show in 1964 that these materials must be tin-tin bonded, tetravalent polymers on the basis of IS values below that for β-tin. The contrary case is presented by the tin-oxygen heterocycles of the general formula

$$\text{Ar} \underset{O}{\overset{O}{<}} Sn$$

which also behave as associated materials. But here tin-tin bonding can be ruled out by higher than β-tin IS values,[18] and the lattice must be held together by O→Sn intermolecular donor-acceptor bonds instead.

Genuine tin(II) compounds containing carbon-tin bonds include the bis($\eta^5$-phenylcarboranyl)tin(II)[19] and bis($\eta^5$-cyclopentadienyl) tin(II) (stannocene)[20] and its ring-substituted derivatives.[21,22] More interesting are the compounds in which the divalent state is of marginal stability. For example, the diphenyltins can be produced as tin(II) or tin-tin bonded, tetravalent derivatives, depending on the substitution pattern.[23] Here the IS value yields at once both the molecularity (monomer or not) and oxidation state [tin(II) for the monomers; tin(IV) for the tin-tin bonded oligomers and polymers]. By this technique it was determined that the 2,6-$(CF_3)_2$, 2,4,6-$(t-C_4H_9)_3$, 2,4,6-$(H_3CO)_3$, 2-$H_3COCH_2$ and 2-$(H_3C)_2CHOCH_2$ groups confer stability on the divalent, monomeric form of diphenyltin(II).

Other, main-group and transition-metal structures can also be predicted using this method. For example, while the species $(C_6H_5)_3SnMBr$ (M = Mg, Zn) and $(C_5H_5)_2C_6H_5SnMgBr$ serve chemically as sources of diorganotin(II), the IS values lower than tin metal rule out the divalent $R_2Sn$→$M(C_6H_5)Br$ structures, at least in the

solid state.[24] We have recently confirmed the tin(IV) nature of
$[(CH_3)_3Si]_3SnMgSi(CH_3)_3 \cdot DME$ by X-ray diffraction which shows the
tetrahedral coordination about the tin atom.[25]

Conventional, directly tin-bonded $M-SnR_nX_{3-n}$ transition metal
systems give IS values characteristic of tin(IV),[26] whereas ionic
derivatives with isolated $SnX_3^-$ groups give values characteristic
of tin(II). Thus Mössbauer spectroscopy can be used to distinguish
these two isomeric forms. Such an example is provided by the red
and green forms of the low-spin cobalt(II) bis(1,2-diphenylphosphino)-
ethane (dpe) complex which can be formulated as either the directly
tin(IV)-Co(II) bonded $Cl(dpe)_2CoSnCl_3$, or as the ionic tin(II)
$[(dpe)_2CoCl]^+[SnCl_3]^-$. As it turns out, the IS values for both are
well within the tin(II) range, as is the IS of the bromine analogue,[27]
and the crystal structures confirm that the $[SnCl_3]^-$ ions are remote
in the lattice, and that the color difference must arise from geometry changes in the cations.[28] Similar application of IS data to the
$Cl_3SnNiLC_5H_5-\eta^5$ and $Cl_3SnNiL_2C_5H_5-\eta^5 \cdot S$ (L = CO, $P(C_6H_5)_3$; S = solvent)
systems serves to distinguish the directly bonded tin(IV)-Ni complexes
$Cl_3SnNiP(C_6H_5)_6C_5H_5-\eta^5 \cdot S$ from the ionic $[Cl_3Sn]^-[NiP(C_6H_5)_3C_5H_5-\eta^5 \cdot S]^+$
salts.[29] Mössbauer IS data also signal non-coordinated
Sn(II) moieties[30,31] in $[Rh(NH_3)_6]^{3+}[Rh(SnCl_3)_4SnCl_4]^{5-}[SnCl_6]^{4-} \cdot 4H_2O$,
whose X-ray structure confirms that these species are not directly
attached by Sn-Rh bonds.[32]

It is appropriate at this juncture to take stock of these systematics. It would seem at first surprising that the tin(IV) and
tin(II) oxidation states are separated in Mosbauer spectra by materials which are formally thin(0). The modern view teaches, however,
that the diamond-structured, $\alpha$-tin allotrope[14] is a tin(IV) material
which behaves chemically[15] as such in its reaction with hydrochloric
acid to yield the tin(IV)-containing $SnCl_4 \cdot 5H_2O$. The tetragonal
$\beta$-tin, on the other hand, behaves structurally[14,15] and chemically
as a tin(II) material showing its VSEPR-expected distortion from
perfect $O_h$ symmetry[14] to accommodate the lone pair present in this
oxidation state and reacting with hydrochloric acid to give
$SnCl_2 \cdot 2H_2O$, with HCl gas to give tin(II) materials[33] and with HCl
in ether[34] to give $H_2SnCl_4 \cdot 2(C_2H_5)_2O$. In this view the crossover
between the two oxidation states must come between the $\alpha$-tin IS
value of 2.10 and the $\beta$- of 2.65 mm s$^{-1}$, but where? The differences
in bonding between the two allotropes as it affects their IS values
have been discussed,[35] and indeed there are a number of tin-bearing
systems whose IS values fall into the range 2.10-2.65 mm s$^{-1}$.

Easiest to consider are the alkaline-earth metal salts of the
$[Sn(OH)_3]^-$ anion.[36] Assuming that no direct group IIA-tin bonding
takes place (no structural data are available), then these tin(II)
materials can help to reassign the velocity dividing the two oxidation states at lower values. For example, while $Ba^{2+}[Sn(OH)_3]_2^-$
exhibits an IS = 2.59, its dihydrate has IS = 2.44 mm s$^{-1}$.

In addition, the structurally characterized $\{[(CH_3)_3Si]_2CH\}_2Sn$ $\rightarrow Cr(CO)_5$ which contains three-coordinated tin[37] exhibits an IS = 2.21 (its Mo analogue gives 2.15 mm s$^{-1}$).[38] Assuming the Mo analogue is in fact isostructural, the dividing line becomes within conventional experimental error from the $\alpha$-tin value of 2.10 mm s$^{-1}$. From the low-velocity side, the directly Sn-Ir bonded (2.642 Å)[39] (cyclo-$C_8H_{12})_2IrSnCl_3$ exhibits[40] an IS = 2.06 mm s$^{-1}$, again within experimental error of the $\alpha$-tin value.

Thus from these systematics it would appear that $\alpha$-tin is the Sn(IV) system with the highest velocity IS value, and that 2.10 mm s$^{-1}$ should now be assigned as the dividing line between the two tin oxidation states, a suggestion put forth by Donaldson[41] in 1979.

Reassignment of this important reference point is useful, but does not serve to remove all the ambiguities presented by some of the nominally tin(II) derivatives. These systems have been categorized into seven classes,[42] with one, embracing tin(II) compounds with electropositive ligands, having no known examples (the claimed $Sn[W(CO)_3C_5H_5-\eta^5]_2$ (IS = 2.08 mm s$^{-1}$)[43] having been shown to be[44] $HSn[W(CO)_3C_5H_5-\eta^5]_3$). Two of the enumerated categories give rise to situations in which it becomes exceedingly difficult, even with both Mössbauer and structural data available, to decide upon the oxidation state of the tin atom. The first involves compounds in which the tin(II) lone-pair electrons may be regarded as taking part in weak donor-acceptor interactions. While the first such claimed example,[45] ($\eta^5$-$C_5H_5)_2Sn \rightarrow BF_3$, has now been shown to be[46] $\{[BF_4]^- (\eta^5$-$C_5H_5)_2Sn[\eta^5$-$C_5H_5Sn]^+ \cdot THF\}_n$, the solid-state, centrosymmetric, diamagnetic, non-planar dimer,[47] $[\{[(CH_3)_3Si]_2CH\}_2Sn]_2$, has a tin-tin distance of 2.764 Å and an IS[38] = 2.16 mm s$^{-1}$. An ethylenic $R_2Sn=SnR_2$ formulation would render the tins as tin(IV), but an alternative tin(II) scheme involving mesomeric $R_2Sn \rightarrow SnR_2 \leftrightarrow R_2Sn \leftarrow SnR_2$ forms or arc-shaped bonds between the tin atoms is available.[47] The weakness of the interaction is reflected in the monomeric nature of the material in the solution and vapor state,[48] which must contain tin(II), where it should exhibit a higher velocity IS. The isoelectronic $\{[(CH_3)_3Si]_2N\}_2Sn$ analogue[49,50] which in the solid,[51] solution[50] and vapor[52] is a monomer, isostructural with the monomeric, gas-phase form[48] of $\{[(CH_3)_3Si]_2CH\}_2Sn$, exhibits[49] an IS = 3.52 mm s$^{-1}$. It is a general rule that a rise in coordination number gives rise to a shift in IS to lower velocities. Without the IS value of a matrix-isolated, monomeric sample of $\{[(CH_3)_3]_2CH\}_2Sn$, it is not possible to assign the oxidation state of the tin atoms in its dimer, or to decide among the various bonding materials presented.[47]

In another related category is found the analogue of the $\{[(CH_3)_3Si]_2CH\}_2SnCr(CO)_5$ compound mentioned above (IS = 2.21 mm s$^{-1}$)[38] with a pyridine (py) molecule coordinated to the tin (II) atom.[53] The compound $(t$-$C_4H_9)_2SnCr(CO)_5 \cdot py$ exhibits an IS = 2.11 mm s$^{-1}$, lowered on increasing the coordination number at tin from three to four. This system could then be formulated as the tin(II) py→Sn→Cr(CO)$_5$ or as a tin(IV), ylide, py$^+$-Sn-Cr(CO)$_5$, with such charge separation helping to rationalize the large QS value found (3.44 mm s$^{-1}$).[54] Again, no definite answer is available.

Other tin-transition metal systems with IS values close to that of α-tin include the acac-type derivative (pbd)$_2$SnW(CO)$_5$ (IS = 2.13),[55] ClPtP(C$_2$H$_5$)$_3$(SnR$_2$)(SnR$_2$Cl) where R = CH[Si(CH$_3$)$_3$]$_2$ (IS = 2.05 for the first and 1.73 for the second),[38] cis-ClFe(CNAr)$_4$SnCl$_3$ (IS = 2.02),[56] the cycloocta-1,5-diene (cod) (IS = 2.16) and norbornadiene (nbd) (IS = 2.23) Rh and analogues[40] of the Ir compound[40] containing a direct bond[39] to tin(IV) discussed above, [(C$_6$H$_5$)$_3$P]$_2$(cod)RhSnCl$_3$ (IS = 2.15), [(C$_6$H$_5$)$_3$P]$_2$(cod)IrSnCl$_3$ (IS = 2.16),[40] [(CH$_3$)$_2$C$_6$H$_5$As]$_2$(cod)IrSnCl$_3$ (IS = 2.17),[40] the 1,2-bis-(diphenylarsino)ethane (dpae) and 1-diphenylarsino-2-diphenylphosphinoethane (dpdae) Ir complexes, (dpae)(cod)IrSnCl$_3$ (IS = 2.15) and (dpdae)(nbd)IrSnCl$_3$ (IS = 2.12),[40] the nbd and cod derivatives of [(CH$_3$O)$_2$C$_6$H$_5$P]$_2$IrSnCl$_3$ (IS = 2.08 and 2.04, respectively),[40] (nbd)$_2$Ir-SnCl$_3$ (IS = 1.98)[40] the tetrafluorobenzylbarralene (tfbb) iridium complex, (tfbb)IrSnCl$_3$ (IS = 1.98),[40] the [(CH$_3$)$_4$N][ClIr(SnCl$_3$)$_5$] salt (IS = 2.08)[57] the o-phenylenebis(dimethylarsine) (diars) Fe complex Cl(diars)$_2$FeSnCl$_3$ (IS = 1.98),[58] (OC)$_4$FeSnCl$_3$ (IS = 2.01)[38] and the Cl(t-C$_4$H$_9$)$_2$AsSnM(CO)$_5$ species (M = Cr, Mo and W) (IS = 2.26, 2.1 and 2.11 mm s$^{-1}$, respectively).[59] Unfortunately, no structural data are available which would tell whether the halotin moieties are directly bonded to the transition metal or lie at a remote distance as an anion, or whether the species listed above are in fact dimers containing four-membered, Sn$_2$M$_2$ rings.

Nevertheless, the rule that the two tin oxidation states fall on opposite sides of the resonance for elemental tin[8] (now taken[41] to be close to α-tin) appears to have no known exceptions authenticated by structural data.[5,6] The method can be used to determine that a compound contains tin in two oxidation states, as in[60] [η$^5$-(CH$_3$)$_3$Sn(IV)C$_5$H$_4$]$_2$Sn(II), or Sn(SO$_3$CF$_3$)$_3$ which was reformulated[61] as Sn(II)[Sn(IV)(SO$_3$CF$_3$)$_6$], or [Sn(II)Sn(IV)O(O$_2$CCF$_3$)$_4$]$_2$[62] or (Sn(II)Sn(IV)O(O$_2$CC$_6$H$_4$NO$_2$)$_4$·THF]$_2$[63] for which X-ray structures are confirmatory, or to show that Sn$_3$F$_8$ is actually [SnF]$_2^+$[SnF$_6$]$^{2-}$, also confirmed structurally.[64] The oxidation of tin(II) species can also be monitored as with the slow precipitation of a tin(IV) species[19] from an ether solution of bis(η$^5$-phenylcarboranyl)tin(II),[18] or the oxidation which accompanies the hydrolysis of tin-oxygen[18] and -sulfur[65] heterocycles, or the conversion of tin(II) chloride on incorporation into a porphyrin ring[66] to give an eight-coordinated tin(IV) sandwich structure,[67] or the oxidation of Sn(II)O(OH)$_2$SO$_4$ on heating[68] above 230°C, or the oxidation of tin(II) chloride on complexation with monohydroxamic acids.[69] A reversible change from tin(II) to tin(IV) can also be monitored[70] in CsSnBr$_3$ at 418K. Reaction product analysis can be aided as in the distinction between SnBr$_2$ diethylacetylenedicarboxylate which is a complex, and other

Table 1.  Selected Compounds with Extreme Tin-119m Mössbauer Parameters, mm s$^{-1}$

| Compound | IS | Ref. |
|---|---|---|
| **Low-Velocity IS:  Inorganic** | | |
| $Na_2SnF_6$ | -0.61 | 75 |
| $SnF_4$ | -0.48 | 75 |
| $SnP_2O_7$ (298K) | -0.40 | 76 |
| $[SnF]^+[Sn(SO_3F)_6]^{2-}$ | -0.37 | 64 |
| $[Cl_2O]^{2+}[Sn(SO_3F)_6]^{2-}$ | -0.30 | 77 |
| $Sn(SO_3F)_4$ | -0.28 | 78 |
| $Cs_2[Sn(SO_3F)_6]^{2-}$ | -0.26 | 79 |
| $Cs[Sn(SO_3F)_5]$ | -0.24 | 79 |
| $Sn^{2+}[Sn(SO_3CF_3)_6]^{2-}$ | -0.24 | 61 |
| **Organotin** | | |
| $(O_2NCH_2)_3SnCl$ | 0.00 | 80 |
| $(CH_3)_2Sn$(kojic acid anion)$_2$ | 0.22 | 81 |
| $[CH_3Sn(OH)]_n$ | 0.40 | 82 |
| $CH_3SnCl$(tropolonate)$_2$ | 0.53 | 81 |
| $C_6H_5SnCl_3 \cdot 4$ morpholine | 0.53 | 83 |
| $C_6H_5SnCl$[diacetylbis-(benzoylhydrazone)] | 0.56 | 84 |
| $C_6H_5Sn(NCS)_3 \cdot 1,10$-phen | 0.57 | 81 |
| $C_6H_5SnCl$(acac)$_2$ | 0.61 | 85 |
| $(C_6H_5)_2Sn$(8-hydroxyquinoline)$_2$ | 0.64 | 86 |
| $(C_6H_5)_2Sn$(tropolonate)$_2$ | 0.78 | 81 |
| **High-Velocity IS:  Inorganic** | | |
| $Sn^{2+}[Sn(SO_3CF_3)_6]^{2-}$ | 4.69 | 61 |
| 3-Sn-1,2-$B_9C_2H_{11}$[3-stanna-1,2-dicarba-closo-dodecarborane(11)] | 4.67 | 87 |
| $Sn^{2+}(SbF_6)_2(AsF_3)_2$ | 4.66 | 88 |
| $3Sn_3Cl_2 \cdot 2$(15-crown-5) | 4.59 | 89 |
| $Sn(SbF_6)_2$ | 4.44 | 90 |
| **Organotin** | | |
| $[\eta^5-(CH_3)_5C_5]_2Sn \cdot Sn(NO_3)_2$ | 4.10 | 91 |
| $[\eta^5-(C_6H_5)_5C_5]_2Sn$ | 4.00 | 92 |
| $[\eta^5-(CH_3)_5C_5]_2Sn$ | 3.83 | 21 |
| $[\eta^5-(CH_3)_5C_5Sn]^+[CF_3SO_3]^-$ | 3.81 | 21 |

| Large QS | QS | Ref |
| --- | --- | --- |
| $(CH_3)_2Sn(SO_3F)_2$ | 5.54 | 78 |
| $(CH_3)_2Sn(SO_3CF_3)_2$ | 5.51 | 78 |
| $(CH_3)_2SnCl_2 \cdot HSO_3F$ | 5.41 | 93 |
| $(CH_3)_2Sn(F_2PO_2)_2$ | 5.13 | 94 |
| $(CH_3)_3SnCl \cdot HSO_3F$ | 4.95 | 93 |
| $(CH_3)_3SnSbF_6$ | 4.75 | 95 |
| $\{[(CH_3)_3Si]_2CH\}_2SnCr(CO)_5$ | 4.43 | 38 |

materials from analogous reactions which are the result of oxidative addition,[71] or in the identification[72] of $(CF_3)_2Sn(II)$. Radiation damage often produces tin(II) species, as on Co-60 irradiation[73] of $(n-C_4H_9)SnSO_4$ to give $SnSO_4$, or in $K_6^{119m}Sn\ IV)(C_2O_4)_7 \cdot 4H_2O$, used as a source, giving photochemical conversion to tin(II) atoms following internal conversion at the tin-119m sites.[74]

Some examples of compounds whose IS values define the extreme range of the tin-119m Mössbauer spectrum are gathered in Table 1. It is noted that the IS data for the two tin sites in the one salt, $Sn^{2+}[Sn(SO_3CF_3)_6]^{2-}$ (-0.24 and 4.69 mm s$^{-1}$),[61] span nearly the whole range of values!

## Fingerprinting

Given the rather broad lines ubiquitous in tin-119m Mössbauer spectroscopy, it is imperative that negative evidence establishing identity be disregarded. Thus the technique should only be applied to demonstrate that two systems are not the same. Even here, interlaboratory comparisons and comparisons with older data may lead to spurious conclusions. In a now classic case, for example, the two tin(IV) atoms in the double salt $[(CH_3)_2SnCl \cdot terpy]^+[(CH_3)_2SnCl_3]^-$, one cationic and six-coordinated, the other anionic and five-coordinated,[96] give identical spectral parameters in a perfectly overlapping doublet.[97] Neither can the two four-coordinated $(\mu-\eta^5-C_5H_5)_3SnX$ (X=O(CH_2)_4, $BF_4^-$)tin(II) sites be distinguished[46] in $\{[BF_4]^-(\mu-\eta^5-C_5H_5)_2Sn[\mu-\eta^5-C_5H_5Sn]^+ \cdot THF\}_n$.

However, restricting the method to cases yielding positive results, examples of decisive contributions abound. For instance, a new, orthorhombic form of $SnF_2$ (IS = 3.30; QS = 2.20) was discovered[90,98] to accompany the known monoclinic material (IS = 3.70; QS = 1.80 mm s$^{-1}$).[90,98,99] Materials containing both tin(II) and tin(IV) atoms have been mentioned above,[60-64,70] and the ability to distinguish tin(II) from tin(IV) bromide after exposure of tin foil to $Br_2$ allows the surface $SnBr_4$ to be detected in the presence of a deeper $SnBr_2$ layer.[100] Two tin(II) sites[101] in $Sn^{2+}[SnEDTA]^{2-}\cdot 2H_2O$ can be seen at 78-226K as two overlapping doublets.[102] Two organo-tin(IV) sites can be detected and rationally assigned in series of N,O-bis(trialkyltin)carbamates[103] and O,S-bis(trialkyltin)mercapto-carboxylates.[104]

Extending this concept to intercompound comparisons encompassing both IS and QS data allows conclusions to be drawn, for example, concerning which compounds in a homologous series are likely to be iso-structural and which are not. Thus were two isomeric forms of $BF_3SnCl_2L$ (L=bipy, TMED) detected and assigned as having the Lewis acid attached to the base nitrogen or to the tin(II) atom.[105] In-tensely colored, air-sensitive $(R_2N)_2Sn(II)$ are monomeric species[51,52] with higher-velocity IS and larger QS values than their more air-stable, presumably associated, colorless analogues ($[(CH_3)_2N]_2Sn$ is a dimer in the solid state[106]). Methyl-substitution at the $\alpha$-carbon of the amine suffices to give rise to the onset of color, air-sensi-tivity and enhanced IS and QS magnitudes.[107] Likewise, among the triorganotin(IV) aziridide (ethylene imino) derivatives, the tri-methyl compound exhibits a QS value (2.37) higher by 0.9 mm $s^{-1}$ than the cyclic compounds bearing three ring carbons, either $R_3SnN(CH_2)_3$ or $R_3SnNCH(CH_3)CH_2N$ (QS = 1.4), or than the compounds bearing four or more carbons attached to the nitrogen (QS = 1.0). The inter-molecular association these data suggest for $(CH_3)_3SnN(CH_2)_2$ is corroborated by solution osmometry which shows dimers, by mass spec-tral fragments containing two tin atoms, by its relative air-stability[108] and by variable-temperature Mössbauer studies[109] dis-cussed below. A similar discontinuity arises among the triorgano-tin(IV) azides, $R_3SnN_3$, when R=$CH_3$ to give higher-velocity IS and larger QS values, interpreted to mean the onset of higher coordina-tion,[110,111] but in this case there are two possible modes of N→Sn donation, utilizing either the $\alpha$- or $\gamma$-(terminal) nitrogen of the pseudohalogen. From the Mössbauer data it is not possible to decide, but the X-ray data show bridging through the $\alpha$-nitrogen.[111,112]

Listing the enumerable examples of compounds suspected of adopt-ing different geometries and coordination numbers at tin and dif-ferent solid-state packing arrangements from their cogeners in homo-logous series is beyond the scope of this essay. Most of the IS and QS results are used to suggest that systems with favorable shapes can form associated solids, as demonstrated above, and these data are taken together with those from other appropriate spectroscopies to argue the brief. Unfortunately, in few cases is corroboration from X-ray diffraction studies[5,6] available. Mössbauer data on the matrix-isolated, monomeric molecules would help to clarify whether the dif-ferences observed in fact arise from association, but such studies are by no means routine.[113] However, preliminary results on tin metal[114] and tin(II) halides[115] are hopeful in this regard.

## Tin-Atom Site Symmetries

The most troublesome phenomenon in tin-119m Mössbauer spectro-scopy is the absence of clearly resolvable quadrupole splitting in compounds possessing obvious bond asymmetry at the tin atom. Many compounds in which tin atoms depart from cubic symmetry fail to yield doublet spectra, and in fact give sharp singlets. This contradiction with basic theory was first observed[116] in 1964 and gave rise to many attempts to rationalize why some $<T_d$ and $<O_h$ compounds give doublets and some only singlets. At the root is of course the rather broad tin-119m Mössbauer resonance lines. Tin(II) compounds because of the presence of the lone pair of electrons in oxidation state nearly always yield doublet, or obviously broadened singlet spectra. In addition, the fac-$SnA_3B_3$ isomer would be expected to have zero electric field gradient at the tin center.

Greenwood generalized[117-119] that when tin is surrounded by atoms, all of which either possess or lack lone pairs of electrons, no QS is resolved, but when some of each kind are bonded, doublet spectra are recorded. Thus derivatives containing only hydrogen or groups IA-IVA elements bonded to tin, or instead only groups VA-VIIA elements, will in general exhibit no resolvable QS, but derivatives in which tin holds elements drawn from each of these two categories will display the expected QS. Listing known data reveals violations to this general rule, most having doublet spectra where none would be expected on the basis of Greenwood's suggestion. These involve $R_nSnR'_{3-n}$ derivatives with severely electron-perturbing groups in R', or severely unbalanced combinations of atoms holding lone pairs or certain transition-metal derivatives.[8]

Tin Mössbauer spectroscopists have by now accommodated to this anomaly, and are careful to apply their models rationalizing QS magnitudes only to those compounds capable of possessing resolvable doublet spectra. Of course, compounds actually showing doublet spectra must possess lower than cubic symmetry at the tin site, and this can aid product analysis, as has been demonstrated innumerable times. Less secure is the distinction between the generally smaller QS values derived from compounds which stand as exceptions to the Greenwood generalization and those which follow it. In an early application to rapid product analysis, the cyclic $R_2SnR'_2$-type 1,1 dimethylstannole which gives no resolvable QS in accord with the Greenwood generalization was quickly distinguished[120] from an open-chain chlorotin(IV), $R_2SnR'Cl$ analogue with an identical melting range in which QS = 2.68 mm s$^{-1}$. Such applications are now typical.

## The $\rho$ (=QS/IS) Value and Coordination Number in Tin(IV) Derivatives

The QS/IS ratio, proposed early on as a guide to tin(IV) coor-dination number,[121] is still effective as a crude indication of higher than four-coordinated state at tin. The original suggestion, that ρ values >2.1 are associated with higher coordination and <1.8 with four coordination, seems to have stood the test of time well.[8]

The systematics are a reflection of the general decrease in IS and increase in QS on moving to higher-coordinated tin(IV) species. Again, as pointed out above, the rule is applied only to those systems in which cleanly resolvable QS split doublets can be observed, i.e., predominately the organotin(IV) systems also holding one or more atoms with lone-pair electrons. Confining the application to these systems, the generalization is a useful first approach to the data.[8]

## The Point-Charge Model, Partial QS Values and the Prediction of <sC-Sn-C in $R_2Sn(IV)$ Systems

The electric-field gradient at the tin nucleus can be treated as arising from the surrounding electrons as distorted from spherical symmetry by the atoms directly bonded to tin. Each atom can be approximated by a point charge, and the molecular point-group symmetries used to calculate the relative field gradients and hence QS values. Equal charges, antishielding factors and tin bond distances are assumed. From such calculations comes the result that for $SnA_2B_4$-type octahedral systems there should be a 2:1 ratio of QS values for the trans- and cis-isomers, that $SnAB_5$ systems should exhibit the same QS as the cis-$SnA_2B_4$ case, and that tetrahedral compounds of the type $SnA_2B_2$ should exhibit a QS 1.6 times larger than that for $SnAB_3$. The first application of this model to rationalize QS data was made by Fitzsimmons[122] in 1968.

Successful use of the model in linking QS data to isomer type is of course limited to those systems giving rise to resolvable doublets. For the organotin(IV) derivatives in which tin is also bonded to atoms holding lone-pair electrons, even the cis-$SnA_2B_4$ isomers generally give QS values of ca. 2 (trans-give $\overline{4}$ mm s$^{-1}$), but for the $SnX_2Y_4$ systems where both X and Y are drawn from groups V-VII, only the trans-isomers yield resolvable doublets,[123] and so the comparison will be unsecured for the cis-isomers.

More serious limitations arise, however. Such a model is too simple for real chemical systems which are covalently bound and in which the tin bonds vary in length and the ligands in charge from ligand to ligand in the same molecule or ion. In addition, few but the most symmetrical systems actually correspond strictly to the classical point-group symmetries employed in the calculations.[5,6] Severe distortions from cis- and trans-geometries and from overall $O_h$ symmetry are legion. In addition, other less well-known isomeric forms involving ionic structures are also possible, and would be missed in treatments limited to considering only $O_h$, $T_d$, etc., geometries. The model seems to yield less-valuable results for five coordination where large distortions from the popular trigonal-bipyramidal geometry are found.[5,6] The QS values for known five-coordinated structures actually span the entire range[124] from four-coordinated $SnA_3B$ to six-coordinated trans-$SnA_2B_4$. Further, assignment of the true coordination number at tin in the solid state is seldom straight forward, and ambiguities abound. Finally, the point-charge model in its simplest form does not yield quantitative predictions of structure.

The first extension of the model was taken by Parish[125] who assigned a set of partial QS values for some of the more common ligands, groups and atoms attached to tin from a parameterization of the known QS data. These values were meant to be universally applied to tin(IV) in any coordination number or geometry based only on the assumption that the QS arises exclusively from intramolecular effects. The goal was to provide a series of numbers which could be used additively to predict QS values. As for other additivity-based methodologies in chemistry (radii, electronegativities, bond dipoles, bond energies, etc.), more efficient parameterization inevitably leads to larger families of numbers appropriate only to narrower classes of systems.

It is apparent that for more accurate results, separate partial QS values are necessary for ligands in different coordination environments, and these were provided by Maddock based upon consideration of a molecular-orbital model.[126] More detailed work by Bancroft[127-129] has improved the accuracy of the calculated QS values possible, but to the same measure limited the applicability of measured QS data to predict structure, since structure must in these treatments be known before the partial QS additivity is applied. In particular, the need to know the coordination number beforehand is a fatal flaw, since the results of X-ray studies would have to be already known. Again, no way to deal with the distortions from ideal geometry in a quantitative way is apparent in this treatment.

Thus it is particularly welcome that in at least one class of tin(IV) compounds, the diorganotin(IV) derivatives, true structural predictions emerge from the recognition by Bancroft[130] that the QS values observed are generated predominantly by the effect of the organic ligands. Since the effect wrought by the other atoms or groups holding lone-pair electrons is so much smaller, it can be ignored. The consequence is that the relationship between the observed QS and the C-Sn-C angle is independent of coordination number and can be calculated from:

$$|QS| = 4\{R\}(1-3\sin^2\theta\cos^2\theta)^{\frac{1}{2}}$$

where $\{R\}$ is the partial QS for R and $<$C-Sn-C $= (180-2\theta)°$. The treatment assumes that there will be no sign inversion in the QS data throughout the systems compared (see below), that the partial QS values for all but the R groups are negligible, and that $\{R\}$ is a constant over the systems compared. Nevertheless, this correlation is supported by the abundant structural data[5,6] for the $(CH_3)_2Sn(IV)$[130,131] and $(C_6H_5)_2Sn(IV)$[132] compounds. Predicted values increase smoothly from ca. 2 for 90° (cis-) to 4 mm s$^{-1}$ for 180° (trans-), and recent crystal structure results lend further confidence.[133-141] However, in certain cases the model breaks down, and QS values well in excess of those calculated for a linear C-Sn-C array are observed.[142-145] A survey of all known $R_2Sn(IV)$ <C-Sn-C data vs. QS values should be undertaken to reveal if the model has real validity.

The point-charge model predicts the signs as well as the relative magnitudes of the QS values: $RSnX_3$(+2); trans-$R_2SnX_4$(+4); cis-$R_2SnX_4$(-2); $R_3SnX_2$(-3 to -4); $R_3SnX$(-2 mm s$^{-1}$) which can be measured by applying a strong magnetic field to a powdered sample, or by observing the intensity ratio of the doublet wings of a single crystal with respect to its principal axis.[8] Sign information can thus potentially help to distinguish isomers and coordination numbers. The first organotin application[146] was to dimethyltin(IV) molybdate and dichloride at 30kG in 1969.

Unfortunately, the early hope of obtaining independent verification of structural predictions from knowledge of the signs of the QS was disappointed in the face of results contradictory with QS magnitudes and X-ray studies when it was learned that sign reversal is a common result of departure from ideal symmetries.[147-149]

## Lattice Dynamics

The magnitude of the Mössbauer resonance effect is related to the recoil-free fraction which depends on the amplitude of thermal motion of in this case the tin-119 nucleus. While spectra have been resolved from viscous molten tin(II) fluoride[150] and smectic liquid crystals,[151] the probability of the Mössbauer effect vanishes for unbounded motion since it is dependent upon the mean-square vibrational amplitude, $<x^2>$ in the direction of $\gamma$-ray emission or absorption. Tin atoms bound anisotropically will have different Mössbauer resonance magnitudes in different directions along crystallographic axes. Hard, refractory tin solids [e.g., $SnO_2$, $SnO$, $SnF_4$, $(CH_3)_2SnO$, $(CH_3)_2Sn(O_2CH)_2$ give strong resonance effects at ambient temperatures which diminish only little when temperature is raised. Soft, molecular solids give weaker resonances which have a sharp temperature dependence. It was an early but still largely unreached goal to relate these factors to the structure and binding of tin materials. Although $<\bar{x}^2>$ should be a function of the tightness of the binding force constants holding the tin atom in the lattice, the real situation appears to be very complex. Organotin compounds generally form molecular solids, i.e., assemblies of molecules which retain their identities in the solid. Two kinds of forces will then be important, the binding of the atom in the molecule and the molecule into the solid, and initial attempts at correlations with melting point temperature, molecular mass, coordination number, solubility, etc., failed.[8]

Resolvable room-temperature spectra signal tight binding, but the attempt to link this phenomenon to association in the lattice in one or more dimensions is thwarted by the observation that tetraphenyltin, a cleanly molecular solid,[5,6] gives such a spectrum.[152] The situation is not measurably improved when area ratios (ambient to 80K) are used instead.[153]

For thin absorbers assuming the Debye model, the recoil-free fraction is linearly related to the area under the resonance, $A_T$, and its temperature dependence is given by

$$A_T\{ f = \exp[- \frac{6E_R T}{k\Theta_M^2}] \text{ for } T \geq \frac{\Theta_M}{2}$$

where $E_R$ is the Mössbauer recoil energy, and $\Theta_M$ is the Mössbauer-probed temperature at the tin site, similar to $\Theta_D$ for a Debye solid. Thus in the high-temperature limit plots of $A_T$ (or $A_T/A_{77K}$ normalized for ease of comparison) vs. T should, where motional anharmonicity can be neglected, be linear. The more tightly bound are the tin atoms, the more gentle the slope. This treatment yields quantitative slope data for comparison.[131,154] Tin(II) oxide gives a particularly low value ($-0.23 \times 10^{-2} K^{-1}$), but solids composed of non-interacting, monomeric molecules exhibit slopes of ca. $-1.8 \times 10^{-2} K^{-1}$, no matter what the coordination number at tin (tetraphenyltin[155,156] gives $-1.60$ to $-1.66 \times 10^{-2} K^{-1}$ and a resolvable spectrum at ambient temperatures[152]). In between are found compounds with strongly hydrogen-bonded lattices, or in which the tin atoms engage in one-, two- or three-dimensional association.

The problem is that the precise lines of demarcation which separate the monomers from the associated species have remained elusive. In addition, while knitting the organotin(IV) molecules into an associated lattice lowers the slope of $A_T/A_{77K}$, as demonstrated for organotin-substituted styrene monomers and polymers,[157] so does hydrogen bonding and other more subtle and generalized lattice forces. For example, the tightly packed lattice of bis(0,0'-diisopropyldithiophosphato)diphenyltin(IV) containing chelated monomers could accommodate its bridged, associated, isomeric form without change in dimensions or density.[158] Its slope of $-1.06 \times 10^{-2} K^{-1}$ reflects a situation which we have termed a "virtual polymer."[132]

Despite the lack of secure interpretation in terms of solid-state structural descriptions, variable-temperature tin-119m Mössbauer data are useful in other regards. For example, slope differences between the ethyl (of known distorted $O_h$ structure) and n-propyl (-1.92 and $-2.19 \times 10^{-2} K^{-1}$, respectively) vs. the isopropyl ($-1.06 \times 10^{-2} K^{-1}$) analogue first signalled[132] that the last named "virtual polymer" was not isostructural[158] (see above). Which tin atom is held more tightly in compounds with multiple sites can be determined, as with the tin(II) crown-ether complexes in which the more tightly held atom was assigned as sandwiched between two crowns.[159] Better structural conclusions will become available when comparisons can be drawn with compounds of analogous formula whose solid-state structures are securely known. However, the slope data are subject to errors inherent in the inclusion of data points collected near the high-temperature limit where the spectra are weakest. These points carry large uncertainties which can introduce significant errors in the calculated least-squares slopes.

When the X-ray structure is also known, the $\langle \bar{x}^2 \rangle_{iso}$ values can be calculated at all temperatures given the temperature dependence and one value of $\langle \bar{x}^2 \rangle_{iso}$. These values can be derived from the derivative with respect to temperature, $d\langle \bar{x}^2 \rangle_{iso}/dT = -\lambda(d\ln A/dT)$, and the one value of $\langle \bar{x}^2 \rangle_{iso}$ is derived from X-ray data.

The anisotropy of the recoil-free fraction reveals itself as an asymmetry in the two wings of doublet spectra. The phenomenon is called the Goldanskii-Karyagin effect,[8] and persists even in a randomly oriented, polycrystalline powder. The effect has been claimed in many one- and two-dimensional organotin(IV) polymers, where the amplitude of motion of the tin atoms would be expected to be anisotropic, but great care must be taken since identical effects would result from spontaneous partial orientation of the sample, and verification is difficult. In a true Goldanskii-Karyagin doublet the more intense wing should also be broader.[160]

Using companion data from X-ray diffraction, the peak-intensity ratios can be made in favorable cases to yield information on the details of the motion of tin atoms in the solid. Where a unique axis can be defined, e.g., the axis of chain propagation in linearly associated organotins or the R-Sn-R axis in trans-$R_2SnX_4$ complexes, the sign of the prinicipal component of the electric field gradient tensor can be assigned and whether the field has a prolate or oblate shape be decided, then the isotropic $\langle \bar{x}^2 \rangle_{iso}$ can be decomposed into its anisotropic $\langle \bar{x}^2 \rangle_{\parallel}$ and $\langle \bar{x}^2 \rangle_{\perp}$ components from the temperature-dependent data. The Mössbauer results can be compared with those from the anisotropic thermal ellipsoids from X-ray studies. For example, in the linear polymer $(CH_3)_3SnO_2CCH_2NH_2$, at 298K the difference between $\langle \bar{x}^2 \rangle_{\parallel}$ and $\langle \bar{x}^2 \rangle_{\perp}$ from X-ray is $4.24 \times 10^{-2}$ vs. $2.96 \times 10^{-2} \overset{\circ}{A}^2$ from Mossbauer.[161] For the distorted ethyl dithiophosphate mentioned above, $\langle \bar{x}^2 \rangle_{\parallel}$ and $\langle \bar{x}^2 \rangle_{\perp}$ at 77K are 1.80 and $2.42 \times 10^{-2}$ vs. 0.65 and $0.81 \times 10^{-2} \overset{\circ}{A}^2$ for the tighter isopropyl "virtual polymer" analogue.[132]

Herber has developed a model which undertakes to answer a fundamental question concerning organotin lattices, namely the molecularity and hence composition of the vibrating mass. From a combination of variable-temperature resonance area and low-energy ($<200$ cm$^{-1}$) lattice-mode absorption data from Raman spectra the treatment allows the effective vibrating mass of a Debye solid to be calculated. Contrarywise, linking the vibrating mass with the Raman frequencies allows the latter to be assigned in terms of the recoiling and vibrating mass in the solid as a hard-sphere particle interacting with the rest of the unit cell masses by van der Waals intermolecular forces.[156,162] For example, in the structurally characterized,[161] hydrogen-bonded $(CH_3)_3SnO_2CCH_2NH_2$ linear polymer, Raman bands are observed at 46, 31 and 23 cm$^{-1}$ which correspond in this treatment to the dimer, tetramer and heptamer, respectively. No band corresponding to the monomer (expected at 61 cm$^{-1}$) is found. These results are interpreted in terms of two, four and seven dimer units of the chain vibrating against the units of an adjacent chain linked to it by hydrogen bonds.[161] For the ethyl dithiophosphate ester derivative discussed above which forms a solid of monomers, the lowest frequency Raman absorption at 28 cm$^{-1}$ corresponds to the monomer (680 vs. 643 daltons calcd.), but for the "virtual polymer" isopropyl analogue the observed bands at 16, 25 and 34 cm$^{-1}$ correspond to 5.4, 2.4 and 1.2 monomer units.

One difficulty lies in the possible absence of low-energy bands important to this treatment if only ambient-temperature Raman data are recorded. The key frequency sought for $Sn[S(CH_2)_2S]_2$ could only be observed at 77K, for example.[162] In addition, for the structurally characterized[163] O,O'-diethyl dithiophosphato triphenyltin(IV) monodentate monomer, Raman frequencies at 34 and 22 cm$^{-1}$ correspond to 1.16 and 2.78 molecular units, respectively, which also seem to be signalling an associated solid.[164] Finally, there is the question of whether the effective vibrating unit as determined in the treatment is capable of independent experimental test. In the most intriguing cases even the results of X-ray studies make the question of the nature of the unit preserving its integrity in the solid a matter of speculation. For example, while the lowest frequency absorption in the Raman spectrum of tetraphenyltin (39 cm corresponds to a vibrating mass of 420 (vs. 427 daltons for the monomer) despite the relatively low slope of its Mössbauer temperature dependence (see above),[155,156] the Raman spectrum of diphenyltin dichloride contains bands at 41 corresponding to the monomer (350 vs. 343 daltons calcd)[155,156] and at 20 cm$^{-1}$ which may correspond to the bridged unit described in the reinterpretation[165] of the original structural data.[166] It is not apparent that anything short of an electron density distribution study could decide this question.

THE NEXT 25 YEARS

Extrapolating the known into the unknown is impossible without knowing the equation of the curve that connects the past with the future. The only common feature linking predictions of our scientific future is that they almost always prove to be wrong. "Predicting is hard," observed one wag, "especially predicting the future." But predict, we will.

The next quarter century should certainly bring about a situation in which tin-119m Mössbauer data will accompany all claims in the literature to the synthesis and characterization of new compositions of matter containing tin. Hotter, cheaper tin-119m source material should encourage the routine recording of spectra to guide synthesis, separation and purification. These, together with user-friendly instrumentation will lead to the collection of much new data, and the redetermination of older values so that the current hodgepodge of randomly collected and scattered parameters on materials of doubtful purity can be replaced by accurate compilations readily accessible by computer which can be applied with confidence.

Hotter, cheaper sources will also encourage the routine collection of variable-temperature data leading to a much better understanding of the systematics of the log $A_T/A_{77K}$ vs. T slopes. Easier availability of X-ray diffractometer time will mean a vastly increased library of known structural data for comparison. In fact, too easy an access to quick molecular structures can threaten to make chemical spectroscopists redundant!

The invention of a routine technique for preparing matrix-isolated organotin species in solid hydrocarbons or other materials which are hard solids at ambient temperatures is needed. The spectra of these samples can then help determine definitively the states of aggregation of organotin solids, as well as aid in developing new chemical reactions of these discrete monomers.

Application of pulsed synchrotron radiation can lead tin-119m Mössbauer spectroscopy into unexpected places and encompass phenomena that no one today has glimpsed.

None of these predictions may come to pass in the next 25 years, or ever. But one thing is certain. The future of tin-Mössbauer spectroscopy depends critically on attracting into its study bright, industrious and imaginative young men and women. Nurturing their development and encouraging their efforts are central tasks of today's practitioners. In this way we ensure a bright future for our discipline.

REFERENCES

1. Barloutard, R., Cotton, E. Picou, J.-L., and Quidort, J., C.R. Hebd. Seances Acad. Sci., 250:319 (1960).

2. Delyagin, N. N., Spinel, V. S., Brukharov, V. A., and Zvenglinski, B., Sov. Phys.-JETP, 12:159 (1960).

3. Zuckerman, J. J., ed., ORGANOTIN COMPOUNDS: New Chemistry and Applications, Advances in Chemistry Series, No. 157, American Chemical Society, Washington, D.C. (1976).

4. Zuckerman, J. J., Reisdorf, R. P., Ellis, H. V., III, and Wilkinson, R. R. in Chemical Problems in the Environment: Occurrence and Fate of the Organoelements, ed. by Bellama, J. M., and Brinckman, F. E., ACS Symposium Series, No. 82, American Chemical Society, Washington, D.C. (1978), p. 388.

5. Zubieta, J. A. and Zuckerman, J. J., Prog. Inorg. Chem., 24:251 (1978).

6. Smith, P. J., J. Organomet. Chem. Library, 12:97 (1981).

7. Petrosyan, V. S., Prog. NMR Spectrosc., 11:115 (1971).

8. Zuckerman, J. J., Adv. Organomet. Chem., 9:21 (1970).

9. Debye, N. W. G. and Zuckerman, J. J., in Determination of Organic Structures by Physical Methods, ed. by Zuckerman, J. J., Vol. 5, Academic Press, New York (1973), p. 235.

10. Parish, R. V., Prog. Inorg. Chem., 15:101 (1972).

11. Mössbauer spectroscopy is surveyed each year in the Specialist Periodical Reports of the Royal Society of Chemistry, London, in the series Spectroscopic Properties of Inorganic and Organo-metallic Compounds, Davidson, G. and Ebsworth, E. A. V., senior reporters (see, for example, Volume 15, 1983), and every other year in the April issue of Analytical Chemistry [see, for example, Bowen, L. H. and Stevens, J. G., Anal. Chem., 56:199R (1984)]. The Mössbauer Effect Data Index, ed. by Stevens, J. G. and Stevens, V. E. (IFI/Plenum Press, New York) compiles the literature data annually. Compilations of tin-119m data appear in ref. 8 as well as in Smith, P. J., Organomet. Chem. Rev., A, 5:373 (1970) and in Ruddick, J. N. R., Rev. Si, Ge, Sn, Pb Compnds., 2:115 (1976).

12. For an exposition of the history of the introduction of new physical methods in chemistry see Zuckerman, J. J., in Mössbauer Spectroscopy and Its Chemical Applications, ed. by Stevens, J. G. and Shenoy, G. K., Advances in Chemistry Series, No. 194, American Chemical Society, Washington, D.C. (1981), p. 619.

13. Karl, R. E. and Zuckerman, J. J., in Mössbauer Spectroscopy and Its Chemical Applications, ed. by Stevens, J. G. and Shenoy, G. K., Advances in Chemistry Series, No. 194, American Chemical Society, Washington, D.C. (1981), p. 221.

14. Donohue, J., Structures of the Elements, Wiley-Interscience, New York (1974), p. 272.

15. Wells, A. F., Structural Inorganic Chemistry, 4th ed., Oxford University Press, Oxford (1975), p.103.

16. Shpinel, V. S., Bryukhanov, V. A., and Delyagin, N. N., Sov. Phys.-JETP, (Engl. Transl.), 14:1256 (1962).

17. Goldanskii, V. I., Rochev, V. Ya., and Khrapov, V. V., Dokl. Akad. Nauk SSSR, 156:909 (1964).

18. Bearden, A. J., Marsh, H. S., and Zuckerman, J. J., _Inorg. Chem._, 5:1260 (1966).
19. Aleksandrov, A. Yu., Bregadze, V. I., Goldanskii, V. I., Zakharkin, L. I., Okhlobystin, O. Yu., and Khrapov, V. V., _Dokl. Akad. Naud SSSR_, 156:804 (1965).
20. Harrison, P. G. and Zuckerman, J. J., _J. Am. Chem. Soc._, 91:6885 (1969).
21. Dory, T. S. and Zuckerman, J. J. _J. Organomet. Chem._, 264:295 (1984).
22. Dory, T. S., Zuckerman, J. J., and Rausch, M. D., _Abstr. 182nd American Chemical Society National Meeting_, New York, Aug., 1981.
23. Bigwood, M. P., Corvan, P. J., and Zuckerman, J. J., _J. Am. Chem. Soc._, 103:7643 (1981).
24. Harrison, P. G., Zuckerman, J. J., and Noltes, J. G., _J. Organomet. Chem._, 31:C23 (1971).
25. Hahn, F. E., Rösch, L., Starke, U., Hossain, M. B., van der Helm, D., and Zuckerman, J. J., unpublished results.
26. Fenton, D. E. and Zuckerman, J. J., _Inorg. Chem._, 8:1771 (1969).
27. Stalick, J. K., Meek, D. W., Ho, B. Y. K., and Zuckerman, J. J., _J. Chem. Soc., Chem. Commun._, 630 (1972).
28. Stalick, J. K., Corfield, P. W. R., and Meek, D. W., _Inorg. Chem._, 12:1668 (1973).
29. Bancroft, G. M. and Butler, K. D., _Can. J. Chem._, 53:307 (1975).
30. Kimura, T., _Sci. Papers I.P.C.R._, 73:31 (1979).
31. Sakurai, T. and Kobayashi, K., _Rep. Inst. Phys. Chem. Res._, 55:69 (1979).
32. Kimura, T. and Sakurai, T., _J. Solid State Chem._, 34:369 (1980).
33. Gela, T., _J. Chem. Phys._, 24:1009 (1956).
34. Bulten, E. J. and Van der Kirk, J. W. G., _J. Organomet. Chem._, 162:161 (1978).
35. Silver, J., Mackay, C. A., and Donaldson, J. D., _J. Mater. Sci._, 11:836 (1976).
36. Davies, C. G. and Donaldson, J. D., _J. Chem. Soc._, A., 946 (1968).
37. Cotton, J. D., Davidson, P. J., and Lappert, M. F., _J. Chem. Soc., Dalton Trans._, 2275 (1976).
38. Cotton, J. D., Davidson, P. J., Lappert, M. F., Donaldson, J. D., and Silver, J., _J. Chem. Soc., Dalton Trans._, 2286 (1976).
39. Porta, R., Powell, H. M., Jr., Mawbry, R. J., and Venanzi, L. M., _J. Chem. Soc._, A, 455 (1967).
40. Mays, M. J. and Sears, P. L., _J. Chem. Soc., Dalton Trans._, 2254 (1974).
41. Donaldson, J. D. and Tricker, M. J., in _Spectroscopic Properties of Inorganic and Organometallic Compounds - A Specialist Periodical Report_, Vol. 11, Royal Society of Chemistry, London (1979), p. 381.
42. Harrison, P. G. and Zuckerman, J. J., _Inorg. Chim. Acta_, 21:L3 (1977).
43. Harrison, P. G. and Stobart, S. R., _J. Chem. Soc., Dalton Trans._, 940 (1973).

44. Dory, T. S., Zuckerman, J. J., Hoff, C. D., and Connolly, J. W., J. Chem. Soc., Chem. Commun., 521 (1981).

45. Harrison, P. G. and Zuckerman, J. J., J. Am. Chem. Soc., 92:2577 (1970).

46. Dory, T. S., Barnes, C. L., van der Helm, D., and Zuckerman, J. J., unpublished results.

47. Davidson, P. J., Harris, D. H., Lappert, M. F., J. Chem. Soc., Dalton Trans., 2268 (1976).

48. Fjeldberg, T., Haaland, A., Lappert, M. F., Schilling, B. E. R., Seip, R., and Thorne, A. J., J. Chem. Soc., Chem. Commun., 1407 (1982).

49. Schaeffer, C. D., Jr. and Zuckerman, J. J., J. Am. Chem. Soc., 96:7160 (1974).

50. Lappert, M. F. and Power, P. P., in ref. 3, p. 70.

51. Fieldberg, T., Hope, H., Lappert, M. F., Power, P. P., and Thorne, A. J., J. Chem. Soc., Chem. Commun., 639 (1983).

52. Lappert, M. F., Power, P. P., Slade, M. J., Hedberg, L., Hedberg, K., and Schomaker, V., J. Chem. Soc., Chem. Commun., 639 (1983).

53. Brice, M. D. and Cotton, F. A., J. Am. Chem. Soc., 95:4529 (1973).

54. Grynkewich, G. W., Ho, B. Y. K., Marks, J. J., Tomaja, D. L., and Zuckerman, J. J., Inorg. Chem., 12:2522 (1973).

55. Cornwell, A. B. and Harrison, P. G., J. Chem. Soc., Dalton Trans., 1486 (1975).

56. Bancroft, G. M. and Butler, K. D., J. Chem. Soc., Dalton Trans., 1209 (1972).

57. Antonov, P. G., Kukushkin, Yu. N., Shkredov, V. F., and Vasilev, L. N., Koord. Khim., 5:561 (1979).

58. Parish, R. V. and Rowbotham, P. J., J. Chem. Soc., Dalton Trans., 37 (1973).

59. du Mont, W.-W., Lefferts, J. L., and Zuckerman, J. J., J. Organomet. Chem., 166:347 (1979).

60. Bulten, E. J. and Budding, E. A., J. Organomet. Chem., 157:C3 (1978).

61. Batchelor, R. J., Ruddick, J. N. R., Sams, J. R., and Aubke, F., Inorg. Chem., 16:1414 (1977).

62. Birchall, T. and Johnson, J. P., J. Chem. Soc., Dalton Trans., 69 (1981).

63. Ewings, P. F. R., Harrison, P. G., Morris, A., and King, T. J., J. Chem. Soc., Dalton Trans., 1602 (1976).

64. Dove, M. F. A., King, R., and King, T. J., J. Chem. Soc., Chem. Commun., 944 (1973).

65. Honnick, W. D. and Zuckerman, J. J., Inorg. Chem., 17:501 (1978).

66. Burnham, B. F. and Zuckerman, J. J., J. Am. Chem. Soc., 92:1547 (1970).

67. Bennett, W. E., Broberg, D. E., and Baenziger, N. C., Inorg. Chem., 12:930 (1973).

68. Davies, C. G. and Donaldson, J. D., J. Chem. Soc., A, 1790 (1967).

69. Das, M. K., Ghosh, M. R., and Zuckerman, J. J., Curr. Sci., 49:428 (1980).

70. Donaldson, J. D., Grimsey, R. M. A., and Clark, S. J., J. Phys.

Colloq., C2 (1979) quoted in Spectroscopic Properties of Inorganic and Organometallic Compounds - A Specialist Periodical Report, Vol. 13, ed. by Adams, D. M. and Ebsworth, E. A. V., Royal Society of Chemistry, London (1981), p. 381.

71. Ewings, P. F. R. and Harrison, P. G., Inorg. Chim. Acta, 18:165 (1976).

72. Hani, R. and Geanangel, R. A., Polyhedron, 1:826 (1982).

73. Aleksandrov, A. Yu., Delyagin, N. N., Mitrofanov, K. P., Polak, L. S., and Shpinel, V. S., Sov. Phys.-JETP (Engl. Transl.), 16:1467 (1963).

74. Sano, H. and Kanno, M., J. Chem. Soc., Chem. Commun., 601 (1969).

75. Zarubin, V. N. and Marinin, Russ. J. Inorg. Chem. (Eng. Transl.), 19:2925 (1974).

76. Yeats, P. A., Sams, J. R., and Aubke, F., Inorg. Chem., 12:328 (1973).

77. Huang, C. H., Knop, O., Othen, D. A., Woodhanes, F. W. D., and Howie, R. A., Can. J. Chem., 53:79 (1975).

78. Yeats, P. A., Poh, B. L., Ford, B. F. E., Sanes, J. R., and Aubke, F., J. Chem. Soc., A, 2188 (1970).

79. Mallela, S. P., Lee, K. C., and Aubke, F., Inorg. Chem., 23:653 (1984).

80. Goldanskii, V. I., Makarov, E. F., Stukan, R. A., Sumarokova, T. N., Trikhtanov, V. A., and Krhapov, V. V., Proc. Acad. Sci. USSR, Phys. Chem. Sect. (Engl. Transl.), 151:598 (1963).

81. Naik,, D. V., May, J. C., and Curran, C., J. Coord. Chem., 2:309 (1973).

82. Davies, A. G., Smith, L., and Smith, P. J., J. Organomet. Chem., 39:279 (1972).

83. Goodman, B. A., Greenwood, N. N., Jaura, K. L., and Sharma, K. K., J. Chem. Soc., A, 1865 (1971).

84. Pellerito, L., Bertazzi, N., and Stocco, G. C., Inorg. Chim. Acta, 10:221 (1974).

85. Bancroft, G. M. and Sham, T. K., Can. J. Chem., 52:1361 (1974).

86. Herber, R. H., in Applications of the Mössbauer Effect in Chemistry and Solid-State Physics, Tech. Rept. Ser. No. 50, I.A.E.A., Vienna (1966), p. 121.

87. Rudolph, R. W. and Chowdhry, V., Inorg. Chem., 13:248 (1974).

88. Edwards, A. J. and Khallow, K. I., J. Chem. Soc., Chem. Commun., 50 (1984).

89. Herber, R. H. and Carrasquilla, C., Inorg. Chem., 20:3693 (1981).

90. Birchall, T., Dean, P. A. W., and Gillespie, R. J., J. Chem. Soc., A, 1777 (1971).

91. Harrison, P. G., Khalil, M. I., and Logan, N., Inorg. Nucl. Chem. Lett., 8:551 (1972).

92. Heeg, M. J., Janiak, C., and Zuckerman, J. J., J. Am. Chem. Soc., 106 (1984), in press.

93. Birchall, T., Chan, P. K. H., Pereire, A. R., J. Chem. Soc., Dalton Trans., 2157 (1974).

94. Tan, T. H., Dalziel, J. R., Yeats, P. A., Sams, J. R., Thompson, R. C., and Aubke, F., Can J. Chem., 50:1843 (1972).

95. Yeats, P. A., Sams, J. R., and Aubke, F., *Inorg. Chem.*, 10:1877 (1971).
96. Einstein, F. W. and Penfold, B. R., *J. Chem. Soc.*, A, 3019 (1968).
97. Debye, N. W. G., Rosenberg, E., and Zuckerman, J. J., *J. Am. Chem. Soc.*, 90:3234 (1968).
98. Donaldson, J. D., Oteng, R., and Senior, J. B., *J. Chem. Soc., Chem. Commun.*, 618 (1965).
99. Donaldson, J. D. and Nicholson, D. G., *J. Chem. Soc.*, A, 145 (1970).
100. Bonchev, Z., Jordanov, A., and Minokova, A., *Nucl. Instr. Methods*, 70:36 (1969).
101. van Remoortere, Flynn, J. J., Boer, F. P., and North, P. P., *Inorg. Chem.*, 10:1511 (1971).
102. Rein, A. J., deVries, J. L. K. F., and Herber, R. H., *J. Inorg. Nucl. Chem.*, 36:825 (1974).
103. Harrison, P. G. and Zuckerman, J. J., *J. Organomet. Chem.*, 55:261 (1973).
104. Stapfer, C. H. and Herber, R. H., *J. Organomet. Chem.*, 56:175 (1973).
105. Hsu, C. C. and Geanangel, R. A., *Inorg. Chem.*, 19:110 (1980).
106. Olmstead, M. M. and Power, P. P., *Inorg. Chem.*, 23:413 (1984).
107. Corvan, P. J. and Zuckerman, J. J., *Inorg. Chim. Acta*, 34:L255 (1979).
108. Bishop, M. E. and Zuckerman, J. J., *Inorg. Chem.*, 16:1749 (1977).
109. Molloy, K. C., Bigwood, M. P., Herber, R. H., and Zuckerman, J. J., *Inorg. Chem.*, 21:3709 (1982).
110. Cheng, H.-S. and Herber, R. H., *Inorg. Chem.*, 9:1686 (1970).
111. Cunningham, D., Molloy, K. C., Hossain, M. B., van der Helm, D., and Zuckerman, J. J., unpublished results.
112. Allman, R., Hohfeld, R., Waskowska, A., and Lorberth, J., *J. Organomet. Chem.*, 192:353 (1980).
113. Power, W. J. and Ozin, G. A., *Adv. Inorg. Chem. Radiochem.*, 23:80 (1980).
114. Bos, A. and Howe, A. J., *J. Chem. Soc., Faraday Trans. II*, 70:451 (1974).
115. Schichl, A., Litterst, F. J., Micklitz,Devort, J. P., and Freidt, J. M., *Chem. Phys.*, 2;)371 (1977).
116. Aleksandrov, A. Yu. Okhlobystin, O. Yu., Polark, L. S., and Shipinel, V. S., *Proc. Acad. Sci. USSR, Phys. Chem. Sect. (Engl. Transl.)*, 157:768 (1964).
117. Gibb, T. C. and Greenwood, N. N., *J. Chem. Soc.*, A, 43 (1966)
118. Greenwood, N. N. and Ruddick, J. N. R., *J. Chem. Soc.*, A, 1679 (1967).
119. Greenwood, N. N., Perkins, P. G., and Wall, D. H., *Sympos. Faraday Soc.*, 1:90 (1968); *Phys. Lett.*, 28A:339 (1968).
120. Zavistoski, J. A. and Zuckerman, J. J., *J. Org. Chem.*, 34:4197 (1969).
121. Herber, R. H., Stöckler, H. A., and Reichle, W. T., *J. Chem. Phys.*, 42:2447 (1965).
122. Fitzsimmons, B. W., Seely, N. J., and Smith, A. W., *J. Chem.*

Commun., 390 (1968); J. Chem. Soc., A, 143 (1969).

123. Harrison, P. G., Lane, B. C., and Zuckerman, J. J., Inorg. Chem.,
     11:1537 (1972).
124. Bancroft, G. M., Kumar Das, V. G., and Sham, T. K., J. Chem.
     Soc., Chem. Commun., 236 (1974).
125. Parish, R. V. and Platt, R. H., Inorg. Chim. Acta, 4:65 (1970).
126. Clark, M. G., Maddock, A. G., and Platt, R. H., J. Chem. Soc.,
     Dalton Trans., 281 (1972).
127. Bancroft, G. M. and Butler, K. D., Inorg. Chim. Acta, 15:57
     (1975).
128. Bancroft, B. M., Kumar Das, V. G., Sham, T. K., and Clark, M.
     G., J. Chem. Soc., Dalton Trans., 634 (1976).
129. Bancroft, G. M. and Sham, T. K., J. Chem. Soc., Dalton Trans.,
     467 (1976).
130. Sham, T. K. and Bancroft, G. M., Inorg. Chem., 14:2281 (1975).
131. Harrison, P. G., in ref. 1, p. 258.
132. Lefferts, J. L., Molloy, K. C., Zuckerman, J. J., Haiduc, I.,
     Curtui, M., Guta, C., and Ruse, D., Inorg. Chem., 19:2861
     (1980).
133. Molloy, K. C., Hossain, M. B., van der Helm, D., Zuckerman, J.
     J., and Mullins, F. P., Inorg. Chem., 20:2172 (1981).
134. Ng, S.-W. and Zuckerman, J. J., J. Chem. Soc., Chem. Commun.,
     415 (1982).
135. Ng, S.-W., Barnes, C. L., Hossain, M. B., van der Helm, D.,
     Zuckerman, J. J., and Kumar Das, V. G., J. Am. Chem. Soc.,
     104:5359 (1982).
136. Nasser, F. A. K., Hossain, M. B., van der Helm, D., and Zucker-
     man, J. J., Inorg. Chem., 22:3107 (1983).
137. Ng, S.-W., Barnes, C. L., van der Helm, D., and Zuckerman, J.
     J., Organometallics, 2:600 (1983).
138. Nasser, F. A. K., Hossain, M. B., van der Helm, D., and Zucker-
     man, J. J., Inorg. Chem., 23:606 (1984).
139. Rheingold, A. L., Ng, S.-W., and Zuckerman, J. J., Organome-
     tallics, 3:233 (1984).
140. Amini, M. M., Rheingold, A. L., Taylor, R. W., and Zuckerman,
     J. J., submitted for publication.
141. Baxter. J. L.. Holt, E. M., and Zuckerman, J. J., Organome-
     tallics, in press.
142. Cunningham, D., Kelly, L. A., Molloy, K. C., and Zuckerman, J.
     J., Inorg. Chem., 21:1416 (1982).
143. Molloy, K. C., Nasser, F. A. K., and Zuckerman, J. J., Inorg.
     Chem., 21:1711 (1982).
144. Molloy, K. C. and Zuckerman, J. J., Acc. Chem. Res., Inorg.
     Chem., 16:386 (1983).
145. Cunningham, D., Firtear, P., Molloy, K. C., and Zuckerman, J.
     J., J. Chem. Soc., Dalton Trans., 1523 (1983).
146. Goodman, B. A. and Greenwood, N. N., J. Chem. Soc., Chem. Com-
     mun., 602 (1969).
147. Goodman, B. A., Greenwood, N. N., Jura, K. L., and Sharma, K.
     K., J. Chem. Soc., A, 1865 (1971).

148. Parish, R. V., Chem. Phys. Lett., 10:224 (1971).
149. Ruddick, J. N. R. and Sams, J. R., J. Chem. Soc., Dalton Trans., 470 (1974).
150. Birchall, T., Denes, G., Ruebenbauer, K., and Pannetier, J., J. Chem. Soc., Dalton Trans., 2296 (1981).
151. Ktorides, P. I., Uhrieh, D.L., Sidocky, R. M. D., and Fishel, D. L., J. Chem. Phys., 77:4188 (1982).
152. Bancroft, G. M., Butler, K. D., and Sham, T, K., J. Chem. Soc., Dalton Trans., 1483 (1975).
153. Poller, R. C., Ruddick, J. N. R., Taylor, B., and Toley, D. L. B., J. Organomet. Chem., 24:341 (1970).
154. Harrison, P. G., Phillips, R. C., and Thornton, E. W., J. Chem. Soc., Chem. Commun., 603 (1977).
155. Herber, R. H. and Leahy, M. F., J. Chem. Phys., 67:2718 (1977).
156. Hazony, Y. and Herber, R. H., Mössbauer Effect Methodology, Vol. 8, ed. by Gruverman, I. and Seidel, C. W., Plenum Press, New York (1978), p. 107.
157. Molloy, K. C., Zuckerman, J. J., Schumann, H., and Rodewald, G., Inorg. Chem., 19:1089 (1980).
158. Molloy, K. C., Hossain, M. B., van der Helm, D., Zuckerman, J. J., and Haiduc, I., Inorg. Chem., 19:2041 (1980).
159. Herber, R. H. and Smelkinson, A. E., Inorg. Chem., 17:1023 (1979).
160. Gibb, T. C., Greatrex, R., and Greenwood, N. N., J. Chem. Soc., A, 890 (1968).
161. Ho, B. Y. K., Molloy, K. C., Zuckerman, J. J., Reidinger, F., and Zubieta, J. A., J. Organomet. Chem., 187:213 (1980).
162. Herber, R. H. and Leahy, M. F., in ref. 3, p. 155.
163. Molloy, K. C., Hossain, M. B., van der Helm, D., and Zuckerman, J. J., Inorg. Chem., 18:3507 (1979).
164. Lefferts, J. L., Molloy, K. C., Zuckerman, J. J., Haiduc, I., Guta, C., and Ruse, D., Inorg. Chem., 19:1662 (1980).
165. Bokii, N. G., Struchkov, Yu. T., and Prok'iev, A. K., J. Struct. Chem. (Engl. Transl.), 13:619 (1972).
166. Greene, P. T. and Bryan, R. F., J. Chem. Soc., A, 2549 (1971).

# MOSSBAUER SPECTROSCOPY OF IODINE

H. de Waard

Laboratorium voor Algemene Natuurkunde
University of Groningen
The Netherlands

## INTRODUCTION

For Mössbauer spectroscopy of iodine, two isotopes are available, stable $^{127}$I and long lived radioactive $^{129}$I ($T_{\frac{1}{2}} = 1.7 \times 10^7$ y). The latter isotope has been used far more than the former (about 250 publications against 50), which reflects the superior properties of the relevant gamma transition in $^{129}$I compared with that in $^{127}$I. To illustrate these, the Mössbauer spectra of equivalent iodine-127 and -129 compounds are compared in Fig. 1. The important differences of the two gamma transitions are: the much longer half life of the 27.7 keV state in $^{129}$I (16.8 ns) than that of the 57.8 keV state in $^{127}$I (1.9 ns), which implies a 9× smaller natural line width; the larger change of nuclear radius between excited and ground state in $^{129}$I than in $^{127}$I, which makes the isomer shift 1.4 times larger, and the lower gamma energy, by which the recoilless fraction is increased to the extent that most $^{129}$I spectroscopy can be carried out at liquid nitrogen temperature, whereas for $^{127}$I spectroscopy liquid helium is almost always required.

On the other hand, the isotope $^{129}$I is an expensive fission product so that highly efficient preparation methods of $^{129}$I-compounds are required. The weak radioactivity of $^{129}$I hardly ever causes any serious background problems. Even so, this same isotope has been used very successfully to provide the source activity in Mössbauer spectroscopy of xenon compounds[3].

In discussing standard source and absorber materials we concentrate on $^{129}$I. There are two different parent activities: 70 min $^{129}$Te and 34 d $^{129m}$Te (the isomeric state of $^{129}$Te). The latter is more convenient because of its longer half life, but the resonant

295

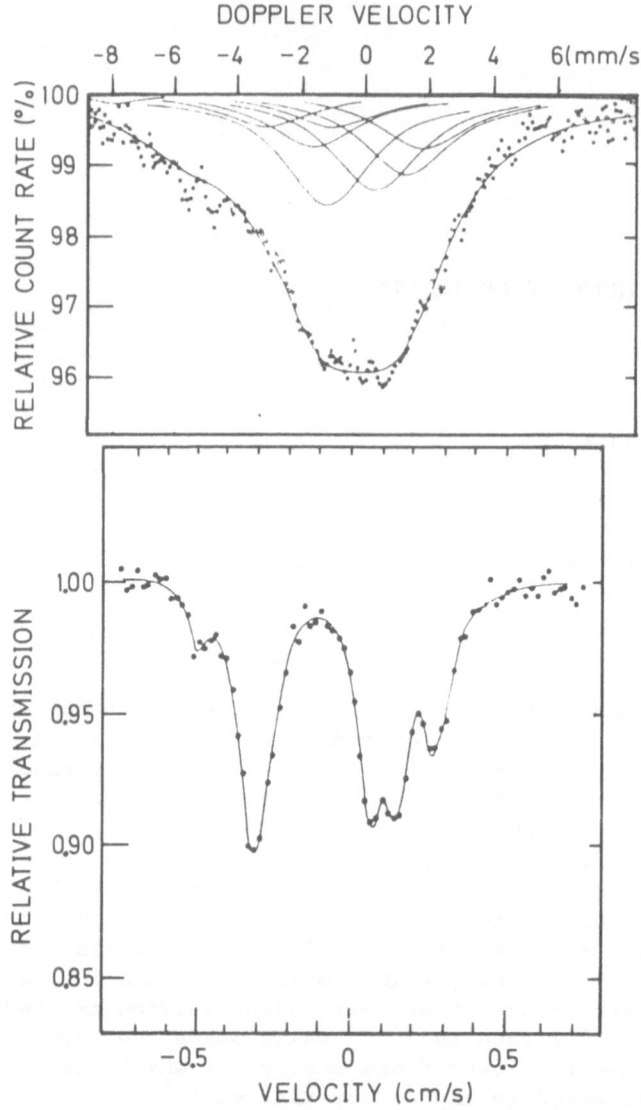

Fig. 1. *Comparison of Mössbauer spectra obtained for equivalent compounds using* $^{127}I$ *(top) and* $^{129}I$ *(bottom). Top: source ZnTe($^{127m}$Te), absorber NaIO$_3$, temp. 4.2K[1]. Bottom: source Zn$^{66}$Te($^{129m}$Te), absorber KIO$_3$, temp. 100K[2].*

effect achieved with it is smaller by a factor of about three because its recoilless resonant gamma ray at 27.73 keV is diluted by the 27.2 and 27.5 keV X-rays generated in the isomeric decay of $^{129m}$Te. The long lived activity is used almost exclusively today.

It is prepared by thermal neutron irradiation of mass separated $^{128}$Te in a high flux reactor ($\cong 2\times10^{14}$ n/cm$^2$s) for about a month, yielding a specific activity of about 0.5 mCi/mg.

For a long time, ZnTe has been used almost exclusively as the source material, so that isomer shifts are generally given with respect to that compound. However, no thin absorber line widths smaller than about 0.9 mm/s can be obtained with it, whereas the natural width is 0.59 mm/s. Better results have been obtained with SnTe: we have found a width of 0.7 mm/s in combination with a thin Cu$^{129}$I absorber. Our analysis shows that the source is still responsible for the larger than natural width[4]. Recently, Pasternak et al.[5] prepared Mg$_3$TeO$_6$ ($^{129m}$Te) and Ca$_3$TeO$_6$ ($^{129m}$Te) sources that can be used at room temperature because of their very high Debye temperatures. With these materials practically the same linewidths are obtained as for SnTe, which shows that the $^{129}$I-atoms resulting after beta decay of $^{129m}$Te are in a nearly perfect cubic environment.

Cu$^{129}$I has been accepted as the standard absorber material. It is very stable chemically and yields close to natural linewidth when the absorber is thin enough. Pasternak et al.[6] point out that Na$_5$IO$_6$ can be used to achieve a large effect at room temperature. There is, however, some line broadening ($\Delta\Gamma = 0.2$ mm/s relative to CuI). Some data on standard sources and absorbers are collected in Table 1.

Clearly, the high f sources and absorbers just mentioned are also useful for $^{127}$I spectroscopy.

Iodine Mössbauer spectroscopy has many applications in chemistry; compounds with iodine in all its possible formal valence

TABLE 1. Standard sources and absorbers for $^{129}$I spectroscopy and values of some of their parameters.

| | isomer shift[a,d] | recoilless fraction (f) 4.2 K | 80 K | 300 K | minimum linewidth[d] | ref. |
|---|---|---|---|---|---|---|
| Zn$^{129}$Te | – | 0.65(5) | 0.42(4) | 0.07(1) | 0.90[b] | 4 |
| Sn$^{129}$Te | 0.43(3) | 0.68(4) | 0.42(3) | 0.05(1) | 0.73(5)[b] | 4 |
| Mg$_3$$^{129}$TeO$_6$ | −3.49(3) | 0.85 | 0.80 | 0.55(2) | 0.78[b] | 6 |
| Cu$^{129}$I | −0.41(2) | 0.58(3) | 0.31(2) | 0.027(2) | 0.73(5)[c] | 4 |
| K$^{129}$I | −0.45 | 0.55(2) | 0.12(2) | <0.01 | 0.96(5)[c] | 4 |
| Na$_5$$^{129}$IO$_6$ | −3.16(2) | 0.80 | 0.67 | 0.25(2) | 0.98(1)[c] | 5 |

a) relative to ZnTe, b) with CuI absorber, c) with SnTe source.
d) in mm/sec.

states, -1 to +7 have been studied. A large majority of the results can be explained in terms of simple 5sp shell electron configurations. Different configurations lead to widely different spectra, often characterized by highly specific combinations of isomer-shift and quadrupole coupling. Some of these cases will be discussed in this paper, while attention will also be paid to a few cases from the realm of physics.

ISOMER SHIFT CALIBRATION

In order to derive quantitative values for the changes $\Delta\rho_e(0)$ of contact electron density at iodine nuclei from isomer shifts S the proportionality constant k in the expression

$$S = k\Delta\rho_e(0) \tag{1}$$

must be determined. This procedure is called isomer shift calibration.

The isomer shift is usually expressed in velocity units, in which case we have

$$k = \frac{c}{E_\gamma} \frac{2}{3} \pi Z e \Delta<r^2> \tag{2}$$

Known factors in this expression are the gamma energy $E_\gamma$ and the atomic number Z of the emitting nucleus, the unknown quantity is the change $\Delta<r^2>$ of the mean square radius between the excited and the ground state of the nucleus. A method that has been used successfully for its determination is the conversion electron method. It uses the proportionality of the decay rate of a nuclear state by emission of electrons from the different atomic shells and the contact density of these electrons at the nucleus. Such electrons are called conversion electrons, they give rise to sharp lines in the electron decay spectrum. In the case of the predominantly magnetic dipole Mössbauer transitions in $^{127}$I and $^{129}$I the above proportionality is well established. The intensity ratio $I_i/I_j$ of conversion electron lines from two different atomic shells (i,j) can be measured quite accurately by beta spectrometry, and therefore yields an accurate value for the ratio of the contact electron densities of these shells $\rho_e^{(i)}(0)/\rho_e^{(j)}(0)$. Of particular interest is the ratio of the electron densities between the valence shell (i=n) and the next inner shell (j=n-1). In iodine compounds, changes $\Delta\rho_e(0)$ due to different chemical environment may be assumed to occur predominantly in the valence shell: $\Delta\rho_e(0) = \Delta\rho_e^{(5)}(0)$. The electron density of the next inner shell (n=4), unaffected by the chemical state of the atom, can be calculated with some confidence by self consistent field methods employed in the computation of free atom wave functions. Thus, we find $\rho_e^{(5)}(0) = (I_5/I_4) \cdot \rho_e^{(4)}(0)$ and, by carrying out the conversion measurements for two or more

TABLE 2. Values of $\Delta\langle r^2 \rangle$ for calibration of the isomer shift in
$^{129}$I Mössbauer spectroscopy obtained by various methods.

| $\Delta\langle r^2\rangle^{129}\times10^3$ fm$^2$ | method | | ref. | |
|---|---|---|---|---|
| 18(4) | conv. spectr. $^{129}$I | Spijkervet and Pleiter | (7) |
| 32.4(7.7) | lifetime $^{125}$I | Müller and Kündig | (8) |
| 18(3) | lifetime $^{123}$I | Ladrière et al. | (9) |
| 21.4(3.0) | theoretical $\rho_e(0)$ | Hartmann and Eifrig | (10) |
| 20.0(3.0) | theoretical $\rho_e(0)$ | Hartmann et al. | (11) |

systems with known isomer shift difference, the calibration constant k. The result of this method for $^{129}$I is given in Table 2.

Another method is the electron capture lifetime method which
uses the circumstance that the electron capture nuclear decay rate
$\lambda$ is proportional to the contact electron density $\rho_e(0)$ if effects
of exchange, overlap and change in electron binding energy are neglected, which is generally justified. For different chemical compounds the relative change of decay rate is then simply given by

$$\Delta\lambda/\lambda = \Delta\rho_e^{(0)}/\rho_e^{(0)} \tag{3}$$

The total contact density $\rho_e(0)$ must again be calculated by a free
atom SCF method. An advantage of this method over the conversion
method is that it does not depend on the assumption that the contact
density change must only occur in the valence shell. On the other
hand, it is a difficult method because the relative changes of $\lambda$
are very small ($\sim10^{-4}$). The results obtained by this method are
also given in Table 2.

The last results in this table stem from theoretical calculations of $\rho_e(0)$. Such approaches have been made from the beginning,
but they were not considered reliable, because essentially, free
atom wave functions were used to calculate contact density changes
for compounds. Recently, a more advanced method was used by Hartmann
et al.[10],[11], who applied the SCF-X$\alpha$MT-SW method, advocated by
Slater[12] for the calculation of localized properties that depend on
chemical surroundings, to the $IO_6^{5-}$, $IO_4^-$, $IO_3^-$ and $ICl_4^-$ groups.
Relativistic effects were included by multiplication with correction
factors.

All results for $\Delta\langle r^2\rangle$ except that of Müller and Kündig lie
within error bars. Possible corrections to Müller and Kündig's
value have been discussed by Makariunas[13] and by Hartmann et al.[11],
but yield no improvement. A $\Delta\lambda/\lambda$ value for $Na_2H_3I(^{125}I)O_6$ and

NaI($^{125}$I)O$_3$ widely different from that of Müller and Kündig, but with much larger error was also measured by Hartmann et al.[11]. In view of these uncertainties we propose to use the weighted average of the other values: $\Delta<r^2>^{129}$ = 19.5(2.0)×10$^{-3}$ fm$^2$ for the time being. The value for $^{127}$I is obtained from this by taking $\Delta<r^2>^{127}$ = -0.72 $\Delta<r^2>^{129}$ (using the isomer shift ratio S$^{127}$/S$^{129}$= -0.345 given by Jones and Warren[14]).

## CHEMICAL BOND INFORMATION OBTAINED FROM ISOMER SHIFT AND QUADRUPOLE COUPLING DATA

The simple semi-empirical formulas that express isomer shift and quadrupole coupling in terms of the numbers of holes $h_{p_x}$, $h_{p_y}$, $h_{p_z}$ and $h_s$ in the full 5($s^2 p^6$) shell of an I$^-$ ion work remarkably well for a large number of iodine compounds. We summarize these formulas here, using the best values of the empirical constants that we could find and referring to other papers for their derivation[15].

The isomer shift for $^{129}$I compounds relative to a ZnTe source is given by

$$S - S_o = -(9\pm1)h_s + (1.5\pm0.1)h_p \quad mm/s \qquad (4)$$

where $S_o$ = -0.54(1) mm/s is the shift between the ZnTe source and the I$^-$ ion and $h_p = h_{p_x} + h_{p_y} + h_{p_z}$. The shifts for $^{127}$I compounds are obtained by multiplying all terms by -0.345[14].

The quadrupole coupling constant for $^{129}$I, measured in mm/s is generally converted to that for $^{127}$I given in MHz, by

$$e^2qQ_{127}/h \text{ (MHz)} = 31.90 \times (\frac{c}{E_\gamma} e^2qQ_{129}) \text{ (mm/s)} \qquad (5)$$

For $^{127}$I an accurate value of the free atom quadrupole coupling is available: $(e^2qQ_{127}/h)_{at}$ = 2292.8(1) MHz. An unbalance in the electron occupation of the 5p shell of a molecule sets up a field gradient. According to Townes and Dailey[16] we may write for this:

$$eq_{zz}^{mol} = \left[ h_{p_z} - \tfrac{1}{2}(h_{p_x} + h_{p_y}) \right] eq^{at} \qquad (6)$$

If the hole numbers $h_{p_x}$ and $h_{p_y}$ are different (loss of axial symmetry) the field gradient has an asymmetry parameter

$$\eta^{mol} = \frac{q_{xx} - q_{yy}}{q_{zz}} = \frac{h_{p_x} - h_{p_y}}{h_{p_z} - \tfrac{1}{3}h_p} \qquad (7)$$

In the equations (6) and (7) the hole numbers may be replaced by occupations numbers $N_{p_{x,y,z}} = 2 - h_{p_{x,y,z}}$. Further, the reduced field gradient

$$U_p = eq_{zz}^{mol}/eq^{at} \tag{8}$$

is frequently used.

There are three standard cases of chemical bonding.

($\sigma$)      The simplest case is the one where only $p_z$ orbitals participate (pure $\sigma$-bonding). Then, combining (4), (6) and (8), we see immediately that (apart from the error bars)

$$S - S_0 = 1.5 \times U_p \text{ mm/s} \tag{9}$$

($\pi$)      If only $p_x$ and $p_y$ participate

$$S - S_0 = -3.0 \times U_p \text{ mm/s} \tag{10}$$

($sp^3$)      If s-electrons participate in the bonding ($h_s \neq 0$), we expect hybridization and for pure $sp^3$ hybridization we expect $h_{p_x} = h_{p_y} = h_{p_z}$ so that $U_p = 0$. As far as I know this is true for all compounds with negative values of $S - S_0$.

The cases ($\sigma$) and ($\pi$) are represented by straight lines in the plot of $S-S_0$ vs $U_p$ given in Fig. 2. A similar figure was given in my review paper in the 1973 Mössbauer Effect Data Index[15]. In the present figure, only points are given that correspond to representative results obtained after 1972. There are no essentially new results on the line ($sp^3$) that would extend vertically downward from the origin.

## Miscellaneous compounds

We briefly review some of the results given in Fig. 2 in order of the iodine oxidation state.

$I^{-1}$. The case of $PbI_2$[17]. The isomer shift yields $h_p \cong 0.5$, indicating some covalency. The absence of quadrupole splitting tells us that the $p_x$, $p_y$ and $p_z$ states are equally populated. This is consistent with the orthogonal coordination of I by 3 Pb atoms in the $CdI_2$ type lattice.

$I^{0-1}$. The plot shows many cases close to the $\sigma$-line which means that the bonding is predominantly due to electrons in $5p_z$ orbitals. The largest deviations from the $\sigma$-line are in the $h_p > U_p$ direction. Such deviations could be caused by unaccounted for contributions of the lattice to the e.f.g., by additional $\pi$-bonding or by general inadequacies of the simple model. As regards the first possibility, we refer to Jones[18] who pointed out that there exists a closely linear relationship between $U_p$ and the Allred-Rochow electronegativity $\chi$ of the atom that is bonded to I. This is shown in Fig. 3 for a number of cases. This should strengthen our belief in the regularity of the e.f.g. values. Thus, the deviations probably

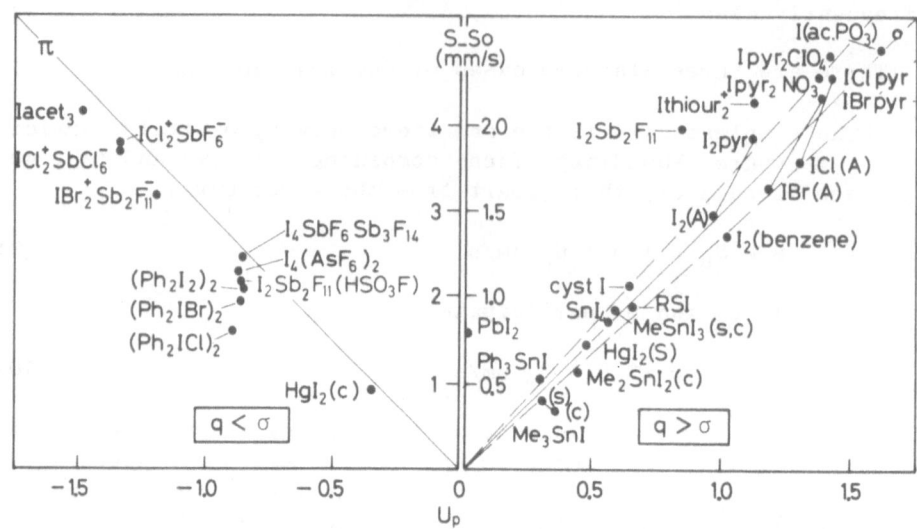

Fig. 2. *Plot of isomer shift vs. reduced quadrupole coupling strength for $^{129}I$ (and some converted $^{127}I$) compounds. Line $\sigma$: pure $p_z$ bonding, line $\pi$: $p_x + p_y$ bonding. Points represent results obtained after 1972 (but by no means all of those). For the earlier results see ref. 15, p. 463. Results for poly-iodide chains are given separately in Fig. 4. Additions in brackets have the following meaning: (s) = in frozen solution; (c) = crystalline; (A) = in argon matrix.*

occur predominantly in the isomer shift, in the direction of larger $h_p$ values. Apparently, the <u>charge unbalance</u> at the iodine atoms $U_p = h_{p_z} - \frac{1}{2}(h_{p_x} + h_{p_y})$ is largely determined by the electron attract- ing power of the electronegative ligand while the <u>total charge</u> on the iodine $h_p = h_{p_x} + h_{p_y} + h_{p_z}$ depends also on $\pi$-bonds with other neighbouring atoms. A further indication that such bonds play a part for the cases of $I_2Sb_2F_{11}$ and $I(thiourea)_2^+$, that show the largest increase of $h_p$ above $U_p$, is the non-zero value of the asymmetry parameter $\eta = (h_{p_x} - h_{p_y})/(h_{p_z} - \frac{1}{3}h_p)$.

A few points in Fig. 2 lie below the $\sigma$-line. It has been argued for some of these cases that a small degree of sp³ hybridization might occur but it is hard to see how this could apply, for instan- ce, for the matrix isolated species IBr(A) and ICl(A). It should be stressed here that the accuracy of the coefficient in the relation

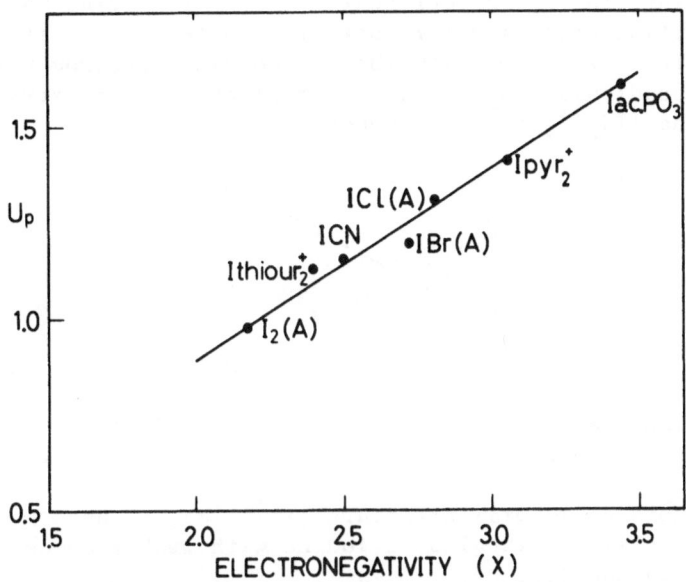

Fig. 3. Plot of reduced field gradient $U_p$ vs Allred-Rochow electro-negativity for some of the results included in Fig. 2.

$h_p = 1.5U_p$ is only of the order of 10%. Those margins, indicated by broken lines in Fig. 2, bring a number of points within the realm of pure $\sigma$ bonding and remind us of the limitations of the model: 16 of the 22 points lie within the margins of error. Moreover, there are experimental errors, mainly in the isomer shift, that should be considered before making statements about bonding anomalies.

One final point on the sigma line merits attention: $HgI_2(s)$[19]. The linear structure of the $HgI_2$ molecule in frozen solution causes regular $\sigma$-bonds with $h_p = U_p$. In crystalline $HgI_2$, however, the I-Hg bonds make angles of 100° by which the $\sigma$-bonding is "switched" to a planar structure where the $p_x$ and $p_y$ orbitals do the bonding, leading to a sign reversal of the field gradient that brings the point $HgI_2(c)$ close to the $\pi$-line, but does not change the total charge on the iodine very much: $h_p = 0.61(3)$ for $HgI_2(c)$ and $0.48(3)$ for $HgI_2(s)$.

This brings us to the cases with planar bonding that have a negative field gradient eq (and positive coupling constant $e^2qQ$ because $Q_{129}$ is negative) and that lie in the second quadrant of figure 2. For such compounds the iodine usually is in the 3$^+$ or 5$^+$ oxidation state.

I$^{3+}$. All points in Fig. 2 represent cases where iodine is in the +3 formal oxydation state and for most of them the iodine atom is rectangularly coordinated with the non bonding electron pairs roughly perpendicular to the molecular plane. The geometry around iodine of the $[IX_2]^+[MY_6]^-$ compounds is:

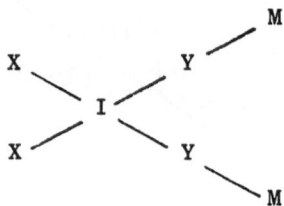

These compounds have moderate asymmetry coefficients $\eta \cong 0.1$-$0.3$ that reflect some degree of asymmetry in the x and y direction[20].

The dimers of general structure $[I_4]^{2+}[MY_6]_2^{2-}$ have been shown[21] to contain rectangles of iodine with weaker bonds to the surrounding anions (Y).

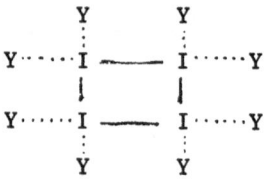

The very close similarity of their Mössbauer parameters S, e$^2$qQ and $\eta$ with those of I$_2$Sb$_2$F$_{11}$ dissolved in HSO$_3$F has led Birchall and Myers[22] to propose that the latter compound is also dimerized in HSO$_3$F solution. It is again worth noting that the total charge on the iodine (h$_p$ $\cong$ 1.3) and the absolute value of the field gradient $|U_p| \cong 1960$ MHz are hardly changed by the dimerization, but that the sign of U$_p$ is opposite to that of the crystalline compounds (see Fig. 2).

Finally, there are the diphenyl iodonium halides, the structures of which were determined by Alcock and Countryman[23]:

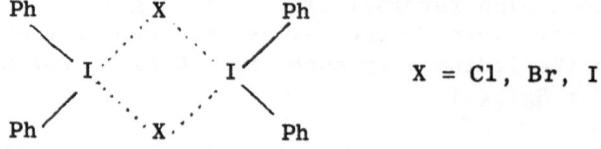

$X = Cl, Br, I$

These have primary $\gtrless$C-I bonds and much weaker bridging halogen bonds. The smaller absolute value of U$_p$ is explained by the smaller electronegativity of the phenylcarbon compared with that of Cl and Br. The not so small value of $\eta$ (= 0.31, 0.36) probably indicates that the phenyl groups are no complanar with the $I_2X_2$ moiety.

## Linear iodine chains

It has been known for a long time that iodine ions can attract iodine molecules to form the linear iodine chains $I_3^-$ and $I_5^-$ in a number of iodine compounds. The first of these to be studied by Mössbauer spectroscopy was $CsI_3$, where the existence of 3 different $S-e^2qQ$ combinations was established[24,25], corresponding to 3 iodine atoms in inequivalent sites (Fig. 4). Each of these is essentially $\sigma$-bonded.

More recently, the formation of iodine chains was established for a variety of organic compounds. These often acquire extraordinary anisotropic properties in the process. For instance, the electrical conductivity may increase by a factor up to $10^{12}$ in the direction of the chains. Mössbauer spectroscopy is an important tool for studying the geometry of the iodine chains because it is very sensitive to bond differences between the atoms of a chain, much more so than, for instance, X-ray structure analysis. It shares this sensitivity with resonant Raman spectroscopy. The two methods have been used together to provide unique answers about the iodine chain structure for a number of compounds. Results of most cases studied so far are summarized in Fig. 4, which is organized in the same way as Fig. 2.

$I_3^-$ chains. These are found in cesiumtriiodide ($CsI_3$), in iodinated nickel phtalocyanine ($NipcI_x$) and in iodinated ruthenium cyclopentadiene iodide ($Ru(cp)_2II_x$). In the latter two compounds the (1,2,3) chain is symmetric, i.e. the bond lengths (1,2) and (2,3) are equal. This results in a spectrum with only 2 components, one for iodine in sites 1 and 3 and one for site 2. The components are identified as such because one has about twice the intensity of the other. This places the iodine atom with the highest $U_p$-value in the middle:

| | 1 | 2 | 3 | $\Sigma U_p - 3$ | ref. |
|---|---|---|---|---|---|
| $Ru(cp)_2I(I_3^-)$: | $U_p$ = 0.50 | 1.07 | 0.50 | -0.93 | 26 |
| $Nipc(I_3^-)$ : | $U_p$ = 0.53 | 1.04 | 0.53 | -0.90 | 27 |

An asymmetric chain is found for

| | 1 | 2 | 3 | $\Sigma U_p - 3$ | ref. |
|---|---|---|---|---|---|
| $C_S(I_3^-)$ : | $U_p$ = 0.64 | 1.10 | 0.36 | -0.90 | 25 |

This chain approaches the combination

| | 1 | 2 | 3 | $\Sigma U_p - 3$ |
|---|---|---|---|---|
| $I^2 + I^-$ : | $U_p$ = 0.98 | 0.98 | 0 | -1.04 |

305

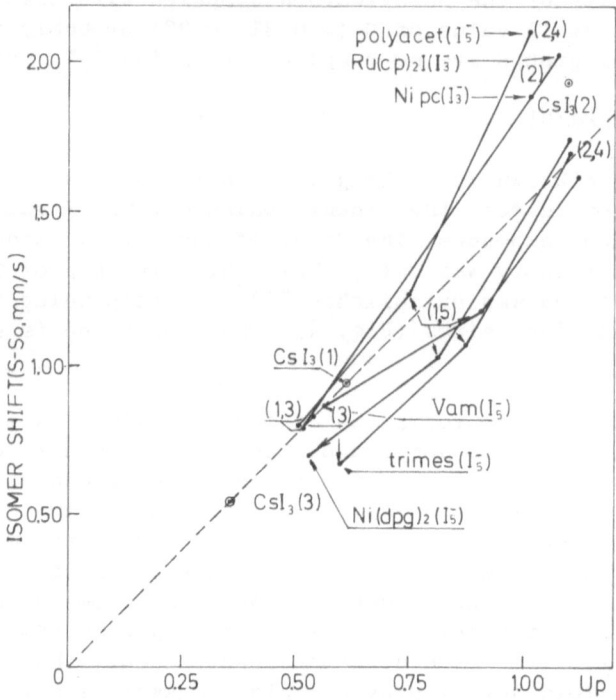

Fig. 4. Plot of isomer shift $S-S_0$ vs reduced quadrupole coupling strength $U_p$ for $^{129}I$ in compounds containing $I_5^-$ and $I_3^-$ chains. Abbreviations are explained in the text. The numbers indicate the positions of the iodine atoms in the chain.

We see that the main influence of the chain formation is on the charge distribution of the terminal iodines (as measured by $U_p$) but that $U_p(1)+U_p(3)$ remains almost constant and equal to $U_p(2)$, keeping the total charge $\Sigma U_p -3$ close to $-1$.

$I_5^-$ chains. These may be formed in stacked arrays of highly polarizable flat molecules. All chains found so far are symmetric, i.e. $(1,2) = (4,5)$ and $(2,3) = (3,4)$. The first example is (benzene 1,3,5 tricarboxylic acid $\cdot$ $H_2O)_{10}H^+I_5^-$ (trimes ($I_5^-$)), that was known to contain linear chains of symmetrical $I_5^-$ ions from crystallographic analysis. Based among others on the similar structure of its Mössbauer spectrum, Treitelbaum et al.[28] were led to ascribe the structure of iodine in the well known blue black "starch-iodine" also to the formation of linear symmetrical $I_5^-$ chains.

306

Very similar chains are found in the iodinated nickel and palladium bisdiphenylglyoximates (Ni,Pd(dpg)$_2$I) as evidenced by the very similar structure of their Mössbauer spectra measured by Cowie et al.[29] (see Fig. 4). The flat molecules of this compound form extended planar structures in which metal atoms alternate with holes. Such structures stack easily in vertical arrays in which the holes lie directly above each other, producing "chimneys" that accomodate the iodine chains.

We may again compare the effective charges on the iodine atoms by inspecting the $U_p$-values. It is clear already from Fig. 4 that for each particular position in the chain the $U_p$-values of the different compounds lie close together, especially for atoms nr. 2 and 4 and for atom nr. 3. The $U_p$ values and chain structure are:

| | 1 | 2 | 3 | 4 | 5 | $\Sigma U_p-3$ | ref. |
|---|---|---|---|---|---|---|---|
| Trimes·I$_5^-$ | $U_p$=0.87 | 1.10 | 0.60 | 1.10 | 0.87 | −0.46 | 28 |
| V·amylose I$_5^-$ | 0.91 | 1.12 | 0.56 | 1.12 | 0.91 | −0.38 | 28 |
| Ni(dpg)$_2$·I$_5^-$ | 0.81 | 1.10 | 0.53 | 1.10 | 0.81 | −0.65 | 29 |
| Polyacet·I$_5^-$ | 0.75 | 1.01 | 0.51 | 1.01 | 0.75 | −0.97 | 30,31 |

It is seen that the total negative charge of the I$_5^-$ chain is lower than that of the I$_3^-$ chain by about 0.5e for the first three cases, pointing to a larger degree of covalency of the bonding to the chain. Chains with more than 5 iodine atoms have not been observed and it should be noted that already the I$_5^-$ chain is not very tightly bound together. In fact, in polyacetylene it competes with I$_3^-$ and the I$_5^-$ chains only predominate at high iodine concentrations[30,31]. Clearly, a symmetric chain can be easily formed from an asymmetric I$_3^-$ unit by attracting an I$_2$-molecule, but we can not see very well how a symmetric I5-chain could attract another I$_2$ molecule and since asymmetric I5-chains have not been found, the possibility of forming I$_5^-$ and longer chains seems remote.

It is seen in Fig. 4 that the I$_3^-$ chains and the I$_5^-$ chain in polyacetylene lie above the $U_p = h_p$ line ($h_p > U_p$) which suggests some π-bonding of the iodines to their environment. Indeed, in the structures with $h_p > U_p$ the distances between the iodines and their neighbours in the host lattice are shorter than for structures with $h_p \lesssim U_p$ where the chains are contained in rather wide cylindrical chimneys.

RADIOGENIC SYSTEMS

In the previous paragraphs Mössbauer spectroscopy was discussed of iodine compounds, that are chemically stable. It is also possible to investigate systems in which the $^{129}$I is formed by beta decay of $^{129}$Te incorporated in tellurium compounds or introduced as an impurity in a wide variety of materials by diffusion or implantation. Such radiogenic systems often can not be produced in bulk by normal iodine chemistry and they are frequently unstable. However, they will usually live longer than the 24 ns lifetime of the 27.7 keV state of $^{129}$I, especially when cooled to liquid helium temperature. Such systems have interesting properties; a few of them will be discussed here.

## Tellurium halides

Compounds with the composition $(NH_4)_2{}^{129m}TeX$ (X = Cl, Br, I), used as Mössbauer sources, were investigated by Jones and Warren[32] and by Johnstone, Jones and Vasudev[33]. It is observed for X = Cl and Br that in a large fraction of the $^{129m}$Te decays the daughter $^{129}$I atoms remain bonded to the chlorine or bromine ligands forming octahedral $ICl_6^-$ or $IBr_6^-$ ions. For II$^-$ the situation is more complicated. However, in each case the Mössbauer spectrum clearly shows one line with large negative shift relative to the $Cu^{129}I$ absorber. This shift corresponds to a large <u>positive</u> regular isomer shift. The measured isomer shift of 6.23(8) mm/s for $ICl_6^-$, for instance, means that there is a positive charge of +3.5 units on the iodine. The absence of quadrupole splitting proves that all bonds are equally strong and perpendicular to each other. Since this excludes hybridization, we can conclude that we have pure p-bonding and that each bond transfers 0.6 units of charge from the iodine to the chlorine in $ICl_6^-$ (0.45 in $IBr_6^-$ and 0.28 in $II_6^-$). No iodine compounds with similar bonding characteristics have ever been made; the $IX_6^-$ unit probably has a very short lifetime. This example illustrates one of the special features of the chemistry of radiogenic systems.

## Heusler alloys

These systems are actually intermetallic (-metalloid) compounds of general formula $X_2YX$ or XYZ, many of which have cubic structures (L2$_1$ or C1$_b$). Of special interest, and widely studied, are the properties of Heusler alloys with Y = Mn that order ferromagnetically. One of the items worth studying is the magnetic hyperfine field at impurities. We have investigated such fields at non magnetic impurities Z' that replace Z in compounds $Pd_{(1,2)}MnZ(Z')$ with Z = Sn, Sb and Z' = In, Sn, Sb, Te, I. It turns out to be possible to substitute a small amount (<1%) of Z by an impurity Z' with atomic number of Z ±1 during preparation of the compound but this is no longer possible is Z' = Z±2. Such impurities can be incorporated, however, by radioactive decay of Z ±1 impurities. We have done this for the following cases: $Pd_{(1,2)}MnSb(^{129m}Te \rightarrow ^{129}I)$[34,35], $Pd_2MnSb(^{125}Sb \rightarrow ^{125}Te)$[36], $Pd_2MnSn(^{111}In \rightarrow ^{111}Cd)$[36], thus extending the impurity hyperfine field systematics. The Mössbauer spectrum of $^{129}$I impurities in $Pd_2MnSb(Te)$ is shown in Fig. 5 as an example.

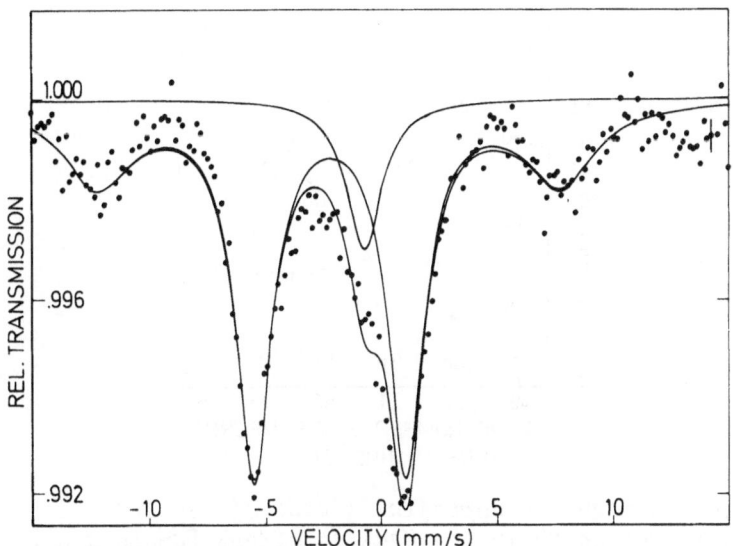

Fig. 5. Mössbauer spectrum of $^{129}I$ impurities in a $Pd_2MnSb_{0.998}$, $Te_{0.002}$ ($^{129m}Te$) source. Absorber: $Cu^{129}I$, temp. 4.2K. The Zeeman splitting of the major component corresponds to a hyperfine field of 51.2(2) T. From ref. 34.

We see the typical Zeeman splitting for most of the iodine and a minor single line component which is probably due to iodine in MnTe. In Fig. 6 the hyperfine fields at (5sp) impurities substituting Sb in PdMnSb and $Pd_2MnSb$ are plotted against their atomic numbers. The downward trend of the field at iodine in $Pd_2MnSb$ and the shift to the right of the curve for PdMnSb can both be understood in terms of a conduction electron polarization model formulated by Blandin and Campbell[37].

## Experiments with implanted sources

Many unusual systems can be prepared by implantation of energetic ions in metals, semiconductors and compounds. Ion implantation in semiconductors has become of vital importance in the semiconductor industry.

In Groningen, we have pursued Mössbauer spectroscopy of implanted sources for almost two decades now with the purpose of studying the impurity hyperfine interaction, of establishing the exact landing sites of the implanted atoms, and of determing the structure and annealing behaviour of associated defects produced by the implantation. A recent review paper on Mössbauer spectroscopy of implanted sources can be found in the Proceedings of the 1981

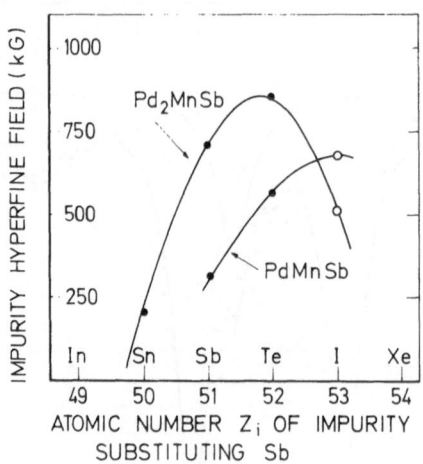

Fig. 6. *Plot of magnetic hyperfine fields of 5sp shell impurities substituting Sb in the Heusler alloys PdMnSb and Pd₂MnSb. From ref. 35.*

International Conference on Applications of the Mössbauer Effect[38].

We shall only discuss here the case of <u>implantation of $^{129m}$Te in silicon</u>, which was studied extensively by Kemerink et al.[39]. As we shall see, this case clearly demonstrates the power of Mössbauer spectroscopy in revealing the electronic and dynamic properties of the implanted impurity.

It should be remarked at the start that we wanted to incorporate the tellurium substitutionally in the silicon lattice. This can not be achieved by alloying or diffusion, since the solubility of Te in Si is extremely low.

Sources were therefore prepared by implantation of 100 keV ions of $^{129m}$Te in phosphorus doped (n$^+$ type), boron doped (p$^+$ type) and compensated silicon single crystals. At this implantation energy, the average depth is 500 Å. The total Te-dose, of the order of $10^{14}$/cm² , has caused heavy damage to the implanted region. Since we wish to start with a sample in which Te is substitutional in single crystal material, this damage must be removed. Foti et al.[40] discovered that in the case of tellurium impurities this could be done very effectively by heating the crystal surface by a short pulse of light such as can be obtained with a Q-switched ruby laser (duration ∿30 ns, energy denstiy 1.5-2 J/cm²). In this way, 80-90% of the Te-atoms become substitutionally distributed over a depth of about 2500 Å in high quality single crystalline host material. The laser annealed sources were used in combination with a Cu$^{129}$I

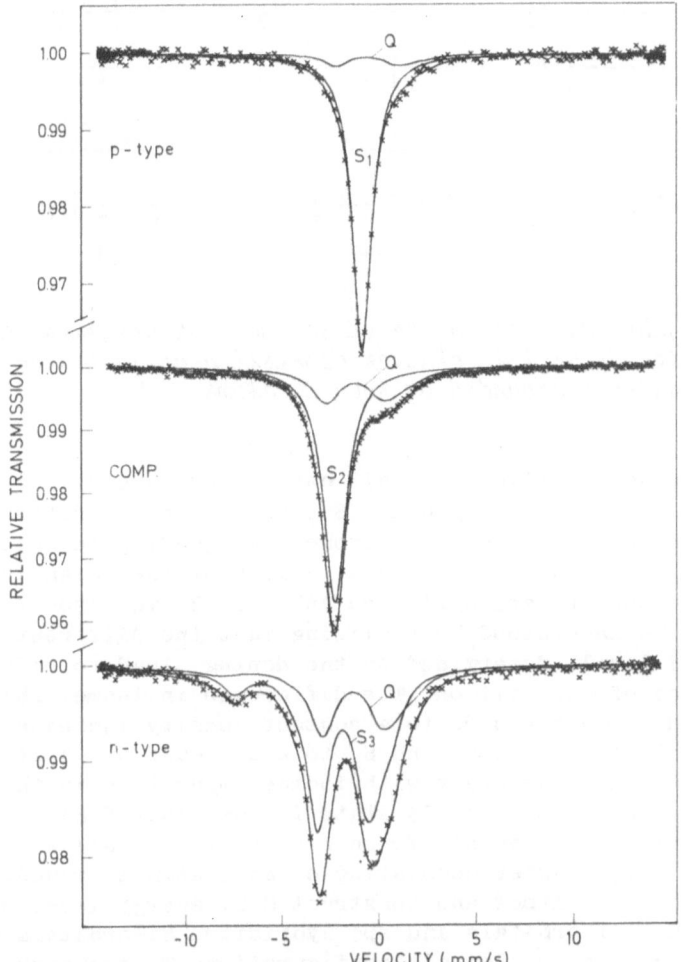

Fig. 7. $^{129}I$ Mössbauer spectra of differently doped Si single crystals implanted with $^{129m}Te$ and subsequently laser annealed. Absorber $Cu^{129}I$, temperature 4.2 K. For doping details see Table 3. From ref. 39.

absorber for Mössbauer spectroscopy. The following quantities proved of importance for the investigation:   (i) the isomer shift which may inform us about the charge state of the impurity, (ii) the quadrupole coupling strength which measures the asymmetry relative to the surrounding charges and (iii) the recoilless fraction, measured as a function of temperature, which gives information about dynamical properties.

Isomer shift. Results of measurements at 4.2 K are shown in Fig. 7. Large differences are observed between spectra obtained for

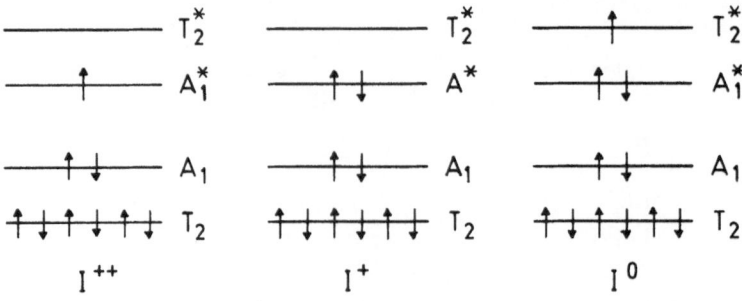

Fig. 8. Possible M.O. energy level schemes for different charge states of an $ISi_4$ cluster consisting of an I atom tetrahedrally surrounded by four Si atoms.

different types of silicon. In all cases there are at least two components. The strong component ($S_1$, $S_2$, $S_3$) has a different isomer shift in each case and a pronounced quadrupole splitting occurs for $^{129}I$ in $n^+$ type silicon. Values of the parameters derived from the spectra are collected in Table 3. The isomer shift results can be understood by realizing that the different positions of the Fermi level, determined by the doping, lead to different charge states of the silicon. The difference in isomer shift between $S_1$ and $S_2$ corresponds to a contact density increase $\Delta\rho_e(0) = 6.8(1)$ $a_0^{-3}$, that between $S_2$ and $S_3$ to a decrease of $-1.1(2)$ $a_0^{-3}$. This behaviour is consistent with iodine impurities in the 2+ charge state in $p^+$ type Si, in the $1\frac{1}{2}$ state in compensated Si and in the neutral state in $n^+$ type Si. To understand this qualitatively we consider an $ISi_4$ cluster consisting of an I atom tetrahedrally surrounded by 4 Si atoms and construct M.O. energy level schemes using 5s and 5p I orbitals and sp$^3$ hybridized Si-orbitals pointing to the central I atom. For all configurations $T_d$ symmetry is assumed and spin-orbit interaction is neglected. This model, given in Fig. 8, can account for the observed isomer shift differences, as follows: the addition of an electron to the $A_1^*$ orbital ($I^{++} \rightarrow I^+$) leads to an increase of s density because this orbital has a partial s character. The next electron ($I^+ \rightarrow I^0$) fills a $T_2^*$ orbital which leads to a (smaller) decrease of the s density because it has p character and screens the nucleus to some extent from the s electrons. The ratio of the contact density charge agrees with that found for many iodine compounds.

A more quantitative approach was also tried. Self consistent Hartree Fock calculations were performed by P. Aerts[41] on $ISi_4H_{12}$ clusters, yielding level energies and contact densities at the I nucleus. Using the level filling scheme presented in Fig. 7, the calculated contact density differences are $\Delta\rho_e(0) = 19$ $a_0^{-3}$ for $I^{++} \rightarrow I^+$ and $-5$ $a_0^{-3}$ for $I^+ \rightarrow I^0$ (after correcting for relativistic effects). Though a factor $\sim 3$ larger than the measured values, the changes are in the right direction.

Table 3. Parameters derived from Mössbauer spectra of $^{129}I$ in $p^+$, compensated and $n^+$ type silicon.

| Te-dose (cm$^{-2}$) | B or P dose (cm$^{-2}$) | Si-type | compo-nent | $S$[a) (mm/s) | $e^2qQ/h$ (MHz) | $\theta'$ (K) |
|---|---|---|---|---|---|---|
| $2.2\times10^{14}$ | $2\times10^{15}$B | $p^+$ | $S_1$ | 0.94(1) | | 196(3) |
| | | | Q | 1.3(1) | 550(20) | |
| $6 \times10^{14}$ | $6\times10^{14}$B | comp. | $S_2$ | 2.34(1) | – | 170(3) |
| | | | Q | 1.7(1) | 550(20) | |
| $1.4\times10^{14}$ | $2\times10^{14}$P | $n^+$ | $S_3$ | 2.15(4) | 452(8)[b) | } 155(7) |
| | | | Q | 1.4(1) | 550(20) | |

a) relative to Cu$^{129}$I absorber; b) at 4.2 K.

_Quadrupole splitting, recoilless fraction_. The component Q has about the same parameters S and e2qQ/h in all three cases (see Table 3). It is thought to be due to I-atoms residing in strained regions in the crystal. The component $S_3$, with $e^2qQ/h = 452(8)$ MHz at 4.2 K is only found in strongly n type silicon (iodine in $I^0$ state). The quadrupole splitting indicates a considerable deviation from cubic symmetry, even though the source atoms ($^{129m}$Te) are known to be in cubic sites. But this is not the only remarkable property of component $S_3$. As shown in Fig. 9, the quadrupole pattern collapses to a single line in the temperature region from 4.2 K to 80 K and further, the intensity of this single line depends strongly on the angle of emission of the gamma rays relative to the crystal axes. It is much higher for emission in the <111> direction than in the <110> direction.

A consistent explanation of these phenomena is provided by the assumption that the iodine-silicon cluster suffers a Jahn-Teller distortion that is static at 4.2 K and has become fully dynamic around 80 K. Such a distortion can indeed exist for an atom with one electron in the orbitally degenerate $T_2^*$ orbital, as assumed in Fig. 8.

The most probable distortion is a trigonal one, where the iodine atom has shifted over a distance $\Delta$ from the center of the tetrahedron of Si-atoms along a <111> axis as shown in Fig. 10a. In the static case, it stays in one of the four possible positions indicated in that figure, but when dynamic behaviour sets in, it starts to jump between these positions with a frequency that rapidly increases with temperature. At temperatures where the jump frequency is much larger than the Larmor precession frequency the collapse of the quadrupole pattern to a single line is complete. The very strong anisotropy of the recoilless fraction observed at these temperatures can also be attributed to the jumping of iodine atoms

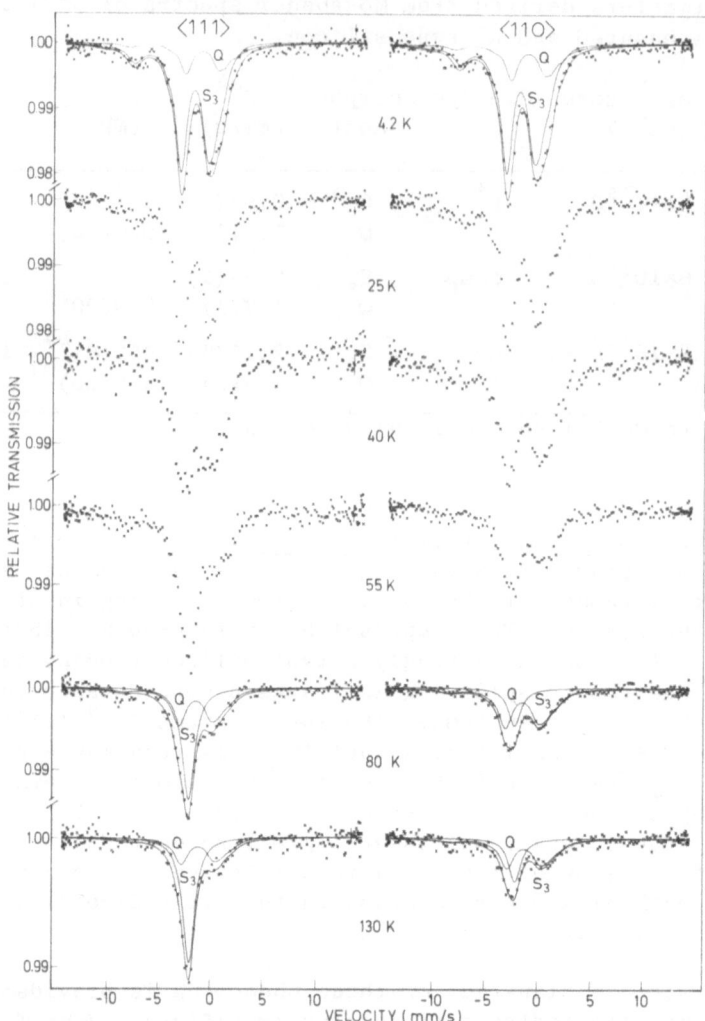

Fig. 9. $^{129}I$ Mössbauer spectra from a laser annealed Si single crystal with $2 \times 10^{14}$ at.P/cm² and $1.4 \times 10^{14}$ at.Te/cm² as a function of temperature. Gamma ray emission along <111> axis (left) and <110> axis (right). Cu$^{129}I$ absorber kept at 4.2 K.

between four equivalent sites in the tetrahedron. The general theory of Singwi  and Sjölander[42] for the effect of jump diffusion on the recoilless fraction, can be applied to the present case of limited diffusion between only four sites. It yields a so called "jump Debye Waller factor" $f_j$, that strongly depends on the angle between crystal axes and direction of gamma emission. This factor

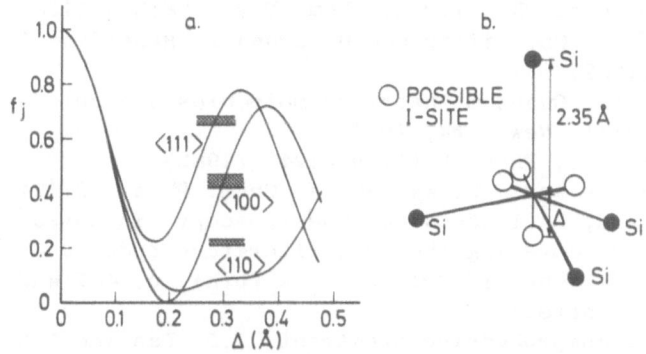

Fig. 10. a. *Jump Debye-Waller factor* $f_j$ *as a function of I displacement* $\Delta$ *for directions of gamma ray observation along* <111>, <110> *and* <100> *crystal axis. Hatched areas indicate experimental* $f_j$ *values. b. I-sites in trigonally distorted tetrahedron.*

is given in Fig. 10a as a function of displacement $\Delta$, for gamma emission along the <111>, <110> and <100> axis. The hatched areas in this figure give the experimental results for $f_j$ in these three directions. Fair agreement is obtained for $\Delta \sim 0.3$ Å.

This ends our review, which was mainly devoted to advances in the Mössbauer spectroscopy of [129]I in the last 10 years. The author wishes to apologize for incompleteness and for the influence of personal bias on the choice of material.

REFERENCES

1.  P. Jung and W. Triftshauser, Phys. Rev. 175:512 (1968).
2.  H. de Waard, G. De Pasquali and D. Hafemeister, Phys. Lett. 5:217 (1963).
3.  H. de Waard, S. Bukshpan, G.J. Schrobilgen, J.H. Holloway and D. Martin, J. Chem. Phys. 70:3247 (1979).
4.  G.J. Kemerink, Internal Report, Nr. IR108, Natuurkundig Laboratorium, Groningen, 1977.
5.  M. Pasternak, M. van der Heyden and G. Langouche, to be published.
6.  M. Pasternak, M. van der Heyden and G. Langouche, J. Chem. Phys. 80:36 (1984).
7.  W.J.J. Spijkervet and F. Pleiter, Hyperfine Interactions 7:285 (1979).
8.  P.E. Müller and W. Kündig, J. de Phys. 41:C1-125 (1980).
9.  J. Ladrière, M. Cogneau and A. Meykens, J. de Phys. 41:C1-131 (1980).

10.  E. Hartmann en Ch. Eifrig, Chem. Phys. 58:283 (1981).

11.  E. Hartmann, Ch. Eifrig and G. Brunner, Hyperfine Interactions 12:219 (1982).

12.  J.C. Slater, Quantum theory of molecules and solids, Vol. 4 (McGraw Hill, New York, 1974).

13.  K. Makariunas, Phys. Lett. 91A:249 (1982).

14.  C.H.W. Jones and J.L. Warren, J. Chem. Phys. 53:1740 (1970).

15.  H. de Waard, $^{129}$I Mössbauer Spectroscopy, in Mössbauer Effect Data Index, covering the 1973 literature (eds. J.G. Stevens and V.E. Stevens, IFI/Plenum, New York), p. 447 and references given therein.

16.  See for a comprehensive treatment: T.P. Das and E.L. Hahn, Nuclear Quadrupole Resonance Spectroscopy (eds. F. Seitz and D. Turnbull, Academic Press, New York, 1958), p. 145.

17.  F.J. Berry, C.H.W. Jones and M. Dombsky, J. Sol. State Chem. 46:41 (1983).

18.  C.H.W. Jones, J. Chem. Phys. 62:4343 (1975).

19.  H. Sakai, Y. Maeda, S. Ichiba and H. Negita, Chem. Phys. Lett. 27:27 (1974).

20.  T. Birchall and R.D. Myers, J. Chem. Soc. Dalton Trans: 885 (1983).

21.  R.J. Gillespie, R. Kapoor, R. Faggiani, C.J.L. Lock, M. Murchee and J. Passmore, J. Chem. Soc. Chem. Commun. 8:422 (1983).

22.  T. Birchall and R.D. Myers, J. Chem. Soc. Commun : 1174 (1982).

23.  N.W. Alcock and R.M. Countryman, J. Chem. Soc. Dalton Trans: 217 (1977).

24.  B.S. Ehrlich and M. Kaplan, J. Chem. Phys. 53:2041 (1969).

25.  H. de Waard and R.L. Spanhoff, unpublished results, referred to in ref. 15.

26.  M.J. Potasek, P.G. Debrunner, W.H. Morrison and D.N. Hendrickson, J. Chem. Phys. 60:2203 (1974).

27.  C.J. Schramm, R.P. Scaringe, D.R. Stojakovic, B.M. Hoffman, J.A. Ibers and J. Marks, J. Am. Chem. Soc. 102:6702 (1980).

28.  R.C. Teitelbaum, S.L. Ruby and T.J. Marks, J. Am. Chem. Soc. 102:3322 (1980).

29.  M. Cowie, A. Gleizes, G.W. Grynkewich, D.Webster Kalina, M.S. McClure, R.P. Scaringe, R.C. Teitelbaum, S.L. Ruby, J.A. Ibers, C.R. Kannewurf and T.J. Marks, J. Am. Chem. Soc. 101:2921 (1979).

30.  G. Kaindl, G. Wortmann, S. Roth and K. Menke, Sol. St. Comm. 41:75 (1982).

31.  T. Matsuyama, H. Sakai, H. Yamaoka, Y. Maeda and H. Shirakawa, J. Phys. Soc. Japan 52:2238 (1983).

32.  C.H.W. Jones and J.L. Warren, J. Chem. Phys. 53:1740 (1970).

33.  J.J. Johnstone, C.H.W. Jones and P. Vasudev, Can. J. Chem. 50:3037 (1972).

34.  H. de Waard, F.Th. ten Broek and N. Teekens, Hyperfine Interactions 4:383 (1978).

35.  H. de Waard and V. Lakshminarayana, Phys. Lett. 67A:219 (1978).

36.  F. Pleiter, R.J.T. Lindgreen and H. de Waard, to be published.

37. A. Blandin and I.A. Campbell, Phys. Rev. Lett. 31:51 (1973).
38. H. de Waard, Proceedings of the Indian National Science Academy, Int. Conf. on the Appl. of the Mössbauer Effect (New Delhi, 1982) p. 5.
39. G.J. Kemerink, H. de Waard, L. Niesen and D.O. Boerma, Hyperfine Interactions 14:53 (1983).
40. G. Foti, S.U. Campisano, E. Rimini and G.Vitaly, J. Appl. Phys. 49:2569 (1978).
41. P. Aerts, Department of Physical Chemistry, University of Groningen; we gratefully acknowledge his collaboration.
42. K.S. Singwi and A. Sjölander, Phys. Rev. 119:863 (1960).

# MÖSSBAUER SPECTROSCOPY OF ANTIMONY COMPOUNDS

John G. Stevens

Department of Chemistry
University of North Carolina at Asheville
Asheville, North Carolina 28814

## INTRODUCTION

After $^{57}$Fe and $^{119}$Sn, one of the most studied Mössbauer isotopes during the last ten years has been $^{121}$Sb. One of the more useful features of this particular Mössbauer transition is that the sign of the quadrupole coupling constant and, in many cases, the asymmetry parameter, can be determined. The addition of data about these two parameters aids in the understanding of the electronic structure of the large variety of antimony compounds that have been synthesized in recent years.

Antimony is an interesting element to study. It has oxidation states of both three and five. The electronic structures of Sb(+3) and Sb(+5) are quite different from each other. For most Sb(III) compounds there is a lone electron pair, usually with a large p character. $^{121}$Sb Mössbauer spectroscopy is a unique instrumental technique that provides information on the population of this particular orbital.

An excellent review of $^{121}$Sb Mössbauer spectroscopy was written by L. H. Bowen covering the literature up to 1972, by which time this area had been well developed.[1] This review, therefore, will primarily concentrate on developments that have taken place since 1972. During the last twelve years there has been an almost fourfold increase in the number of published papers, the major emphasis being on the applications of the technique to problems in chemistry. Some of the applications in recent years have included studies on catalysts, glasses and amorphous substances, semiconductors, and a number of magnetic systems especially Heusler alloys.

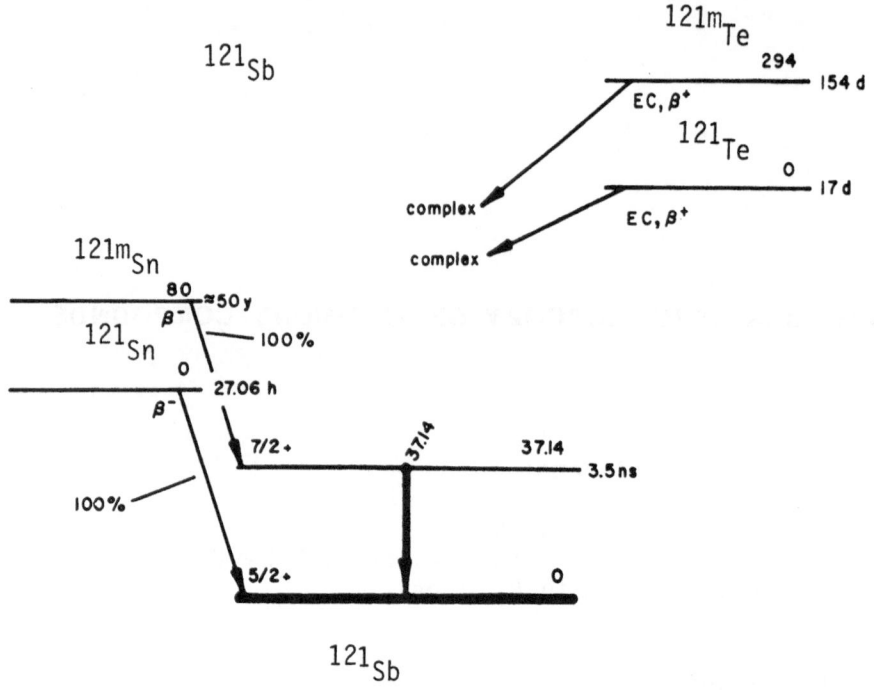

Figure 1. Simplified decay scheme of $^{121}$Sb.

A simplified decay scheme diagram on antimony is shown in Figure 1. The gamma-ray of interest is the 37.15 keV, which is between the nuclear spin 7/2 and 5/2 states. The 3.5 nanosecond lifetime of the excited state results in a somewhat broadened linewidth compared to some of the other Mössbauer isotopes. The narrowest possible linewidth is 2.10 mm/s. There is no problem with the isotopic abundance of this particular isotope since antimony contains 57.25% of $^{121}_{2}$Sb. The cross section for nuclear resonance is $1.95 \times 10^{-19}$ cm$^2$. This compares very favorably with the other Mössbauer transitions. The higher spin states of the 7/2 and 5/2 result in a much more complex spectrum. It is unfortunate that, because of the relatively large linewidth, the transitions which are caused by interactions of the nucleus with its environment cannot be fully resolved.

This overlapping can be seen in Figure 2 showing eight peaks. In Figure 3 the splittings of the 7/2 and 5/2 levels by a pure quadrupole interaction result in eight possible transitions.

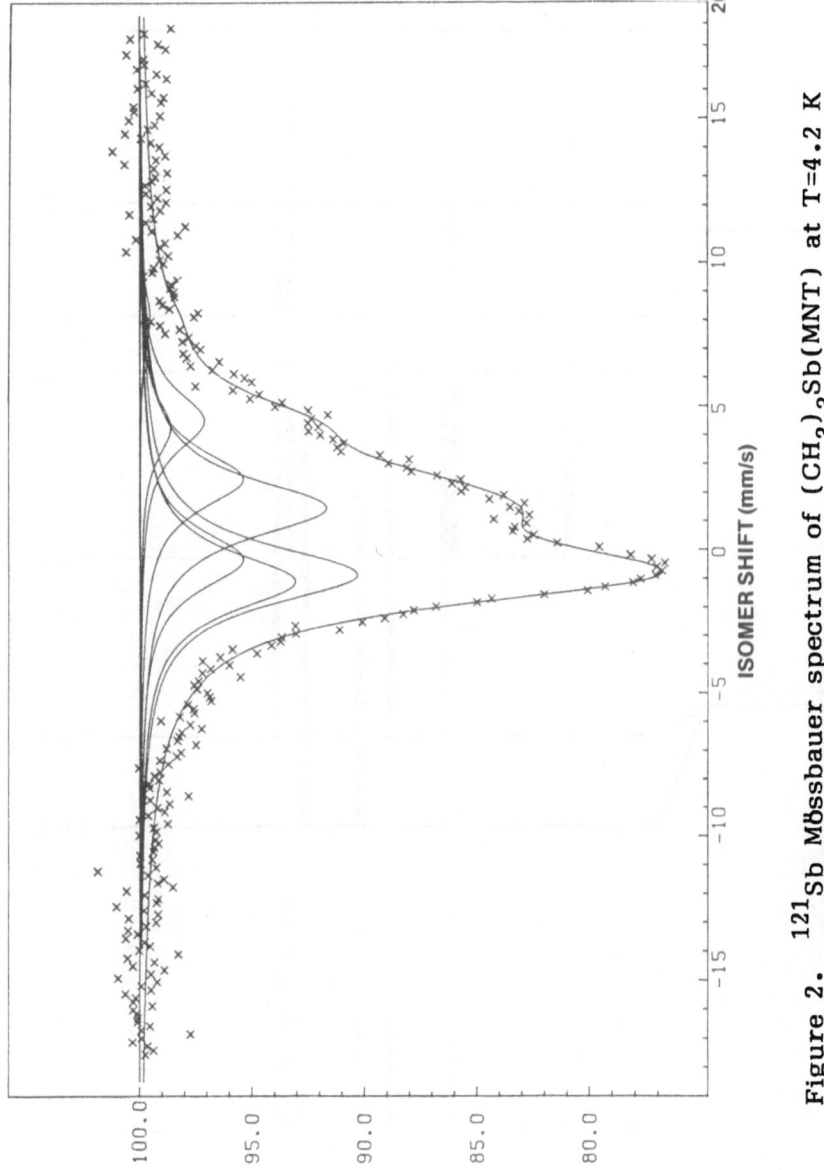

Figure 2. $^{121}$Sb Mössbauer spectrum of $(CH_3)_3Sb(MNT)$ at T=4.2 K (MNT=maleonitriledithiolate).

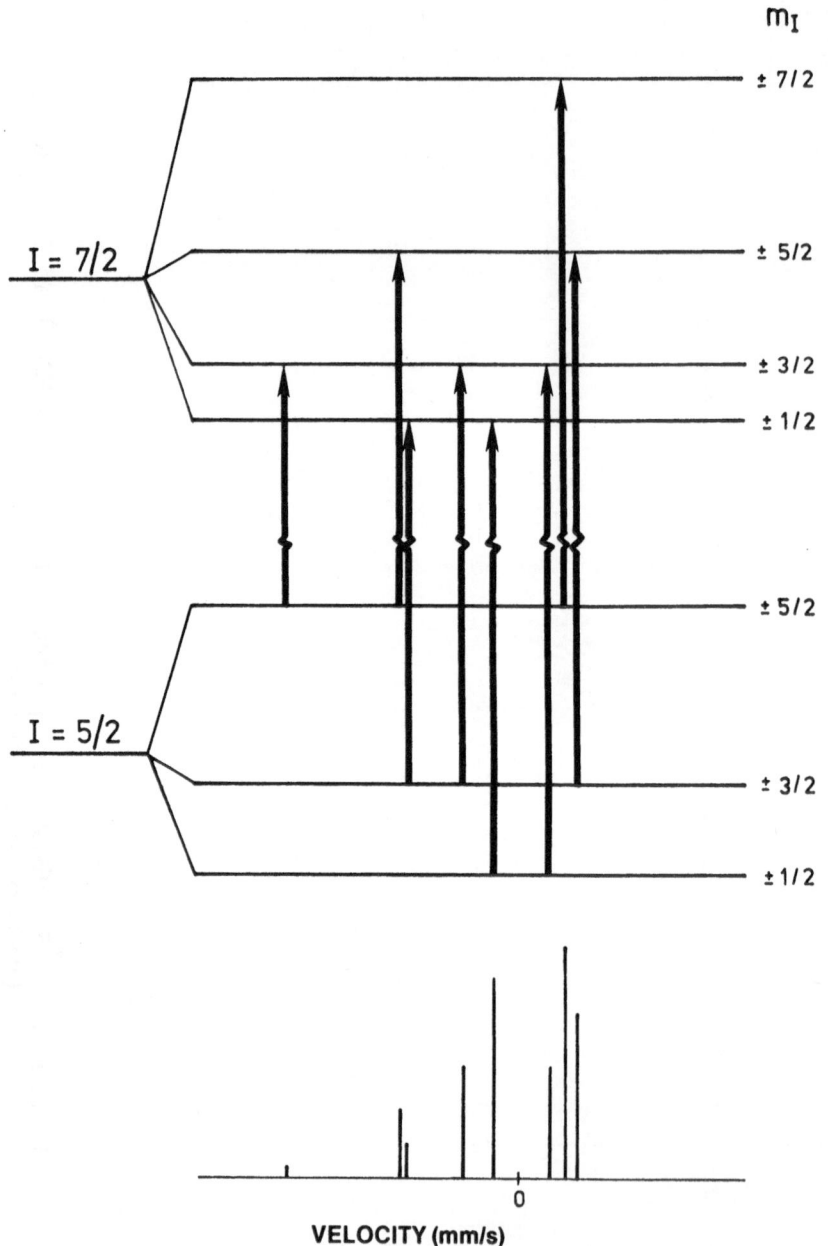

Figure 3. $^{121}$Sb Mössbauer energy level diagram and bar spectrum for pure quadrupole interaction.

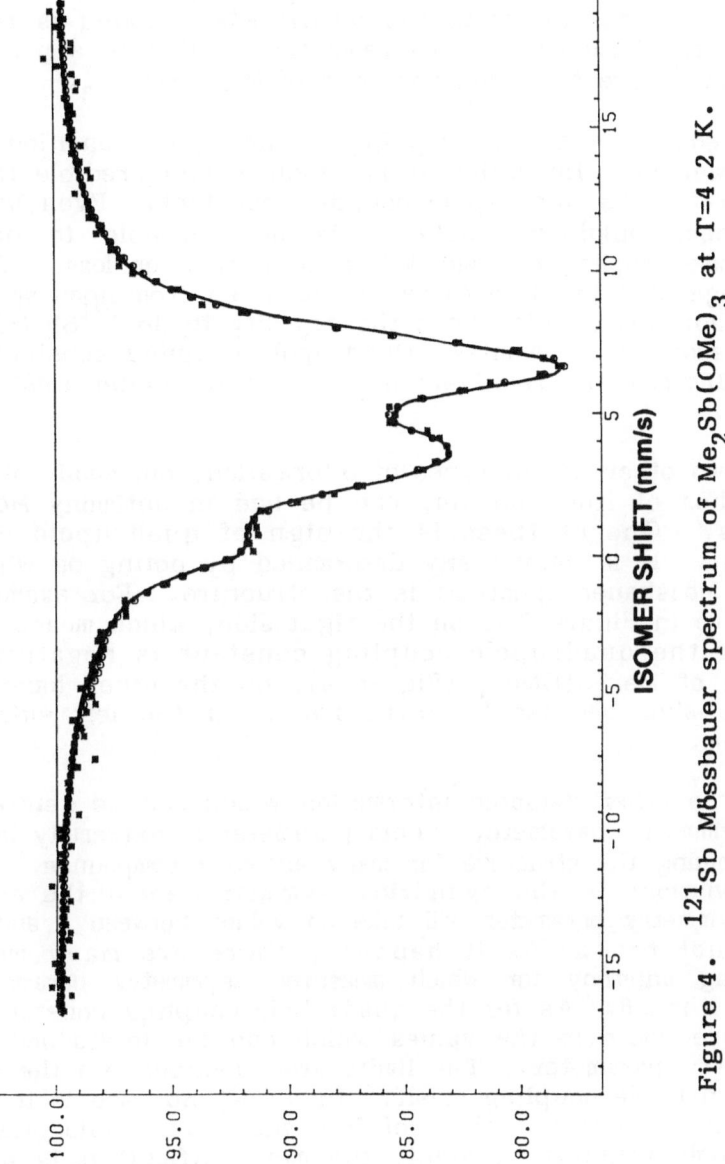

Figure 4. $^{121}Sb$ Mössbauer spectrum of $Me_2Sb(OMe)_3$ at $T=4.2$ K.

The relative positions and intensities of these transitions are shown at the bottom of Figure 3. Because of the relatively broad linewidth compared to the magnitude of the interaction, the resultant spectra consist of an envelope containing eight single lines. While Figure 2 is a more typical spectrum, there are a number of cases in which the quadrupole coupling is relatively large. In these cases some resolution begins to appear, as is shown in Figure 4 for the spectrum of $Me_2Sb(OMe)_3$.

Because of the overlapping of the eight transitions, there is a lower limit for which it is practical and credible to assign a value for the quadrupole coupling constant. Even using the very best obtainable data it is not possible to determine quadrupole coupling constant values of 5 mm/s or less. Values in the range of 5 to 10 mm/s can be obtained from good spectra by most laboratories which have the capacity to do $^{121}Sb$ Mössbauer spectroscopy experiments. Quadrupole coupling constant values above 10mm/s are relatively easy to obtain under most circumstances.

Two other useful types of information, not easily obtainable in studies of iron and tin, can be had in antimony Mössbauer studies. One of these is the sign of quadrupole coupling constant. It is very easily determined by noting on which side of the Mössbauer spectrum is the structure. For example, the structure in Figure 2 is on the right side, which means that the sign of the quadrupole coupling constant is negative. The spectra of $Me_2Sb(OMe)_3$ (Figure 4), on the other hand, has a positive sign because its structure is on the left side of the spectrum.

The other valuable information which can be determined is the asymmetry parameter. This parameter is extremely important in examining the structure for many antimony compounds. If there are deviations in the cylindrical symmetry about the antimony, the asymmetry parameter will take on values between 0 and 1, but not equal to 0. As it happens, there are many molecules containing antimony for which measured asymmetry parameters are greater than 0. As for the quadrupole coupling constant value, there are limits to the values which can be determined for the asymmetry parameter. The limits are dependent on the value of the quadrupole coupling itself. In almost no case is it possible to obtain asymmetry values of less than 0.2. The smaller the quadrupole coupling constant, the more difficult it is to obtain the smaller values of the asymmetry parameter. It is even more important than for determining quadrupole coupling constant to have data of good quality.

Electronic configurations for different kinds of antimony bonding situations are given in Table I. It is important to note

## Table I

### Electronic Configurations for Antimony

| | |
|---|---|
| Free antimony atom | $[Kr]4d^{10}5s^25p^3$ |
| $Sb^{3+}$ | $[Kr]4d^{10}5s^25p^0$ |
| Sb(III) | $[Kr]4d^{10}5s^{2-m}5p^n$ |
| Sb(V) | $[Kr]4d^{10}5s^{1-m}5p^{3-n}$ |
| $Sb^{5+}$ | $[Kr]4d^{10}5s^05p^0$ |

that a large change is expected in the isomer shift when going from ionic $Sb^{+3}$ to ionic $Sb^{+5}$. Since the difference between these states is two 5s electrons, and the magnitude of the nuclear parameter, $\delta r/r$, is relatively large for $^{121}$Sb, there is a definite effect on the values of the isomer shifts. Isomer shift data reported relative to indium antimonide, the preferred material for reporting antimony isomer shifts, is positive for ionic or covalent Sb(V) and negative for ionic or covalent Sb(III).

It is also possible to observe magnetic hyperfine interactions in antimony Mössbauer spectroscopy. However, overlapping lines also create problems with evaluating magnetic hyperfine spectra.

With antimony Mössbauer spectroscopy one can obtain isomer shift values, quadrupole coupling constants with their signs, asymmetry parameters, and magnetic hyperfine interactions. This has made the use of antimony Mössbauer spectroscopy extremely valuable in the study of structures and bondings for many different antimony compounds. Antimony has an interesting chemistry because it has a variety of structural and bonding possibilities. Some applications of Mössbauer spectroscopy to the study of structure and bonding are discussed below.

**EXPERIMENTATION**

The most commonly used detector for $^{121}$Sb Mössbauer spectroscopy is the xenon proportional counter. The pulse height spectrum which results from this particular type of detector is unusual. The peak corresponding to the 37 keV transition is relatively weak compared to the other observed peaks and is not

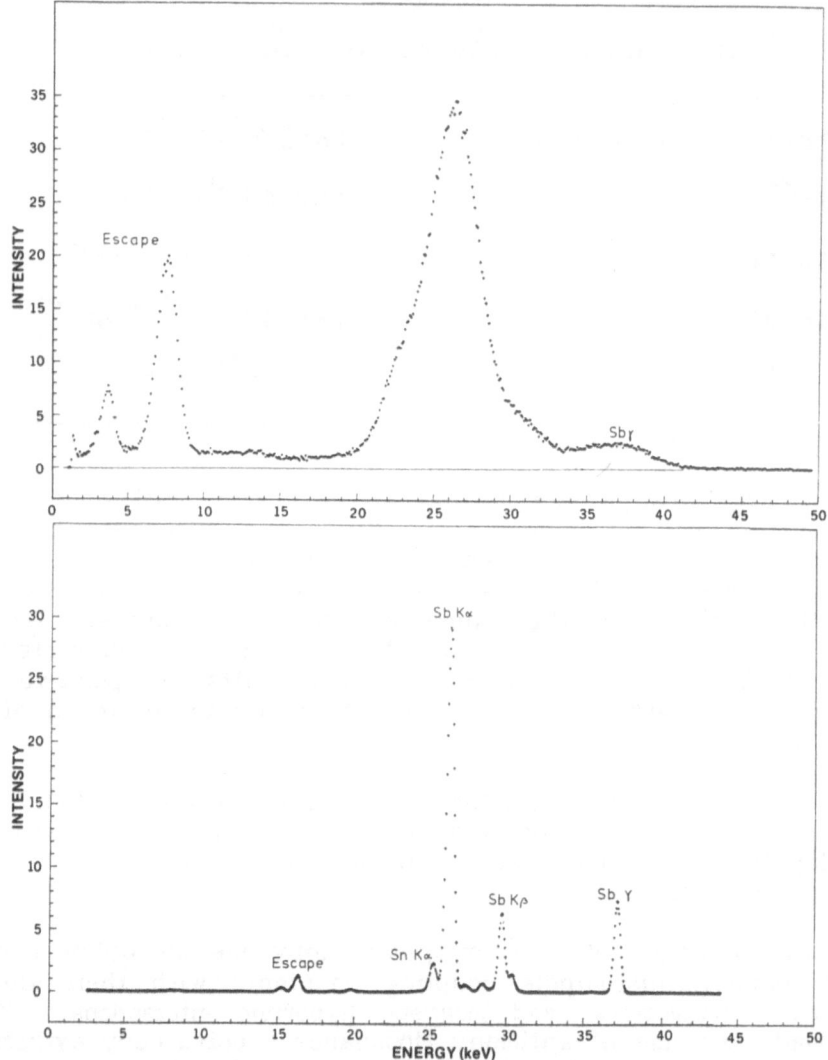

Figure 5. Pulse height spectra of the [121m]Sn source. The top spectrum is from a xenon filled proportional detector, and the bottom spectrum is from an intrinsic germanium detector cooled to a liquid nitrogen temperatue.

itself usually used for counting. One of the primary peaks is the escape peak which corresponds to a difference between the 37 keV gamma-ray and the binding energy of the K-electron in the xenon gas (See Figure 5). It is this well-resolved peak which is used when counting during the collection of data of a Mösssbauer spectrum. In some cases, additional, but not many, counts can be obtained by setting the single channel analyzer on both the escape peak and the weak 37 keV peak. The typical count rates for these weak $^{121m}$Sn sources are on the order of 50 to 300 cps. Despite the relatively low count rates, it is quite easy to obtain spectra because large absoprtions are possible. These large effects are due primarily to the small amount of background radiation that is under the escape peak itself and the large nuclear resonance cross section. In practice it is not too difficult to obtain spectra with absorptions up to 50%.

A pulse height spectrum from an intrinsic germanium detector is also shown in Figure 5. In this case the single channel analyzer is set on the actual antimony peak at 37 keV. There appears to be little advantage of the intrinsic detector over the xenon proportional counter. There is very little difference in the count rate and the background counts for either of the two types of counters. This makes the xenon proportional counter preferable since it is much less expensive.

In order to maximize the count rate, it is important that the distance between the source and detector be minimized. This is sometimes difficult to achieve because most of the experiments are performed using either a liquid nitrogen or a liquid helium Dewar. Conversely, the source and the detector should not be too close. Most antimony sources have a large area. The result of having a large source area and a short source to detector distance is both a lowering of and a smearing of the velocities in the Mössbauer spectrometer. Close source-to-detector distance and large source diameters can result in symmetric lines taking on an asymmetric shape and, therefore, being incorrectly interpreted as a small quadrupole coupling spectrum.

Most of the isomer shift data reported in the literature is given relative to InSb. In other cases, isomer shift values are given relative to tin dioxide, calcium stannate, barium stannate, and a few other materials. In order to compare isomer shift data among the various laboratories, it is important to have reliable conversions. Table 2 contains such conversions.[2]

The 37 keV nuclear level of $^{121}$Sb is populated by the decay of $^{121m}$Sn. The $^{121m}$Sn is produced by irradiating $^{120}$Sn for a long period of time in a high flux nuclear reactor. Because of the long half-life of the $^{121m}$Sn, the activity of these sources is quite weak. It is very unusual to have sources with activities much greater than 100 $\mu$c. Because of the weak activity there has been interest in producing a different source for populating the 37 keV nuclear level. One candidate is $^{121}$Te. Bukshspan et al. at Groningen produced the $^{121}$Te through a (p,n) reaction with a $^{121}$Sb target.[3] Even though the group was successful in obtaining a Mössbauer spectrum, the spectrum

Table II

Isomer Shift Reference Table for $^{121}$Sb

|  | InSb | $Ni_{21}Sn_2B_6$ | $SnO_2$ |
|---|---|---|---|
| InSb | 0.0 | 1.666±0.014 | 8.55±0.03 |
| $Ni_{21}Sn_2B_6$ | -1.666±0.014 | 0.0 | 6.88±0.03 |
| $SnO_2$ | -8.55±0.03 | -6.88±0.03 | 0.0 |

Note: The isomer shifts for $SnO_2$, $CaSnO_3$, and $BaSnO_3$ are experimentally identical.

produced had a broadened linewidth and weakened intensity, which resulted in a very poor signal.

At Leningrad State University a group made another attempt at producing $^{121}$Te, this time by radiating $^{120}$Te with neutrons.[4] The spectra they obtained also resulted in a broadened linewidth and extremely small intensities.

Even though the $^{121}$Sn source has undesirably low activity, it is a source that can be used for many decades. Also, as mentioned above, one is routinely able to obtain intensities on the order of 10%.

A plot of the recoil-free fraction according to the Debye model versus temperature is given in Figure 6. As this plot clearly shows, it is very difficult to obtain spectra at room temperature. Only in the cases of materials having extremely high Debye temperatures can useful spectra be obtained. Most of the materials that have been studied and are of interest to the chemists during the last decade have been organo-antimony compounds. These compounds have Debye temperatures usually below 100K which means that even at the liquid nitrogen temperature it is very difficult to obtain good spectra. However, at liquid helium temperature excellent spectra can be obtained without much difficulty.

Care must be taken when fitting experimental data to either the quadrupole coupling constants or the magnetic hyperfine interactions. Shenoy et al. have pointed out instances in which data should not be fitted with the normal sum of Lorenztians but

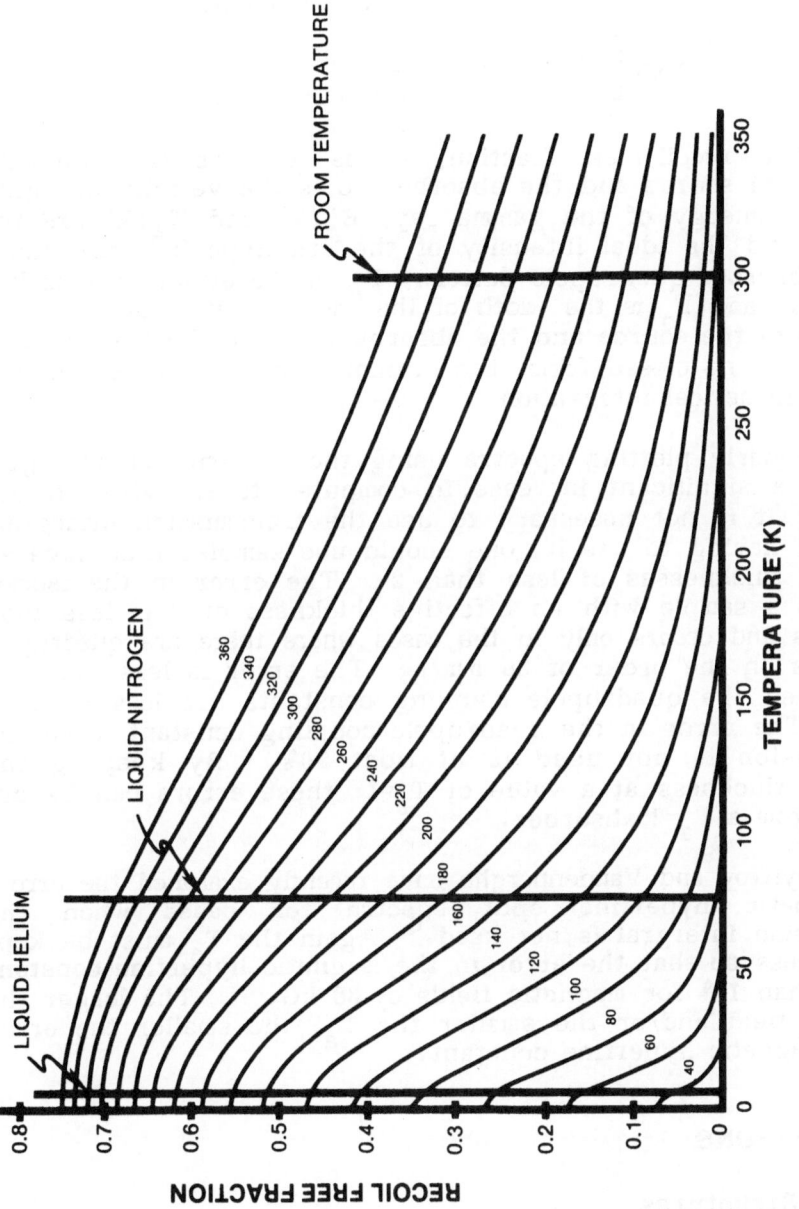

Figure 6.  Plot of recoil-free fraction according to the Debye
model vs. temperature.  Curve lines for Debye temperatures are drawn for values of 40 K to 360 K in
steps of 20 K.

329

with the transmission intergral.[5]   The transmission intergral is given as follows:

$$I_{TI}(v) = I_0 \left\{ (1-f_s) + \frac{f_s}{\pi} \int_{-\infty}^{+\infty} \frac{\frac{1}{2}\Gamma_0 \, dE}{(E + vE_0/c)^2 + (\frac{1}{2}\Gamma_0)^2} \times \right.$$

$$\left. \times \exp\left[ -\sum_k W_A(k)\, T_A \frac{(\frac{1}{2}\Gamma_0)^2}{[E - E_A(k)]^2 + (\frac{1}{2}\Gamma_0)^2} \right] \right\}$$

$f_s$ is the recoil-free fraction, $v$ is the relative velocity between the source and the absorber, c is the velocity of light, $E_0$ is the energy of the gamma ray, $E_A(k)$ and $W_A(k)$ are the position and the ideal intensity of the $k$th hyperfine component of the absorber quadrupole pattern, $T_A$ is the effective absorber thickness, and $\Gamma_0$ is the width of the excited state assumed to be the same in the source and the absorber.  Since the transmission integral has no closed form, the evaluation of the above equation involves numerical integration.

Properly plotting spectra using the transmission intergral requires a significant increase in computer time.  Many times, however, it is not necessary to use the transmission intergral. To avoid having to use it, one should use samples that have an effective thicknesses of less than 2.  The error in the isomer shift for a sample with an effective thickness of 2 is less than 0.2 mm/s and occurs only in the cases where there are quadrupole couplings on the order of 30 mm/s.  The error is less than 0.1 mm/s when the quadrupole coupling constants are less than 15 mm/s.  The error in the quadrupole coupling constant when the transmission is not used is at most 10%.  By keeping the effective thickness at a value of $T_a=1$, these errors can be cut in half from a $T_a=2$ absorber.

Gryffroy and Vandenberghe have recently examined the errors in magnetic hyperfine split spectra for cases when the transmission intergral is not used.[6]  Again the $T_a$ must be kept at 2 or less so that the error in the magnetic hyperfine constant is less than 10% for magnetic fields of 80 kG.    The larger the magnetic field and/or the smaller the $T_a$, the smaller the error in the magnetic hyperfine constant.

## APPLICATIONS

### Ph$_4$SbX Structures

Organoantimony compounds with the general formula Ph$_4$SbX have been known for quite some time.  A variety of experimental techniques including infrared spectroscopy and electrical conductivity measurements have been used to gain information on

330

their structure.[7]  The two most likely structures are shown in Figure 7.  Structure A has the antimony tetrahedrally surrounded by the four phenyls.  For this structure a $Ph_4Sb^+$ ion exists in the solid state, which, when dissolved in certain solvents, results in a solution with measurable electroconductivities. Structure B consists of four phenyls and the X group in a trigonal bipyramid configuration.  This structure is the same as that of $Ph_3SbX_2$ compounds for which the general structure is fairly well known and is shown as Structure C.  Note especially that the X ions are in the axial positions of the trigonal bipyramid structure.

If antimony is tetrahedrally surrounded by the phenyls (Structure A), the quadrupole coupling constant is expected to be 0.  Also, the isomer shift is not expected to change very much as the X ion/group changes, because the X ion/group is not bonded directly to the antimony.  However, according to Mössbauer data

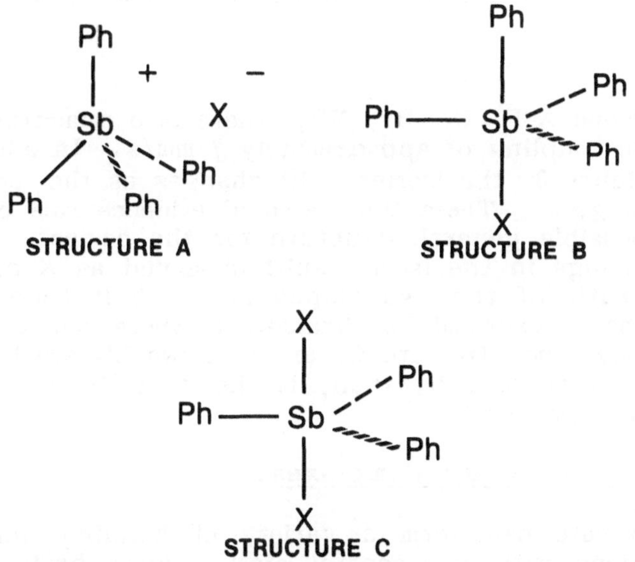

Figure 7.  Structure diagrams for $Ph_4SbX$ and $Ph_3SbX_2$ where X is an anion.

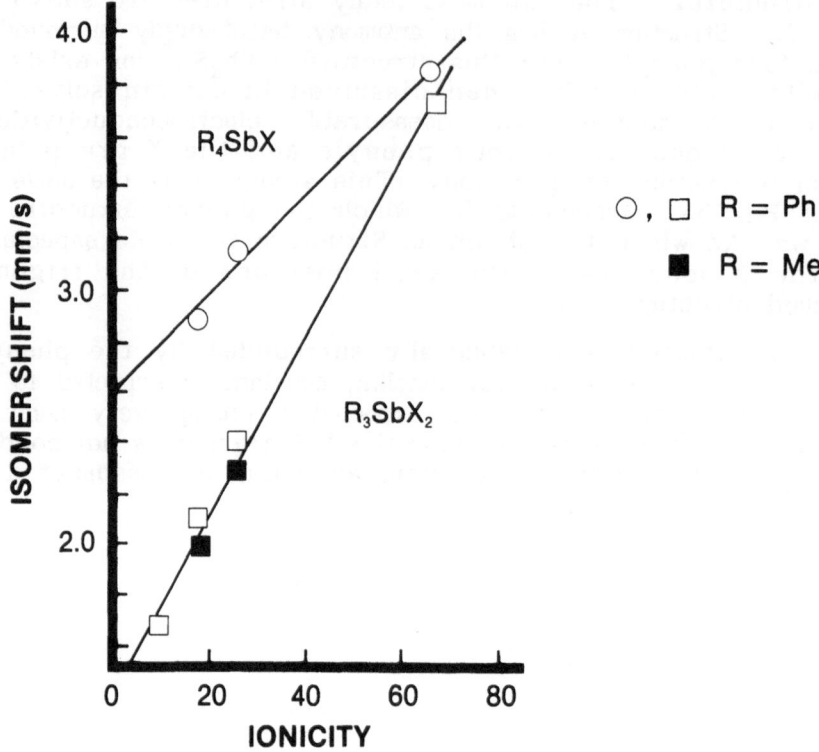

Figure 8. Isomer shift vs. Pauling bonding ionicity (of the SbX bond) for some organoantimony compounds.

obtained when X=F, Cl, Br, $NO_3$, there is a measurable negative quadrupole coupling of approximately 7 mm/s. In addition, as is seen in Figure 8, the isomer shift changes as the ionicity of the X ion changes. These two pieces of evidence rule out Structure A as a possible general structure for the series. Not only is there a change in the isomer shift observed as X changes, but also the ratio of the two slopes is one-half (see Figure 8). This is what is expected for Structure B where one X is bonded to the antimony, and Structure C, in which two X's are bonded to the antimony. Structure B, then, is the most likely structure for the $Ph_4SbX$ compounds.

## Dithiocarbamate (dtc) Complexes.

Dithiocarbamate can form a variety of bonding and structure configurations with the central atom. Four basic possibilities are shown in Figure 9.[8] For bidentate, both the sulfurs of the dithiocarbamate are covalently bonded to the central ion, shown in Figure 8 as M. For an isobidentate two bonds are also formed with the central ion, but one is a covalent bond, and the other a very weak bond. Only one of the sulfurs is strongly covalently bonded to the central ion in the monodentate, the third candidate. The fourth possibility is formed when the dithiocarbamate complex itself is ionically bonded to the central ion. Antimony Mössbauer spectroscopy, aided by other experimental data, has been useful in identifying structures for many of the antimony dithiocarbamate complexes. Data for a selected number of dtc complexes are given in Table 3. Four are discussed below.

**RSb(Et$_2$dtc)$_2$.** Similar Mössbauer parameters were obtained for two RSb(Et$_2$dtc)$_2$ compounds (R=Me, Ph). As noted in Table 3 these two similar compounds have large positive quadrupole coupling constants, indicating an excess of electrons in the axial positions and electron withdrawal in the equatorial positions. The structure with these characteristics would have the electronegative atoms (the sulfur in this case) in the equatorial plane. Also, as shown in the Table, the asymmetry parameter is 0, indicating near cylindrical or cylindrical symmetry. The most obvious choice for a structure is octahedral with a sterically active pair of electrons in one of the axial positions. Interpretation of infrared spectra also confirms bidentate as the

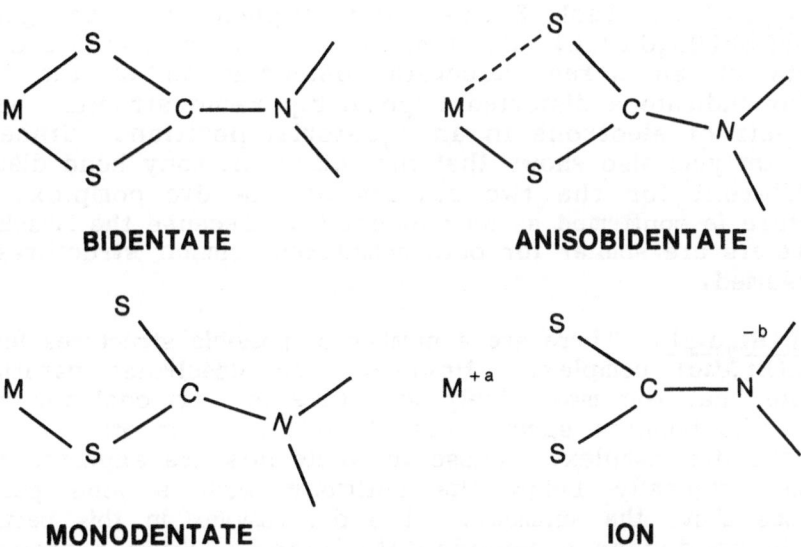

Figure 9.  Diagrams of four types of bonding and structure configurations for dithiocarbamate complexes.

333

# Table III

$^{121}$Sb Mössbauer Parameters for Miscellaneous
Dithiocarbamates at 4.2 K. Isomer Shifts are
Relative to InSb

| Complex | $\delta$ (mm/s) | $e^2qQ$ (mm/s) | $\eta$ |
|---------|-----------------|----------------|--------|
| $MeSb(Et_2dtc)_2$ | −4.0 | 25.1 | 0.0 |
| $PhSb(Et_2dtc)_2$ | −4.1 | 22.2 | 0.0 |
| $Sb(Bu_2dtc)_2I_3$ | −5.5 | 9.4 | 0.5 |
| $Sb(Bu_2dtc)_2 \cdot 0.5Cd_2I_6$ | −5.1 | 10.3 | 0.0 |
| $Ph_2Sb(Et_2dtc)$ | −2.5 | 20.0 | 1.0 |
| $Me_4Sb(Et_2dtc)$ | 3.0 | 0.0 | 0.0 |

likely structure. Consequently, the dithiocarbamate in this particular case is a bidentate.

**$Sb(Bu_2dtc)_2L$.** Table 3 shows two complexes with the general formula $Sb(Bu_2dtc)_2L$. For these L is equal to $I_3$ or 0.5 $Cd_2I_6$. Results of an x-ray structural determination[3] of the latter complex indicate a distorted trigonal bipyramid structure with a lone pair of electrons in an equatorial position. Since the x-ray analysis also shows that the sulfur-antimony bond distance is different for the two sulfurs of the dtc complex, this structure is confirmed as anisobidentate. Because the Mössbauer parameters are similar for both complexes, similar structures can be assumed.

**$Ph_2Sb(Et_2dtc)$.** There are a number of possible structures for the $Ph_2Sb(Et_2dtc)$ complex. However, the Mössbauer parameters indicate that the most likely structure is trigonal for which two of the bonding species are phenyls and the other a sulfur from the dtc complex. These three ligands are expected to be bonded trigonally below the antimony with a lone pair of electrons above the antimony. The dtc complex in this particular case is bonded as a monodentate ligand. Such a structure explains the large value for the asymmetry parameter, the recorded values of the quadrupole coupling constant, and the isomer shift. A fuller explanation is given in the paper by Stevens and Trooster.[8]

**Me$_4$Sb(Et$_2$dtc).** Mössbauer parameters for the complex Me$_4$Sb(Et$_2$dtc) are also given in Table 3. It should immediately be noted that for this particular complex there is no quadrupole coupling. As mentioned in an earlier example, a proposed structure for cases in which there is no quadrupole coupling for the antimony is one in which the antimony is tetrahedrally surrounded by four of the same ligands. For this particular compound, one can conclude that the antimony is tetrahedrally surrounded by four methyls and that the dtc complex is ionically bonded to the Me$_4$Sb$^+$ ion.

A few applications have been mentioned above; another will be described in the next section. However, there are many examples of the application of antimony Mössbauer spectroscopy to the understanding of the bonding and the structure of antimony compounds. In the examples that have been mentioned, use has been made not only of the isomer shift and the quadrupole coupling constant, but also of the sign of the quadrupole coupling constant and the value of the asymmetry parameter. Having these additional two parameters helps greatly in understanding the bonding and the structure of many complexes. While examples mentioned in this review are organoantimony, many other materials have been studied, including inorganic compounds, antimony minerals, magnetic materials, glasses, and catalysts.

## INTERPRETATION OF MÖSSBAUER PARAMETERS

One of the major benefits of interpreting Mössbauer parameters is discovering the relationships they have with the electronic structures of the atoms examined. In the interpretation $^{121}$Sb has two advantages over that of $^{57}$Fe and $^{119}$Sn. The first, already mentioned, is that additional information can be obtained from the quadrupole interactions in the antimony compounds. Secondly, the electronic structure is almost always much simpler because d electrons can be disregarded. Consequently, only the s and the p electrons need be considered. For antimony these are the 5 s's and the 5 p's.

### Self Consistent Atomic Orbital Calculations

The Hartree-Fock consistent field calculation is one of the first theoretical approaches to the interpretation of antimony isomer shifts. These calculations have been refined several times through the years. One of the major refinements has been correction for relativistic effects. The relationship between the electronic populations and the isomer shift is shown in Figure 10.[9] Even though there are differences in interpretation on the positioning of the isomer shifts of the antimony compounds (shown on the right-hand scale of this figure), every interpretation recognizes a strong relationship between the

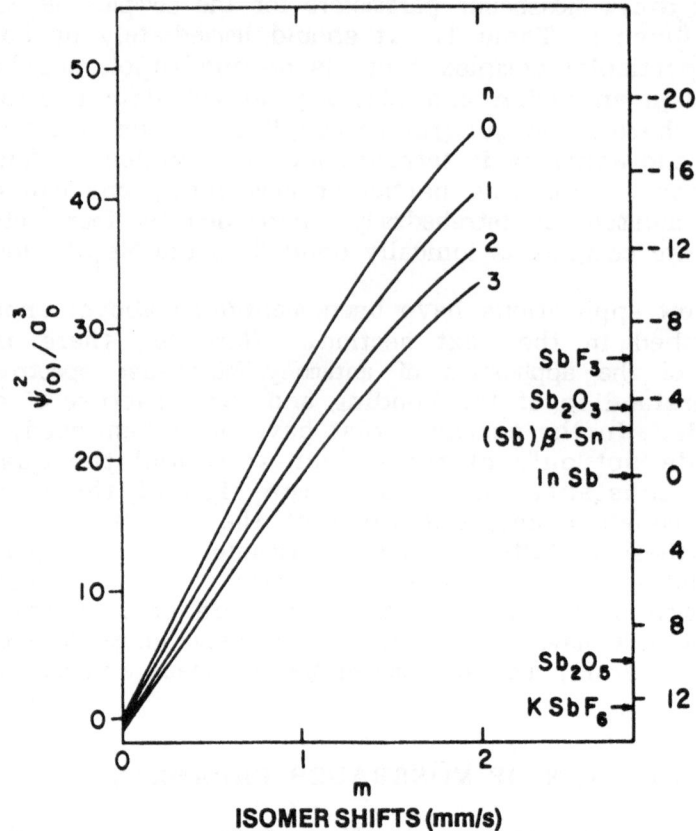

Figure 10. Correlation between electron density $\Psi^2(o)/a_o^3$ at the nucleus and the isomer shift for $^{121}$Sb as a function of the number m of 5s electrons. The parameters n is the number of 5p electrons.

isomer shift and the s electron population. There is a second order effect due to the p electron population. Because there are similarities in the electronic structures of Sn and Sb compounds, plots similar to Figure 10 can be drawn for $^{119}$Sn compounds and comparisons made between $^{119}$Sn and $^{121}$Sb Mössbauer spectroscopic data.

Since isomer shift data can be obtained with the precision of $\pm 0.1$ mm/s, it is easy to observe small changes in the electronic structure, particularly to distinguish between the two major oxidation states, III and V, of antimony. It is also possible to use the isomer shift data to interpret the ionic contributions, electronegativity effects, overlap contributions, hybridization of the valence shell orbitals, polarization of non-bonding orbitals, and bond multiplicity.

## Point Charge Models

Using the point charge model to interpret $^{57}$Fe and $^{119}$Sn Mössbauer quadrupole splittings has had moderate success. For $^{121}$Sb, as already mentioned, there is an additional parameter, the asymmetry parameter. This parameter can also be calculated using the point charge model.

An example of an application of the point charge model demonstrated using the data of a series of methylchlorostibines (Me$_x$SbCl$_{3-x}$).[10] Mössbauer data for this particular series is given in Table IV.[10] In a simplified interpretation, the structure consists of antimony tetrahedrally surrounded by four point charges, the three ions (methyls and/or chlorides) and the lone electron pair. If one starts with the SbCl$_3$ data and assigns a point charge

Table IV

Mössbauer Spectroscopic Experimental
Parameters for Methylchlorostibines
at 4.2 K

| Compound | $\delta$ (mm/s) | $e^2qQ$ (mm/s) | $\eta$ |
|---|---|---|---|
| SbCl$_3$ | −5.9 | 13.3 | 0.0 |
| MeSbCl$_2$ | −4.2 | 31.0 | 0.35 |
| Me$_2$SbCl | −2.6 | −30.8 | 0.80 |
| Me$_3$Sb | −0.2 | 16.3 | 0.0 |

Isomer shift relative to InSb.

value of 5.0 to the chloride ion, the value for the lone pair (E) is calculated from the quadrupole coupling data as 11.6. With the Me$_2$SbCl Mössbauer data, a point charge value for methyl of 22.3 is calculated. Using these three point charge values (Cl=5.0, E=11.6, and Me=22.3), the remaining quadrupole coupling interaction values (the quadrupole coupling constant and the asymmetry parameter) can be calculated. The value for the

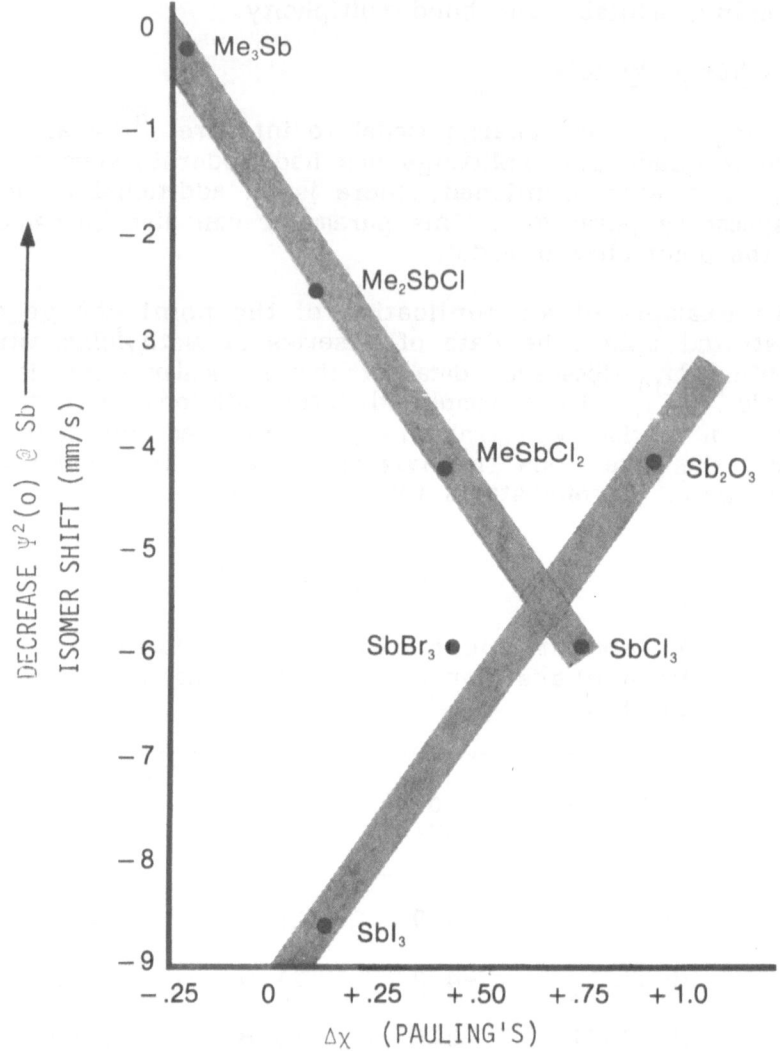

Figure 11. Isomer shift vs. average of Pauling's bonding
ionicity (of SbX bonds) for the series (1)methyl-
chlorostibines and (2)$SbX_3$ (where X=I, Br, Cl, O).

asymmetry parameter for the $Me_2SbCl$ is calculated to be equal to
0.8 which compares with 0.8 obtained experimentally. Likewise,
the quadrupole coupling constant calculated is 31.3 mm/s versus
the experimental value of 31.0 mm/s for $MeSbCl_2$. The asymmetry
parameter for this compound is calculated as 0.5 compared to the
experimental value of 0.35. Such results lead to the conclusion

that this approach provides a good explanation for data obtained by Mössbauer spectroscopy.

But when one applies these point charge values to the $Me_3Sb$, the calculated value of the quadrupole coupling constant is -21.4 while the experimental value is +16.3, a difference not only in magnitude, but also in sign. A second difficulty with the data is seen upon examination of the isomer shift values given in Table IV and plotted in Figure 11. The isomer shift increases as a more electronegative chloride is replaced by a methyl. An increase in the isomer shift for $^{121}Sb$ Mössbauer is caused by a decrease in the amount of s electron density. The experimental results are opposite from what is expected. In Figure 11 the data for both the $SbX_3(X=F,Br, Cl, O)$ series and the methylchlorostibine series are plotted. It can easily be seen that their slopes are opposite to each other, the $SbX_3$ data's being what is expected and the methylchlorostibine's not. Structural data for Sb(III) compounds generally give bond angles between $90-95^O$ for the ligands bonded to the antimony rather than the $109.5^O$ tetrahedral bond angle. If the smaller bond angle is taken into consideration in the point charge model and the quadrupole coupling data recalculated, reasonable results are obtained. These are given in Table V. In this case, in which a bond angle of $90^O$ is assumed, the quadrupole coupling data, including that of $Me_3Sb$, can be explained.

It should be particularly noted that the importance of the contribution to the quadrupole coupling constants comes from the

Table V

Mössbauer Spectroscopic Calculated Parameters for Methylchlorostibines at 4.2 K

| Compound | $e^2qQ$ (mm/s) | $\eta$ |
|---|---|---|
| $SbCl_3$ | 15 | 0.0 |
| $MeSbCl_2$ | 29 | 0.3 |
| $Me_2SbCl$ | -28 | 0.6 |
| $Me_3Sb$ | 15 | 0.0 |

Isomer shift relative to InSb.

lone pair electrons in these Sb(III) complexes. For any simple trigonal compounds of $SbM_3$, which have bond angles of $90^O$, all the contribution to the quadrupole coupling constant comes from the lone pair electrons. These electrons populate a molecular orbital that is projected in space. That is, the lone pair electrons have a high $p_z$ character. The difficulty is that the point charge model interpretation conflicts with the hybrid atomic orbital method of explaining bonding, which concludes that a bond angle of $90^O$ implies entirely s, not p, character for the lone pair.

## Molecular Orbital Calculations

One way to resolve the apparent conflict in the interpretations of the methylchlorostibine data between the point charge and the hybrid atomic orbital models is to perform molecular orbital calculations and determine the populations of each of the orbitals. This would allow one to determine if there is any $p_z$ character contribution to the lone pair of electrons when the bond angles are of the order of $90^O$. This has been done using Extended Hückel calculations. The orbital population results are given in Table VI for two molecules, $SbCl_3$ and $MeSbCl_2$. Note from these calculations that even at $90^O$ there is a substantial contribution to the molecular orbital from the 5 p orbital. It is also interesting to note from these calculations that, as expected, the replacement of a methyl by the more electronegative chloride results in an overall decrease in electron density, but that there is a redistribution of electronic populations. Specifically, there is an increase in s electron density. This explains the reverse in the slope of the $Me_xSbX_{3-x}$ line in Figure 11.

There are more sophisticated methods of calculation that can be performed beside the Extended Hückel calculations. One of these, Hartree-Fock-Slater ($X\alpha$) LCAO calculations, has been done on a series of thirteen Sb compounds.[11] This technique is often simply referred to as ($X\alpha$) calculations. These are very precise and have resulted in both isomer shifts and quadrupole coupling calculations that agree very well with the experimental data. ($X\alpha$) calculations on the methylchlorostibines give the same results as the Extended Hückel calculations.

## Other Models

There have been several other models successfully used in the interpretation of [121]Sb Mössbauer parameters. One such approach has been noting the relationship of the ligand ionicities to the Sb isomer shifts. These relationships have been summarized in the review by L. H. Bowen.[1] They have also been noted for most of the other Mössbauer isotopes.

Table VI

Orbital Populations of $SbCl_3$ and $MeSbCl_2$ Based
on Extended Hückel Molecular Calculations

| Bond Angle | SbCl$_3$ | | | |
|---|---|---|---|---|
| | 5s | $5p_z$ | $5p_x + 5p_y$ | total |
| $90°$ | 1.49 | 1.24 | 1.05 | 3.78 |
| $100°$ | 1.42 | 1.35 | 1.08 | 3.85 |
| $109.5°$ | 1.31 | 1.51 | 1.10 | 3.92 |

| Bond Angle | MeSbCl$_2$ | | | |
|---|---|---|---|---|
| | 5s | $5p_z$ | $5p_x + 5p_y$ | total |
| $90°$ | 1.42 | 1.30 | 1.12 | 3.84 |
| $100°$ | 1.36 | 1.42 | 1.12 | 3.90 |
| $109.5$ | 1.26 | 1.59 | 1.13 | 3.98 |

Another approach combines the Mössbauer isomer shifts and the quadrupole coupling constants to obtain electron populations. This technique is referred to as orbital population analysis and has been successful in examining several different series of Sb(III) and Sb(V) compounds.[12,13] The shortcoming of this approach is its use of hybrid atomic orbital theory.

In the future the areas of Mössbauer spectroscopy that show the most promise for understanding the electronic structures of antimony compounds are orbital population analysis, Hartree-Fock-Slater (Xα)LCAO, and Extended Hückel calculations, all of which combine the isomer shift and quadrupole coupling parameters. The orbital populations analysis uses hybrid orbital for which there are shortcomings. The (Xα) calculations are the best, but they entail sophisticated computer calculations. The Extended Hückel calculations give very similar results to those of the (Xα), but are considerably easier to perform.[11]

# REFERENCES

1. L. H. Bowen, $^{121}$Sb Mössbauer Spectroscopy, in: "Mössbauer Effect Data Index," J. G. Stevens and V. E. Stevens, ed., IFI/Plenum, New York (1973).
2. J. G. Stevens, Isomer Shift Reference Scales, Hyp. Int. 13: 221 (1983).
3. S. Bukshspan, F. T. ten Broek, W. Hilbrants, S. R. Reintsema, and H. de Waard, New Cyclotron Produced Sources for Mössbauer Spectroscopy, Int. J. Appl. Radiat. Isot. 27: 15 (1976).
4. A. K. Avenirov, S. I. Bondarevskii, and S. A. Timofeev, Technique of Mössbauer Spectroscopy of $^{121}$Sb with a $^{121}$Te Source, Prib. Tekh. Eksp. 2: 49 (1981)/Instrum. Exp. Tech. (Engl. Transl.) 24: 337 (1981).
5. G. K. Shenoy and J. M. Friedt, Influence of Absorber Thickness on the Mössbauer Quadrupole Spectrum of $^{121}$Sb, Nucl. Instrum. Methods 116: 573 (1974).
6. D. Gryffroy and R. E. Vandenberghe, Determination of Hyperfine Interaction Parameters from Poorly Resolved Mössbauer Spectra with the Transmission Integral, Nucl. Instrum. Methods Phys. Res. 207: 455 (1983).
7. G. G. Long, J. G. Stevens, R. J. Tullbane, and L. H. Bowen, Antimony-121 Mössbauer of Some Organoantimony Compounds, J. Amer. Chem. Soc. 92: 4230 (1970).
8. J. G. Stevens and J. M. Trooster, Antimony-121 Mössbauer Spectroscopic Studies of Antimony Dithiocarbamate Complexes, J. Chem. Soc., Dalton Trans. 740 (1979).
9. S. L. Ruby, G. M. Kalvius, G. B. Beard, and R. E. Snyder, Interpretation of Mössbauer Measurements in Tin and Antimony, Phys. Rev. 159: 239 (1967).
10. J. G. Stevens, J. M. Trooster, H. A. Meinema, and J. G. Noltes, Preparation and $^{121}$Sb Mössbauer Spectroscopy of Methylchlorostibines, Inorg. Chem. 20: 801 (1981).
11. W. Rutter, J. W. M. Jacobs, and A. van der Avoird, Hartree-Fock-Slater LCAO Calculation of the Mössbauer Parameters of Some Antimony Compounds, Chem. Phys. 78: 391 (1983).
12. L. H. Bowen and G. G. Long, Antimony(V) Orbital Populations from Antimony-121 Mössbauer Data, Inorg. Chem. 15: 1039 (1976).
13. S. W. Hedges and L. H. Bowen, Antimony(III) Orbital Population Analysis Using Antimony-121 Mössbauer Data, J. Chem. Phys. 67: 4706 (1977).

# RARE-EARTH MÖSSBAUER STUDIES OF CHEMICAL PROBLEMS

G. K. Shenoy

Materials Science and Technology Division

Argonne National Laboratory, Argonne, IL 60439, USA

## INTRODUCTION

Mössbauer spectroscopy has greatly increased our understanding of the microscopic properties of rare-earth ions in numerous chemical systems. There are as many as 46 rare-earth Mössbauer transitions in 14 rare-earth atoms. Some of these transitions have excellent characteristics appropriate for the investigation of the chemical problems. The hyperfine interactions, which provide us with the chemical information, are large in many cases and are easily measurable.

Much of the early work on Mössbauer spectroscopy of rare-earth based systems has concentrated on magnetic studies. There are excellent reviews[1,2] and data compilations[3] on this subject. In this review we will confine our attention to some of the other chemical studies conducted during the past 25 years. In particular, we will present the investigations of chemical bond, electron hopping, and lattice anisotropy. Finally, conversion electron Mössbauer spectroscopy with rare-earth isotopes will be discussed pointing to its future in resolving chemical problems.

In the next section we will briefly remind the reader of the salient aspects which make the rare-earth Mössbauer transitions suited for chemical problems. In the subsequent sections of this article we will deal with each of the above mentioned applications. This article should in no way be considered a comprehensive survey of this vast subject. The author considers the topics discussed here to be of interest to chemists. There are excellent opportunities to make advances in this field.

Table 1.  Characteristics of Useful Rare-Earth Mössbauer
Transitions

| Isotope | $E_0$(keV) | $W_0$(mm/s) | Free-ion Values (mm/s) | |
|---|---|---|---|---|
| | | | $\Delta S(3+ - 2+)$ | A |
| La-139 | 166 | 1.1 | -- | -- |
| Pr-141 | 145 | 1.0 | 1.2 | 9.19 |
| Nd-145 | 72 | 5.2 | 0.38 | 2.27 |
| Sm-149 | 23 | 1.7 | 0.92 | 7.58 |
| Eu-151 | 22 | 1.3 | 14.10 | -- |
| Gd-155 | 87 | 0.5 | -1.0 | -- |
| Dy-161 | 26 | 0.4 | 5.8 | 5.83 |
| Er-166 | 81 | 1.9 | -- | 3.87 |
| Tm-169 | 8 | 8.3 | -- | 6.00 |
| Yb-170 | 84 | 2.0 | 0.31 | 4.50 |

RARE-EARTH MÖSSBAUER SPECTROSCOPY

All the four windows of the Mössbauer effect, viz., the isomer
shift, the quadrupole interaction, the magnetic hyperfine interaction, and the resonance fraction, play an important role in understanding chemical problems.  However, there are only a few Mössbauer
transitions which are best suited for chemical research (see Table 1).

Useful isomer shifts can be measured with Pr-141, Sm-149,
Eu-151, Eu-153, Gd-155,and Dy-151.  Although the minimum observable
width ($W_0$) is not often achieved, in many cases the isomer shift
difference between trivalent and divalent rare-earth ionic
compounds ($\Delta S$) is larger than the experimental width.  The magnetic
hyperfine interactions are large in most of the rare-earths and so
also the quadrupole interactions (with the exception of Nd-145,
Sm-149, Eu-151, and Eu-153).  In Table 1, the values of free-ion $\Delta S$
and hyperfine magnetic coupling constants, A, for the trivalent
ions have been given for comparison.

In most of the rare-earths, the Mössbauer studies can be conducted only at low tempratures, preferably at helium temperatures;
the exceptions being Sm-149, Eu-151, Dy-151 and Tm-169.  The latter
group can be investigated even above room-temperature.  The con-
straint on temperature significantly reduces the versatality of
high-energy rare-earth Mössbauer transitions because of the small
resonance fractions at higher temperatures.  However, as will be
discussed later, this greatly enhances the possibility to observe
the Goldanskii-Karyagin effect in some cases using which one can
investigate lattice dynamical anisotropies in detail in favorable
situations.

Amongst the rare-earth transitions, Mössbauer conversion-electron spectroscopic studies have been reported in Eu-151 and Tm-169.[4,5] It is clear that Eu-151 is a potential candidate for investigating surfaces and diluted samples, and this aspect will be discussed in detail later.

## CHEMICAL BONDING IN RARE-EARTH COMPOUNDS

Usually, one derives much of the chemical information from the values of the isomer shift.[6] Many of the rare-earth nuclei are strongly deformed, and for the Mössbauer transitions involving rotational excitations (particularly the 0 ---> 2 transitions), the single nucleon states are not altered. This results in a small value of $\Delta R/R$ on excitation and hence a small isomer shift. Conversely, the transitions in Pr-141, Eu-151 and Dy-151 can be described as single-particle excitations having large $\Delta R/R$ and this results in their large isomer shifts.

From the point of view of chemical bonding, the rare-earth atoms are considerably different from the d-transition series atoms. In the rare-earth series the 4f orbitals are progressively occupied while the outer 5s, 5p and 6s orbitals have already been filled. The electron configuration can hence be described as $[Xe]4f^n6s^2$ for free-atoms with n=0 for La and n=14 for Yb (with the exception of Ce and Tb). The 4f orbitls are above the Xe-core states on an energy scale, but their radial extension is much less than those of the 5s, 5p, 5d and 6s orbitals. These facts are of great consequence in understanding the isomer shift variation in chemical systems containing rare-earths.

The trivalent state is the most common oxidation state for the rare-earths and is usually derived by the removal of two 6s and one 4f electrons from the atom. Because of the localized nature of the 4f electrons, they do not participate in chemical bonding and hence the variation in the isomer shift within the trivalent oxidation state is rather small. In a chemical system, Sm, Eu, Tm and Yb can also be observed in their divalent state which is usually derived by removing only two 6s electrons from the free-atom. Again, because of the small radial extent of 4f electrons, they do not usually involve in the chemical bonding. The divalent componds of rare-earth thus have a small spread in their isomer shifts. The difference of one 4f electron between the divalent and the trivalent produces a large change in the electron density at the nucleus, primarily through the shielding of 5s and 6s electrons. This results in a sizeable difference in the isomer shifts for divalent and trivalent rare-earth ions (given in Table 1).

Some of the unusual charge states such as the tetravalent state have been observed in the Mössbauer emission spectrum[7] of Dy-161. While this state is formed in the nuclear decay and is

likely to be short lived, one can stablize[8] tetravalent state in
the compounds of Pr. Similarly the divalent state of Dy has been
identified from the isomer shift measurements of its halides.[9] In
Fig. 1 we present the systematics of the isomer shifts for Pr-141,
and Dy-161.

One of the largest values of $\Delta R/R$ in the Mössbauer periodic
table is for the transition in Eu-151. The isomer shift difference
between the divalent and trivalent Eu compounds is approximately 6
times the experimental linewidth.

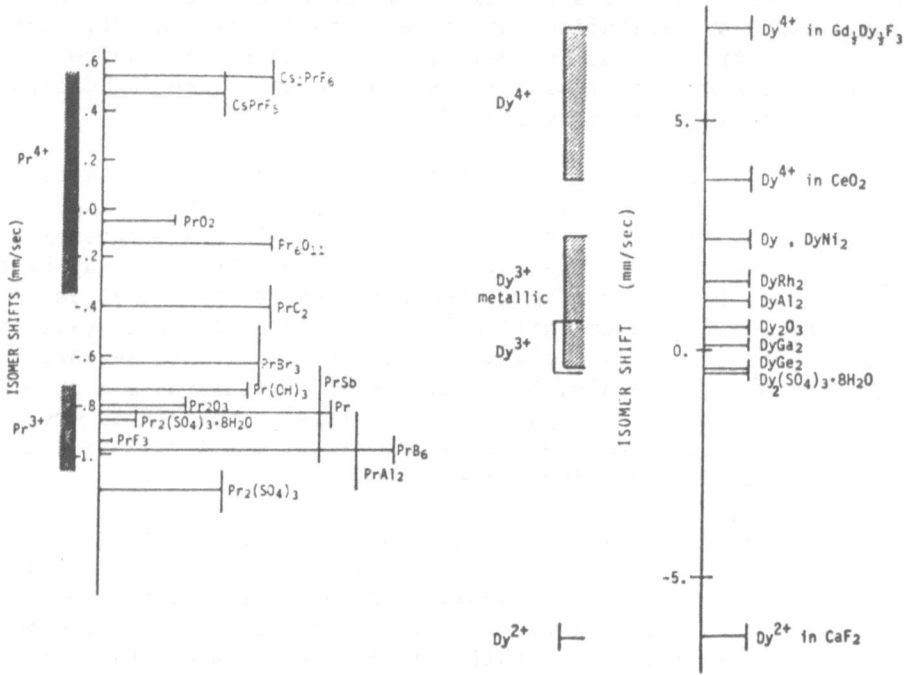

Fig. 1. Isomer shift scales for Pr-141 and Dy-161 (Ref. 6).

In this case, for various compounds even within an oxidation state
there is a measurable variation in the shift. Gerth et al.[10] have
used this fact to correlate the isomer shift of isostructural
divalent and trivalent Eu-compounds with the ionicity. This is
shown in Fig.2. Within an oxidation state, the decrease in the
shift (and also the decrease in the electron density at the
nucleus) reflects the participation of the 6s electrons in the
covalent bonding. This trend is qualitatively seen for the
halides. The sharing of the 5p electrons in the chemical bond, on
the other hand, has two influences on the isomer shift. First, the
decrease in the 5p population on Eu will decrease the shielding of

the s-like electrons which in turn will increase the electron
density at the nucleus.  Second, the decrease of·5p will reduce the
direct contribution to the electron density from $5p_{1/2}$ electrons
arising from relativistic effects.  The isomer shift variation for
the mono-chalcogenides is opposite to that in halides and might
reflect the complex nature of bonding involving both the 5p and 6s
electrons.  A detailed calculation to understand these diverse
trends is appropriate.

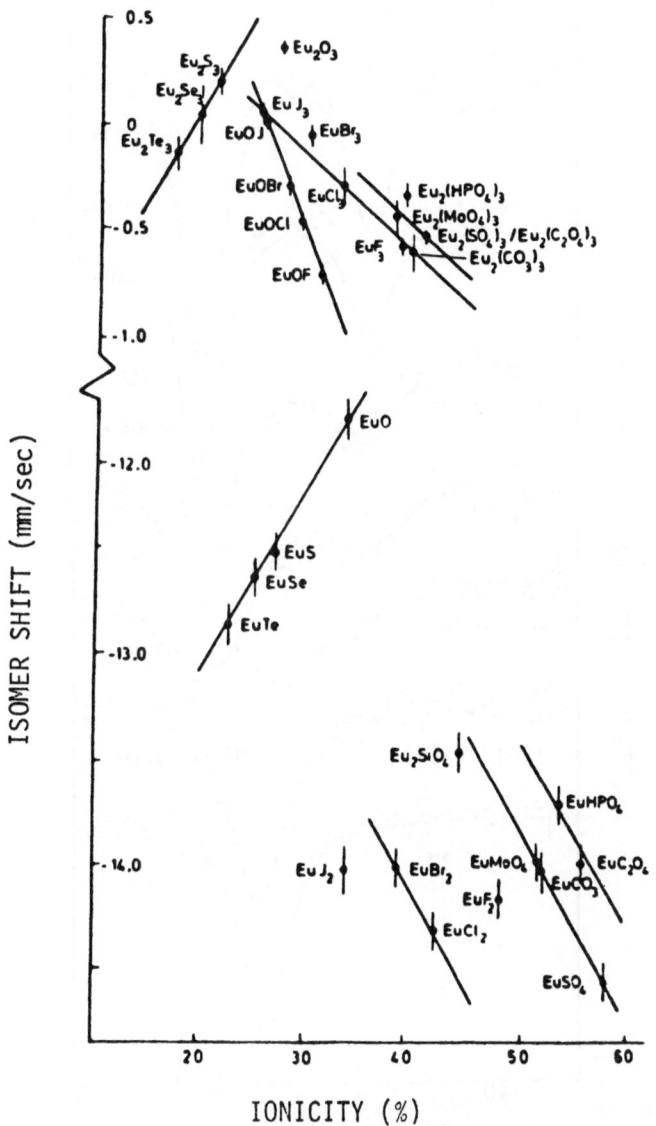

Fig. 2.   Eu-151 isomer shifts for Eu compounds vs ionicity.  The
          upper and lower parts contain trivalent and divalent
          compounds (Ref. 10).

347

In compounds like $Eu_3S_4$ one observes two absorption peaks representing the divalent and the trivalent states of Eu (Fig. 3).[11,12] At low temperatures the relative intensity of the two peaks (2:1) represents the ratio of the two crystallographic sites occupied by Eu ions. As the temperature is raised the lines broaden and finally merge into one broad line. The center of gravity of the two peaks remains unaltered over the entire temperature range

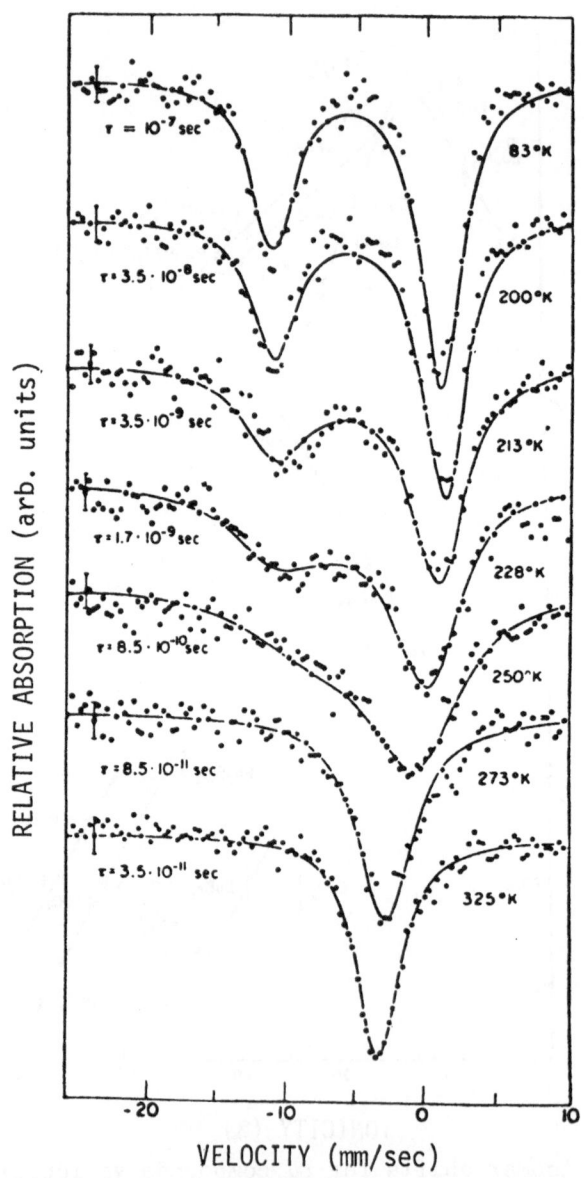

Fig. 3. Eu-151 resonance spectra of $Eu_3S_4$ at various temperatures showing electron hopping (Ref. 11).

investigated. This behavior is due to hopping of valence electron. The jump-rate, $\tau$, is characterized by an activation energy, $E_p$. With increasing temperatures the rate of hopping will rise following an Arrhenius behavior. The spectra have been analyzed to deduce the rate of hopping (given in Fig.3) using the lineshape theory similar to the one used for from the paramagnetic relaxation.[13] The activation energy is found to be 0.24 eV in this case.

Somewhat similar phenomenon is also observed in conducting systems[14] where the fluctuation rate usually remains large, but the population of the two valance state changes with temperature. The resulting motionally-narrowed single-line spectrum consequently shows a strongly temperature-dependent isomer shift.

LATTICE VIBRATIONAL ANISOTROPY

In an anisotropic solid, the Mössbauer resonance fraction, $f(\theta,\phi)$, is dependent on the angle $(\theta,\phi)$ between the direction of the gammaray absorption and the principal axes system of the ellipsoid of lattice vibrations. The anisotropy in $f(\theta,\phi)$ can be readily observed in the hyperfine pattern of a powder sample since it modifies the intensities of the hyperfine transitions in comparison to those in an isotropic crystal. Thus, in a powder, a transition with an angular distribution $P_J(\theta,\phi)$ will have an intensity

$$I = I_J (CG)^2$$

with

$$I_J = \int_0^{2\pi} \int_0^{\pi} f(\theta,\phi)\, P_J(\theta,\phi)\, \sin\theta\, d\theta d\phi$$

and where CG is the appropriate Clebsh-Gordan coefficient. This is the Goldanskii-Karyagin effect.[15] In the case of an axial symmetry, the resonance fraction becomes

$$f(\theta,\phi) = f(\theta) = A \exp(-\varepsilon \cos\theta)$$

where

$$A = \exp(-k\langle x^2\rangle)$$
$$\varepsilon = k(\langle z^2\rangle - \langle x^2\rangle)$$

Here, k is the wave-vector of Mössbauer gamma-ray and $\langle z^2\rangle$ and $\langle x^2\rangle$ are the mean-squared displacements of resonant nuclei along the z and x directions, respectively.

In the rare-earth Mössbauer transitions two aspects makes it worthwhile to investigate lattice dynamical anisotropies. First, many of the rotational transitions (for example, in Sm-152, Gd-156,

Dy-160, Er-166, and Yb-170,174) involve large energies. This will increase k; consequently one can detect considerably smaller values of $(\langle z^2 \rangle - \langle x^2 \rangle)$. For example, with the Yb-170 resonance one can measure a vibrational anisotropy which is 30 times smaller than that measureable with Fe-57 resonance. Second, the rotational transitions have the E2 (0→2) radiation pattern which because of their higher order harmonics are more sensitive to the vibrational anisotropies.[16,17]

As an example, in Fig.4, the data obtained on $Gd_2Ti_2O_7$ measured with 89 keV resonance in Gd-156 is shown.[16] The solid lines are fits to the data which gave $(\langle z^2 \rangle - \langle x^2 \rangle)$ of $-0.00072$ $\AA^2$ at 4.2K. The dashed curves in Fig. 4 are obtained with an isotropic vibrational model. The anisotropy increases at 70 K.

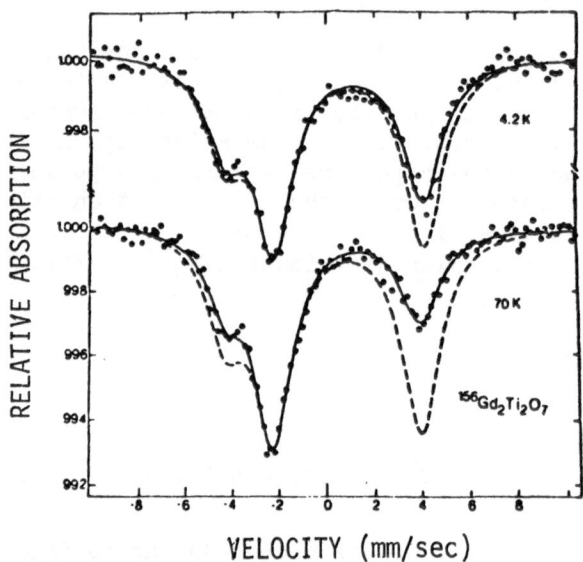

Fig. 4.   Mössbauer spectra of $Gd_2Ti_2O_7$ at 4.2 and 70 K (measured with 89 keV transition in Gd-156) showing large Goldanskii-Karyagin effect (Ref. 16).

Using the 84 keV Yb-170 resonance in Chevrel-phase $YbMo_6S_8$, the anisotropy was found[15] to be $-0.0015$ $\AA^2$ at 4.2K. This is about 6 times smaller than the disorder found from the neutron diffraction studies. The Mössbauer effect measures the dynamic disorder which is smaller than the total disorder measured from diffraction studies. This implies a delocalization (static disorder) of Yb atoms in this lattice closely related to the superconducting behavior of $YbMo_6S_8$.

CONVERSION ELECTRON MÖSSBAUER SPECTROSCOPY

A recent phenomenon is the use of Fe-57 and Sn-119 conversion electron Mössbauer spectroscopy (CEMS) for evaluating chemical and

Table 2. Partial Internal Conversion for the 21.5 keV Transition in Eu-151.

| Shell | Binding Energy (keV) | Conversion % | Electron Energy (keV) |
|-------|----------------------|--------------|-----------------------|
| K | 48.5 | -- | -- |
| $L_I$ | 8.05 | 71 | 13.5 |
| $L_{II}$ | 7.62 | 9 | 13.9 |
| $L_{III}$ | 6.98 | 4 | 14.6 |
| $M_I - M_{IV}$ | ~1.4 | 16 | ~20 |

metallurgical problems. There are some good candidates for such studies in the rare-earth series.[4,5] The 21.5 keV transition in Eu-151 is one of the best for CEMS since it has a total conversion coefficient of 28.6 (which is 3 times that in Fe-57). In Table 2, the partial internal conversion coefficients are given for this case. The majority of the observed signal is expected from the $L_I$ electrons with energy of 13.5 keV. These would have a penetration depth of about 1500 Å suitable for many surface investigations.

In Fig. 5 the conversion spectra for three compounds of Eu are compared with their gamma-ray absorption spectra measured in the transmission geometry.[4] There is a 6- to 10-fold enhancement of the signal in the conversion spectra which is not surprising considering the large conversion coefficient. This enhancement represents an increase in the sensitivity of the technique. For example the conversion spectra of $Eu_2SiO_4$ (Fig. 5) shows a weak resonance corresponding to a trivalent Eu which arises primarily in the surface layer probed by CEMS. The increased sensitivity has permitted us to detect 170 ppm of Eu in Mg.[16] The half-thickness of the L-conversion electrons in Mg is about 4500 Å and this result demonstrates the ability to detect and provide chemical charge assignment for very small dilutions normally encountered in implantation experiments.[5]

CONCLUSIONS

In this article we have presented some of the chemical applications of rare-earth Mössbauer spectroscopy. In understanding chemical bonding in rare-earth systems the Mössbauer isomer shifts have played a role. However the effort has limited to the identification of oxidation states. This is primarily because unlike the d-electrons, the 4f electrons are highly localized and do not participate in the bond in any major way. There are however situations where within an oxidation state, the isomer shift variation for different chemical systems is large enough to warrant detailed electronic structure calculations.[19,20]

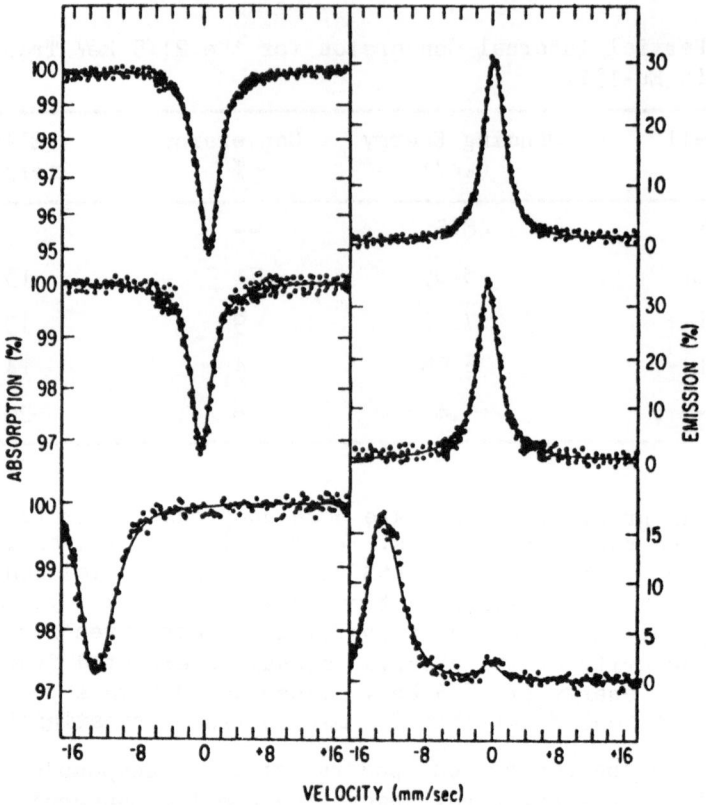

Fig. 5.  Comparison of absorption (left) and conversion electron
         (right) spectra of $Eu_2O_3$ (top), $EuF_3$ (middle) and $Eu_2SiO_4$
         (bottom) (Ref. 4).

    The ability to observe the electron hopping in rare-earth
systems is perhaps one of the important applications.  There are
great potentials for such studies as also for those showing charge
fluctuation in conducting systems.

    One of the least explored areas of rare-earth Mössbauer spec-
troscopy is the investigation of lattice dynamical anisotropy.  As
has been demonstrated the opportunities in this area are immense
and the information derivable is most valuable.

    The transition in Eu-151 is ideally suited for CEMS and many
newer applications of this technique will have far reaching effects
on chemical problems in the near future.

    This article would be incomplete without gratefully acknow-
ledging the pioneering and most significant contributions to the
rare-earth Mössbauer spectroscopy by the late Professor Solly Cohen

and the late Professor Shimon Ofer, both from the Racah Institute of the Hebrew University in Jerusalem. We lost both of them during the past 6 months.

I am grateful to Dr. S. K. Malik for a critical reading of this manuscript. This work was supported by the U. S. Department of Energy.

REFERENCES

1. S. Ofer, I. Nowik, and S. G. Cohen in "Chemical Applications of Mössbauer Spectroscopy", V. I. Goldanskii and R. H. Herber, eds., Academic Press, New York, 1968, p. 428.
2. R. G. Barnes in "Rare Earths", Vol. 2, K. A. Gschneidner, Jr. and L. R. Eyring, eds., North-Holland Publ. Co., Amsterdam, 1979, p. 387.
3. J. G. Stevens and V. E. Stevens, "Mössbauer Effect Data Index", Plenum Press, New York, 1968-77.
4. G. K. Shenoy, D. Niarchos, P. J. Viccaro, and B. D. Dunlap in "Advances in Chemistry Series, No. 194", J. G. Stevens and G. K. Shenoy, eds., American Chemical Society, Washington D. C., 1981, p. 117.
5. J. A. Sawicki, T. Tyliszezak, B. D. Sawicka, and J. Kowalski, Phys. Lett. 91A, 414 (1982).
6. E. R. Bauminger, G. M. Kalvius, and I. Nowik, in "Mössbauer Isomer Shifts", G. K. Shenoy and F. E. Wagner, eds., North-Holland Publ. Co., Amsterdam, 1978, p. 661.
7. A. Almog, E. R. Bauminger, A. Levy, I. Nowik and S. Ofer, Solid State Commun. 12 ,673 (1973)
8. W. Kapfhammer, W. W. Maurer, F. E. Wagner, and P. Kienle, Z. Naturf. 26A, 357 (1971)
9. J. M. Friedt, J. MacCordick, and J. P. Sanchez, Inorg. Chem. (in press).
10. G. Gerth, P. Kienle, and K. Luchner, Phys. Lett. 27A, 557 (1968).
11. O. Berkooz, M. Malamud, and S. Shtrikman, Solid State Commun. 6, 185 (1968).
12. M. Eibschutz, R. L. Cohen, E. Buehler, and J. H. Wernick, Phys. Rev. B 6,18 (1972).
13. H. H. Wickman and G. K. Wertheim in "Chemical Applications of Mössbauer Spectroscopy", V. I. Goldanskii and R. H. Herber, eds., Academic Press, New York, 1968, p. 604.
14. E. R. Bauminger, I. Felner, D. Froindlich, D. Levron, I. Nowik, S. Ofer and R. Yanovsky, J. Physique, Suppl. 35, C6-6 (1974).
15. V. I. Goldanskii, G. M. Gorodinskii, S. R. Karyagin, L. A. Korytko, L. M. Krizhanskii, E. F. Makarov, I. P. Suzdalev, and V. V. Khrapov, Akad. Nauk. SSSR 147, 127 (1962).
16. H. Armon, E. R. Bauminger, A. Diamant, I. Nowik and S. Ofer, Solid State Commun. 15, 543 (1974).

17. G. K. Shenoy and J. M. Friedt, Nucl. Instrum. Methods 136, 569 (1976).
18. J. D. Jorgensen, D. G. Hinks, D. R. Noakes, P. J. Viccaro, and G. K. Shenoy, Phys. Rev. B 27, 1465 (1983).
19. J. D. Cahion, M. A. Couthard, and D. B. Prowse., J. Phys. C 7, 3620 (1972).
20. A. J. Freeman and D. E. Ellis, in "Mössbauer Isomer Shifts", G. K. Shenoy and F. E. Wagner, eds., North-Holland Publ. Co., Amsterdam, 1978, p. 111.

# CHEMICAL ASPECTS OF $^{237}$Np MÖSSBAUER SPECTROSCOPY

D. G. Karraker

Savannah River Laboratory
E. I. du Pont de Nemours & Co.
Aiken, SC 29808

## INTRODUCTION

The $^{237}$Np Mössbauer effect has been especially useful in studies of neptunium chemistry, by virtue of its excellent resolution and straightforward experimental techniques. Neptunium can have valences from +3 to +7, and a broad range of compounds can be prepared that are analogous to those of other actinide elements. Studies on neptunium compounds, for example, have a ready application to the analogous compounds of uranium and plutonium. The emphasis in this paper will be on the application of the $^{237}$Np Mössbauer effect to problems in neptunium chemistry.

## Description of $^{237}$Np Mössbauer Effect

The Mössbauer effect of $^{237}$Np was discovered by Stone and Pillinger[1] at the Savannah River Laboratory, and the experimental techniques were further developed by G. M. Kalvius and co-workers at the Argonne National Laboratory.[2] The fundamentals of $^{237}$Np Mössbauer spectroscopy have been discussed in three excellent reviews[3-5] and will be reviewed very briefly here.

The gamma ray used for the $^{237}$Np Mössbauer effect results from the 59.5 keV, $5/2^- \rightarrow 5/2^+$ El transition in $^{237}$Np, and has a $t_{1/2}$ of 63 ns. The 59.5 keV level is accessible from the $\alpha$-decay of $^{241}$Am, $\beta$-decay of $^{237}$U or the electron-capture decay of $^{237}$Pu. The 433-y $^{241}$Am is the most convenient source, and the minimum line-width is obtained with 5% Am metal alloyed with a cubic thorium metal.[6]

The transitions between the 5/2⁻ excited state and 5/2⁺ ground state of $^{237}$Np are illustrated in Fig. 1. Both the excited and ground state are split by electric and magnetic fields. Fig. 1 shows the appearance of the spectra for (bottom Fig. 1, left to right) for a quadrupole-split spectrum, an unsplit spectrum, a magnetically-split spectrum, and a spectrum that has a combination of quadrupole and magnetic splitting. Pillinger and Stone[3] have discussed the analysis of $^{237}$Np Mössbauer spectra.

For chemical effects, the isomer shift is the most useful hyperfine interaction. The isomer shift is determined by the Coulomb interaction of the electrons with the nucleus. The central field interactions that cause the isomer shift are spherically symmetric and result primarily from s orbitals. For neptunium ions, the 6s orbitals are shielded from the nucleus by the inner 5f orbitals (Fig. 2). An increase in the number of 5f electrons increases the shielding and decreases the isomer shift.

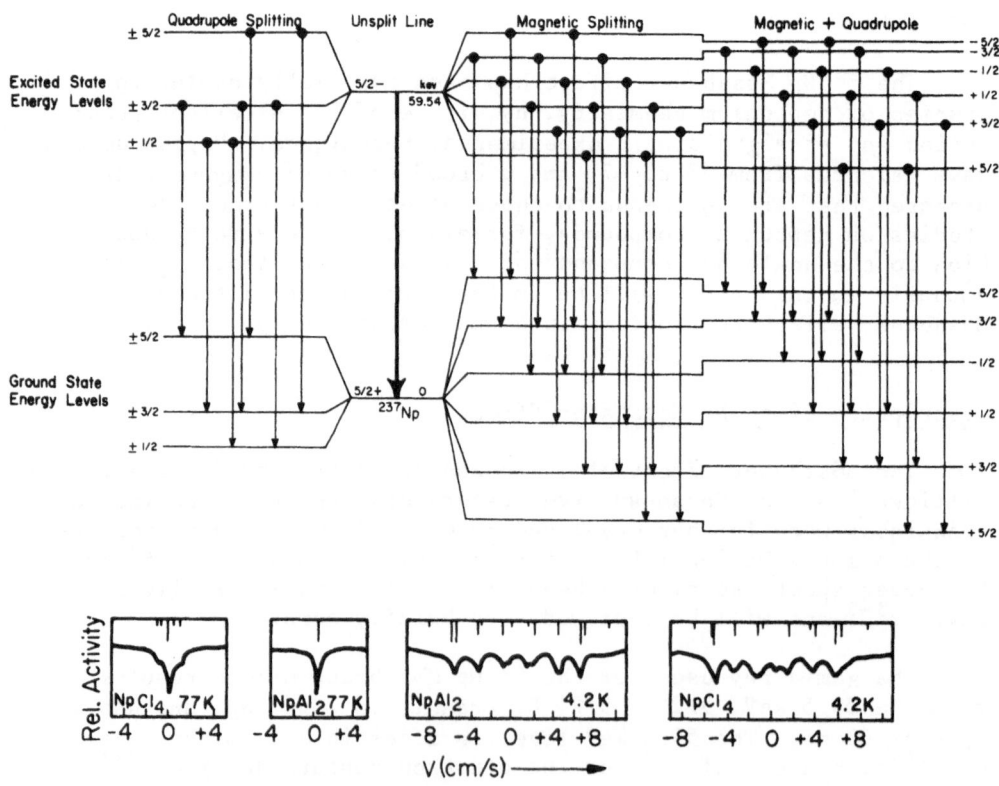

Fig. 1.   Splitting of the ground state and 59.5 keV level of $^{237}$Np in magnetic and electric fields

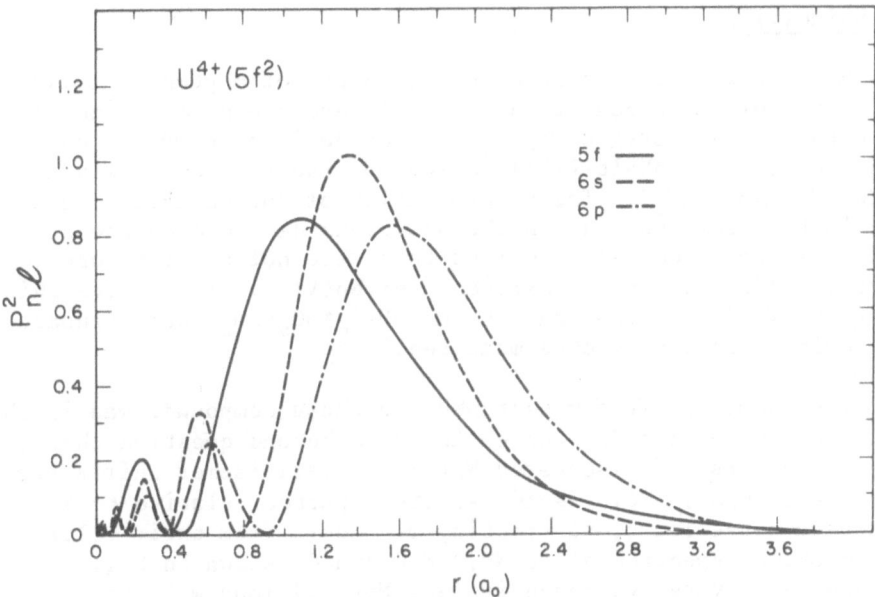

Fig. 2. Radial charge density for $U^{4+}$ $\left(\text{(——) } 5f; \text{ (---) } 6s;\right.$ $\left.\text{(— • —) } 6p\right)$ (courtesy of N. M. Edelstein, Lawrence Berkeley Laboratory, Berkeley, CA)

The isomer shifts of $^{237}$Np in its different valences (+3 to +7) cover a range from -6.9 to +3.5 cm/s, and provide an easy identification of the neptunium valence in a neptunium compound. The isomer shifts of the neptunium valence states are shown in Fig. 3 for fluorides and oxides. The Np(VI) and Np(V) isomer shifts for the oxides are more positive than the isomer shifts of the fluoride, because the covalent bonding of the oxygen atoms in the $NpO_2^{2+}$ and $NpO_2^+$ contributes electron density to the shielding 5f orbitals.

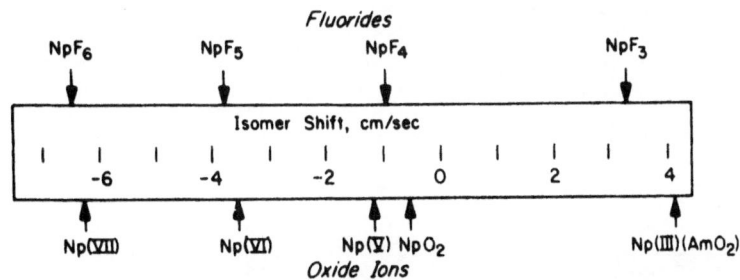

Fig. 3. Isomer shifts of Np fluorides and Np oxide ions

## Neptunium(VII)

Russian workers reported the discovery of neptunium (VII) in 1967.[7],[8] Solutions containing Np(VII) were prepared by bubbling ozone through a slurry of Np(VI) hydroxide in 3M NaOH. A compound with the hexaminecobalt (III) cation was isolated and analyzed to be $Co(NH_3)_6NpO_5 \cdot xH_2O$. The determination of the valence of neptunium in this compound and in the solutions depended on the stoichiometry of some chemical reactions that were not straightforward, so there was doubt about the existence of Np(VII). To resolve this doubt, Np(VII) solutions and the $Co(NH_3)_6NpO_5 \cdot xH_2O$ were prepared and their Mössbauer spectra measured.[9]

The isomer shift for neptunium in these compounds was in the range -6.5 to -6.8 cm/s, and established beyond question that the Russian workers had discovered Np(VII). (All isomer shifts are relative to $NpAl_2 = 0$. Isomer shifts reported relative to $NpO_2$ have been converted by subtracting 0.53 cm/s from their values.) The Mössbauer spectrum of $Co(NH_3)_6NpO_5 \cdot xH_2O$, shown in Fig. 4, was analyzed as having two nonequivalent Np(VII) ions with the same isomer shift -6.81 cm/s, but different quadrupole splittings

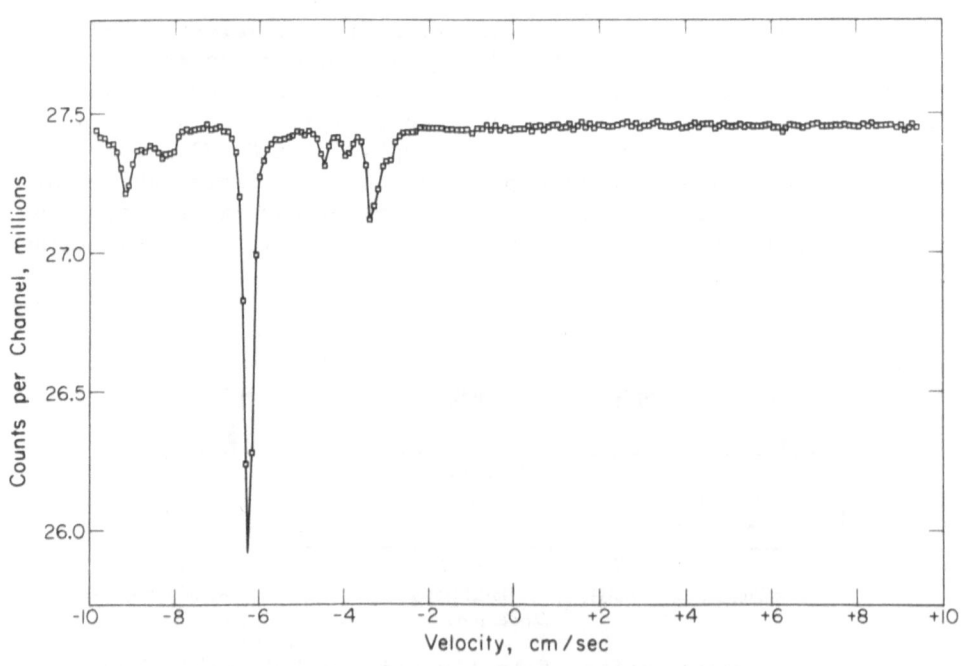

Fig. 4. Mössbauer spectrum of $Co(NH_3)_6NpO_5 \cdot xH_2O$ at 4.2°K

(eqQ/4), 2.1 cm/s for the A site, and 3.1 cm/s for the B site.[9] Later studies by Frohlich, et. al.,[10] reported isomer shifts of -6.64 cm/s for Np(VII) in $Co(en)_3NpO_5 \cdot xH_2O$ (eqQ/4 = 2.45 cm/s) and -7.42 cm/s for Np(VII) in $Li_5NpO_6$ (eqQ/4 = 0.85 cm/s).

## Proposed $NpOCl_5^{-2}$ Compounds

Virtually all Np(V) compounds prepared in aqueous media have the neptunium as an $NpO_2^+$ ion. The neptunyl (V) ion has an ellipsoidal shape, with the two oxygen ions covalently bonded to the central Np(V) ion. Reported preparations[11] of the compounds $Cs_2NpOCl_5$ and $[(C_6H_5)_4As]_2NpOCl_5$ were quite surprising, since these compounds must have one oxygen and one to three chloride ions bonded to the Np(V) ion. As this would surely be reflected by an unusual Mössbauer spectrum, the Mössbauer spectrum of $[(C_6H_5)_4As]_2NpOCl_5$ was investigated.[12]

The $^{237}$Np Mössbauer spectrum of $[(C_6H_5)_4As]_2NpOCl_5$ (Fig. 5a) was blurred by intermediate relaxation effects, but the center of gravity of the magnetically split spectrum has an isomer shift that corresponded to a Np(IV) ion, with some extra absorbances added.[12] Comparison of the $[(C_6H_5)_4As]_2NpOCl_5$ spectrum with the spectra of Np(IV) in $(TPAs)_2NpCl_6$ [TPAs = $(C_6H_5)_4As$] (Fig. 5b) and Np(VI) in $(TPAs)_nNpO_2Cl_{n+2}$ (Fig. 5c) showed that the spectrum of $(TPAs)_2NpOCl_5$ could be reproduced by a superposition of the major absorptions from the Np(IV) and Np(VI) compounds, and that the $NpOCl_5^{-2}$ ion was not present in $(TPAs)_2NpOCl_5$. This conclusion was later corroborated by absorption and infrared spectra. Obviously, in the preparation of $(TPAs)_2NpOCl_5$, the $NpO_2^+$ ion had disproportionated in strong HCl, as:

$$4H^+ + 2NpO_2^+ \rightarrow Np^{4+} + NpO_2^{2+} + 2H_2O$$

and the mixed Np(IV) and Np(VI) compounds were precipitated from solution by TPAsCl.

## $Np(C_8H_8)I \cdot xTHF$

$Np(C_8H_8)I \cdot xTHF$ is an unstable compound that could not be characterized by standard measurements but whose preparation was established through $^{237}$Np Mössbauer spectra.[13] $Np(C_8H_8)I \cdot xTHF$ was prepared as green-brown solid by the reaction of $NpI_3$ and $K_2C_8H_8$ in THF solution at -20°C. The reaction mixture was warmed to room temperature, filtered and the product recovered from the filtrate by vacuum evaporation to a damp-dry solid. Further evaporation caused a color change that was an apparent sign of decomposition. Chemical analysis established that the ratio of I/Np was about 1, but the degree of solvation varied; neither x-ray or absorption spectra yielded any useful results.

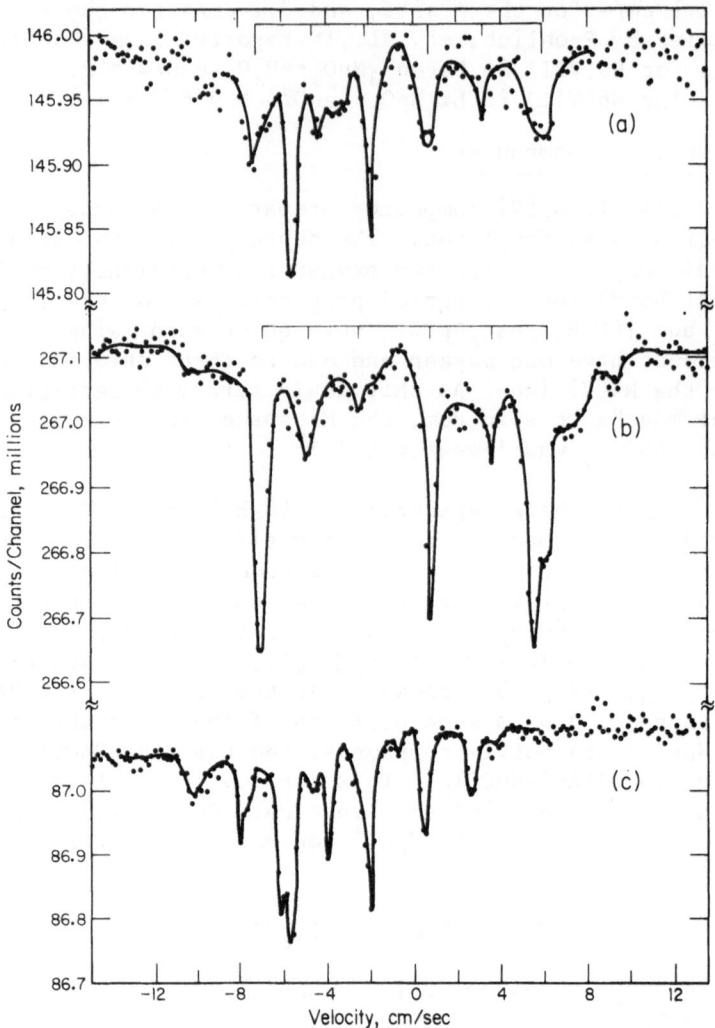

Fig. 5. Mössbauer spectra of (a) "$(TPAs)_2NpOCl_5$," (b) $(TPAs)_2NpCl_6$, and (c) $(TPAs)_nNpOCl_{n+2}$

[237]Np Mössbauer spectra for the reaction reactants and products are shown in Fig. 6. The top spectrum is a quadrupole-split of $NpI_3$ ($\delta$ = 3.33 cm/s, eqQ/4 = 0.81 cm/sec), the next lower spectrum is the THF solvate of $NpI_3$ ($\delta$ = 3.32 cm/s, eqQ/4 = 0.74 cm/s, asymmetry parameter, $\eta \cong 1$), and the next lower spectrum shows a strong single absorption with $\delta$=3.83 cm/s. This line is not the compound $KNp(C_8H_8)_2 \cdot 2THF$ ($\delta$ = 3.92 cm/s, eqQ/4 = 0.74 cm/sec, $\eta \approx 0$), and the pronounced isomer shift from $NpI_3 \cdot 6THF$ indicates a species with a small covalent contribution to its bonding. This line at $\delta$ = 3.83 cm/s is considered to demonstrate a

$Np(C_8H_8)I \cdot xTHF$ species.  The bottom spectrum in Fig. 6 shows the spectrum of the decomposition products that results when the $Np(C_8H_8)I \cdot xTHF-NpI_3 \cdot 6THF$ mixture is thoroughly dried in vacuum – probably the spectrum of a mixture of Np(IV) products.

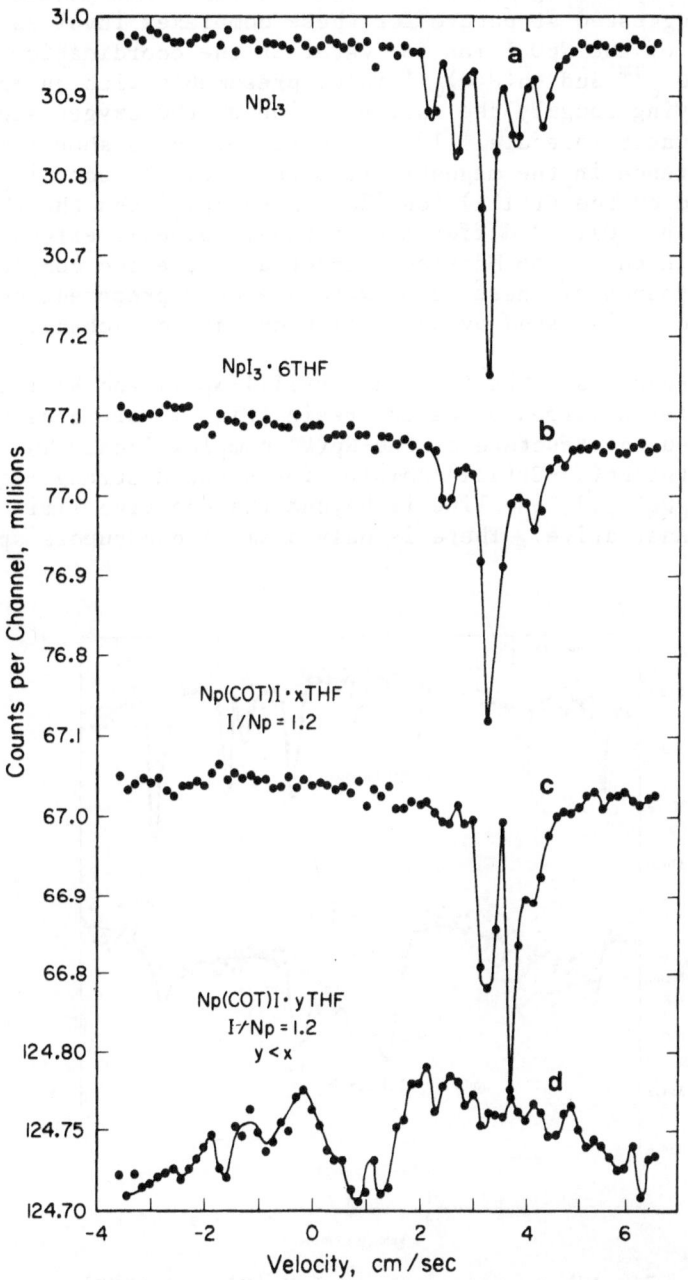

Fig. 6.  Mössbauer spectra of (a) $NpI_3$, (b) $NpI_3 \cdot 6THF$, (c) $Np(COT)I \cdot xTHF$, and (d) $Np(COT)I \cdot yTHF$

## Np(V) - Transition Metal Ion  Complexes

Sullivan[14] discovered from kinetic measurements that the aqueous Np(V) ion forms complexes with a number of other cations, and particularly strong complexes with Cr(III)[15] and Rh(III)[16] ions.  The suggested structure for these complexes involves the substitution of the $NpO_2^+$ ion for water in the coordination sphere of the $Cr(H_2O)_6^{3+}$ and $Rh(H_2O)_6^{3+}$ ions, presumably with an $NpO_2^+$ oxygen occupying roughly the same position as the oxygen atom of a coordinated water molecule.  If this were so, there should be a marked difference in the magnetic properties of the coordinated $NpO_2^+$ induced by the Cr(III) ion ($3d^3$, high spin) and the Rh(III) ion ($4d^6$, spin = 0).  A difference in these magnetic effects should be manifest in the $^{237}$Np Mössbauer spectra of the two complexes. No solid compounds of these complexes have been prepared, but the complexes can be isolated by absorption on cation exchange resin.[14-16]

The Mössbauer spectra[17] of the Cr(III)-Np(V) and Rh(III)-Np(V) complexes absorbed on cation resin (Fig. 7) are consistent with the presumed structure of the Np(V) complex ions.  The spectrum of the Np(V)-Cr(III) complex ion shows a strong magnetic splitting ($g_0 \mu_N H_{eff}$); one line is beyond the negative limit of the Mössbauer drive.  There is only a small quadrupole split-

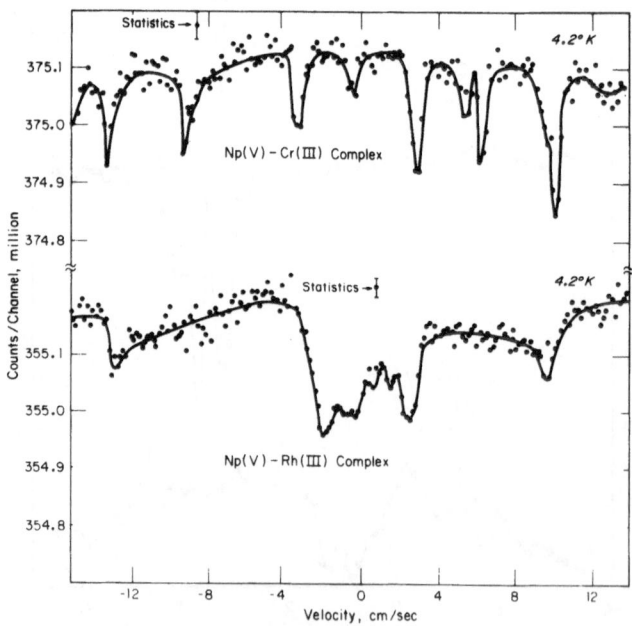

Fig. 7.  Mössbauer spectra of Np(V) - Cr(III) and Np(V) - Rh(III) complexes

ting. The Mössbauer parameters for the Cr(III)-Np(V) spectrum are $\delta$ = 1.69 cm/s, $g_0\mu_N H_{eff}$ = 9.91 cm/s and eqQ/4 = 2.9 cm/s.

The spectrum of the Rh(III)-Np(V) complex shows extreme resonances that correspond to the strongest lines of a magnetically split spectrum, but the strong mixture of resonances in the center of the spectrum are due to the bulk of the Np(V) ions in the complex. These resonances are interpreted as dominated by quadrupole splitting and complicated by intermediate relaxation effects. In the fast relaxation limit, the spectrum will show single or quadrupole-split lines; at long relaxation times, the spectrum will be magnetically split. Intermediate relaxation times produce broadened resonances, and with unequal relaxation times in the different directions along the solid lattice, an extremely strange spectrum may result. However, the conclusion from this spectrum is that the major part of the Np(V) ions are in a non-magnetic environment, and that the spectrum is consistent with the suggested structure for the Rh(III)-Np(V) ion. The comparison of Mössbauer spectra for the Cr(III)-Np(V) and the Rh(III)-Np(V) ions agrees with a structure of these complex ions with an $NpO_2^+$ ion occupying a position in the octahedral coordination sphere of the Cr(III) and Rh(III) ions.

## $[(CH_3)_4N]_2NpCl_6$, $[(C_2H_5)_4N]_2NpCl_6$

These compounds have the $Np^{4+}$ ion located at the body center of the octahedron of six $Cl^-$ ions. $TMA_2NpCl_6$ (TMA = $(CH_3)_4N$) has a face-centered cubic cell ($O_h$) with the $Np^{4+}$ ion at a point of full cubic symmetry. $(TEA)_2NpCl_6$ has an orthorhombic cell ($D_{3d}$) that distorts the field on the $Np^{4+}$ sites. The ground level of the $Np^{4+}$ ion ($5f^3$) is a $\Gamma_8$ quartet[18] in cubic symmetry; the complicated splitting[19] of the $\Gamma_8$ level has been extensively studied in optical spectra,[20-23] and EPR.[24,25] In general, the $\Gamma_8$ level splitting can be described by two splitting factors, g and f, which may be unequal and are anisotropic except in special cases.

The Mössbauer spectra of $(TMA)_2NpCl_6$ and $(TEA)_2NpCl_6$ (Fig. 8) are both characterized by magnetic splitting with an additional quadrupole contribution.[26] The Mössbauer parameters for $(TMA)_2NpCl_6$ are $\delta$ = -0.66 cm/s, $g_0\mu_N H_{eff}$ = 7.85 cm/s and eqQ/4 = 0.091 cm/s. The spectra, however, show marked deviation from an isotropic splitting pattern. The hyperfine pattern expected for pure isotropic splitting is shown at the top of Fig. 8; both the line spacings and intensities are inconsistent with the hyperfine pattern expected for isotropic splitting. The explanation may be that the paramagnetic field has split the $\Gamma_8$ level, and unequal splitting has caused the anomalies in the $(TMA)_2NpCl_6$ Mössbauer spectrum.

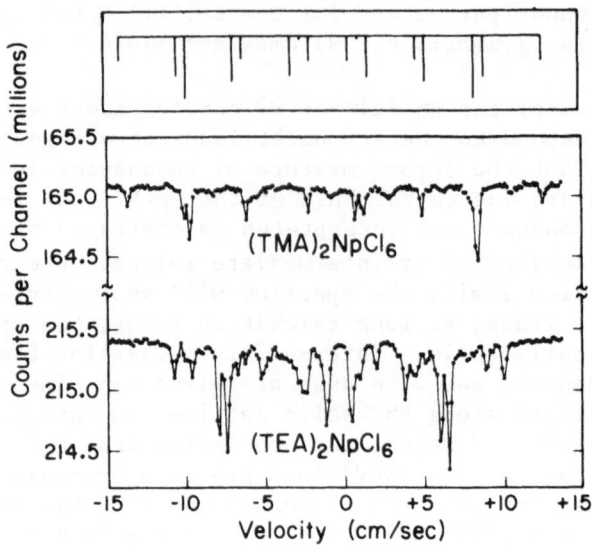

Fig. 8. Mössbauer spectra of $Cs_2NpCl_6$, $(TMA)_2NpCl_6$, and $(TEA)_2NpCl_6$

The Mössbauer spectrum of $(TEA)_2NpCl_6$ (Fig. 8) is quite complex, with about twice the number of lines expected for a magnetically split spectrum. The optical absorption spectrum[21] at 4.2°K was interpreted as resulting from $Np^{4+}$ ions in two unequal sites. The Mössbauer spectrum confirms this interpretation; the parameters for the two sites are (A) $\delta = 0.75$ cm/s, $g_0\mu_N H_{eff} = 5.78$ cm/s, $eqQ/4 = 0.35$ cm/s and (B) $\delta = 0.71$ cm/s, $g_0\mu_N H_{eff} = 6.26$ cm/s, $eqQ/4 = 0.27$ cm/s. As with the $(TMA)_2NpCl_6$ spectrum, the line spacings and intensities are not consistent with an isotropic splitting pattern. However, the spectrum does confirm the presence of two $Np^{4+}$ sites found by optical studies.

## Covalency of Np(IV) Organometallic Compounds

The wide range of isomer shifts in $^{237}Np$ Mössbauer spectra, besides identifying the valence of Np in compound, can also show more subtle effects within compounds of the same valence. This is of particular interest for organometallic compounds, where the orbitals of highly Π-bonding ligands contribute appreciable electron density to the shielding 5f orbitals. The isomer shift of $Np^{4+}$ in covalency-bonded compounds is shifted toward more positive values, sometimes quite dramatically. Table I compares the isomer shift of some $Np^{4+}$ organometallics, with the isomer shifts of $NpCl_4$ and $NpCl_3$ included for reference. As this work has been both reported[27-29] and reviewed,[30,31] this discussion is necessarily brief.

Table 1. Isomer Shifts of Np(IV) Organometallic Compounds

| Compound | Isomer Shift, cm/s | Hyperfine Splitting |
|---|---|---|
| $NpCl_4$ | $-0.35 \pm 0.05$ | MAG |
| $NpCp_3Cl*$ | $1.4 \pm 1.0$ | – |
| $NpCp_3C_4H_9$ | $0.27 \pm 0.07$ | MAG |
| $NpCp_3OC(CH_3)_3$ | $0.86 \pm 0.3$ | MAG |
| $NpCp_4$ | $0.72 \pm 0.02$ | QUAD |
| $Np(C_8H_8)_2$ | $1.94 \pm 0.05$ | MAG |
| $NpCl_3$ | $3.54 \pm 0.05$ | QUAD |
| $KNp(C_8H_8)_2$ | $3.92 \pm 0.10$ | QUAD |

* $Cp = C_5H_5^-$, cyclopentadienyl anion.

## Future

Mössbauer studies[32,33] of the charge states of $^{237}$Np following alpha-decay of the $^{241}$Am source have shown a general persistence of the same valence in the $^{237}$Np daughter as in the $^{241}$Am parent. Experiments of the same nature with the $^{237}$U or $^{237}$Pu as the Mössbauer sources are considerably more difficult, but may lead to the application of the $^{237}$Np Mössbauer effect to uranium or plutonium chemical studies.

## ACKNOWLEDGMENT

The information contained in this article was developed during the course of work under Contract No. DE-AC09-76SR00001 with the U.S. Department of Energy.

## REFERENCES

1. J. A. Stone and W. L. Pillinger, Phys. Rev. Lett. 13:200 (1964).

2. B. D. Dunlap, M. B. Brodsky, G. M. Kalvius, G. K. Shenoy, and D. J. Lam. J. Appl. Phys. 40:1495 (1969).

3. W. L. Pillinger and J. A. Stone. "Mössbauer Effect Methodology." I. J. Gruverman, Ed., Plenum Press, New York, Vol. 4, pp. 217-236 (1968).

4. G. M. Kalvius. "Plutonium 1970 and Other Actinides." Proceedings of the 4th International Conference on Plutonium and Other Actinides, Santa Fe, NM., October 5-9, 1970. W. N. Miner, Ed., Metallurgical Society of American Institute of Mining, Metallurgical and Petroleum Engineers, Inc., New York, NY, pp. 296-330 (1970).

5. B. D. Dunlap and G. M. Kalvius. In "Actinides Electronic Structure and Related Properties." A. J. Freeman, and J. B. Darby, Jr., Eds., Academic Press, New York, NY, p. 237 (1974).

6. B. D. Dunlap, G. M. Kalvius, S. L. Ruby, M. B. Brodsky, and D. Cohen, Phys. Rev. 171:316 (1968).

7. N. N. Krot and A. D. Gelman. "Dokl. Akad. Naak USSR," 177:124, ANL-trans-574, Argonne National Laboratory, Argonne, IL (1968).

8. N. M. Krot, M. P. Mefodyeva, T. V. Smirnova, and A. D. Gelman, "Radiokhimiya," 10:412. ANL-trans-678, Argonne National Laboratory, Argonne, IL (1968).

9. J. A. Stone, W. L. Pillinger, and D. G. Karraker. Inorg. Chem. 8:2519 (1969).

10. K. Frohlich, P. Gutlich, and C. Keller. J. Chem. Soc. Dalton Trans. 971 (1972).

11. K. W. Bagnall and J. B. Laidler. J. Chem. Soc. (A) 516 (1966).

12. D. G. Karraker and J. A. Stone. J. Inorg. Nucl. Chem. 41:1153 (1979).

13. D. G. Karraker and J. A. Stone. J. Inorg. Nucl. Chem. 39:2215 (1977).

14. J. C. Sullivan. J. Am. Chem. Soc. 84:4256 (1962).

15. J. C. Sullivan. Inorg. Chem. 3:315 (1964).

16. R. K. Murmann and J. C. Sullivan. Inorg. Chem. 6:892 (1967).

17. D. G. Karraker and J. A. Stone. Inorg. Chem. 16:2979 (1977).

18. K. R. Lea, M. J. M. Leask, and W. P. Wolfe. J. Chem. Phys. Solids, 25:1381 (1962).

19. B. Bleaney. Proc. Roy. Soc. (London) 73:939 (1959).

20. J. B. Gruber and E. R. Menzel. J. Chem. Phys. 50:3773 (1964).

21. E. R. Menzel, J. B. Gruber, and J. L. Ryan. J. Chem. Phys. 57:4387 (1972).

22. E. R. Menzel and J. B. Gruber. J. Chem. Phys. 54:3857 (1971).

23. R. P. Richardson and J. B. Gruber. J. Chem. Phys. 56:256 (1972).

24. J. E. Bray. Phys. Rev. B, 18:2973 (1978).

25. N. Edelstein, W. Kolbe, and J. E. Bray. Phys. Rev. B, 21:338 (1980).

26. D. G. Karraker and J. A. Stone. Phys. Rev. B, 22:111 (1980).

27. D. G. Karraker and J. A. Stone. Inorg. Chem. 11:1742 (1972).

28. D. G. Karraker and J. A. Stone. Inorg. Chem. 18:2205 (1979).

29. D. G. Karraker. Inorg. Chem. 22:503 (1983).

30. D. G. Karraker. In "Organometallics of the f-Elements." T. J. Marks and R. D. Fischer, Eds., D. Reidel: Dordrecht, Holland, p. 395 (1979).

31. D. G. Karraker. In "Recent Chemical Applications of the Mössbauer Effect." G. Shenoy, J. G. Stevens, Eds., American Chemical Society: Washington, DC, Adv. Chem. Ser. 194:347 (1981).

32. B. D. Dunlap, D. J. Lam, G. M. Kalvius, and G. K. Shenoy. J. Appl. Phys. 42:1719 (1971).

33. J. Gal, Z. Hadari, E. R. Bauminger, I. Nowik and S. Ofer. Solid State Communication, 15:1805 (1974).

# INDEX

Pulse height spectra, of $^{121}$Sn, 326

Quadrupole coupling constant, in antimony compounds, 339
  sign of, 324
  in iodine compounds, 300
Quadrupole Interaction, definition of, 4
  magnetic field effects on, 68
Quadrupole splitting, definition of, 5
  of iodine in silicon, 313
  in iron spectra, 4
  magnetic field effects on, 68
  partial, 6
    additivity of, 7
Radiogenic compounds, of iodine, 307
Radiolysis, local, in emission spectra, 188
Radiation damage, local, 189
Radicals, produced in nuclear decay, 184
Ramsdellite, lattice dynamics of, 206
    iron and tin probe atoms in, 208
Rare earth compounds, bonding in, 345
  Mossbauer spectroscopy of, 343
  trivalent charge state in, 345
Recoil free fraction, 200
  absolute, 201
  angular dependence, 251
  in antimony spectra, 328
  determination of, 204
  from Debye model, 158
  of iodine absorbers, 297
  of iodine implanted in silicon, 313
  in membranes, 263
  of tellurium sources, 297
  temperature dependence of, 253
Relaxation, paramagnetic, 72
  slow paramagnetic, 74
Relaxation time, magnetic, of sublattice, 115
Rho ($\rho$) value, for tin compounds, 279

Second order Doppler shift, 3
  relationship to mass, 203
Sediments, deep sea, 227
  glacial, 226
  Mossbauer studies of, 217
Smectic liquid crystal phase, 243
Silicon, $^{129m}$Te implanted, 310
  isomer shifts of, 311
  quadrupole splitting of, 313
  recoil free fraction of, 313
Single crystal studies, of magnetic interactions, 76
Six coordination, in organo iron complexes, 7